Higher Education and the Challenge of Sustainability

Higher Education and the Challenge of Sustainability

Problematics, Promise, and Practice

Edited by

Peter Blaze Corcoran
Florida Gulf Coast University,
Florida, U.S.A.

and

Arjen E.J. Wals
Wageningen University,
Wageningen, The Netherlands

KLUWER ACADEMIC PUBLISHERS
DORDRECHT / BOSTON / LONDON

Library of Congress Cataloging-in-Publication Data

ISBN 1-4020-2134-8 (PB)
ISBN 1-4020-2026-0 (HB)
ISBN 0-306-48515-X (e-book)

Published by Kluwer Academic Publishers,
P.O. Box 17, 3300 AA Dordrecht, The Netherlands.

Sold and distributed in North, Central and South America
by Kluwer Academic Publishers,
101 Philip Drive, Norwell, MA 02061, U.S.A.

In all other countries, sold and distributed
by Kluwer Academic Publishers,
P.O. Box 322, 3300 AH Dordrecht, The Netherlands.

*LB
2324
H54
2004*

Printed on acid-free paper

TABLE OF CONTENTS

FOREWORD

The challenge of sustainability is to create a new approach to social and economic development and global security which integrates concerns for short term economic gain with concerns for future generations, cultural and biological diversity, and social well-being. The precise meaning and balance between these concerns is debated. Some argue that sustainable development requires only improvements in ecoefficiency and freer global markets. Others argue for a profound shift in worldviews in which the intrinsic worth of nature and the spiritual meaning of life should guide development.

The academy has always played a major role in the theoretical debates and practical experimentation concerning the best route to the good life for all. Higher education is vested by society with the mission of discerning truth, imparting knowledge, skills and values and preparing responsible citizens and competent workers who will contribute to an improving world. While colleges and universities increasingly serve the globalizing market economy, they are also increasingly sources of innovation in sustainability.

Definitions of sustainability or sustainable development are contested, but most agree they involve recalibrating economic and social policies and practices to support economy, ecology, and equity. A university's approach to sustainability will vary widely depending on its cultural and political context and the level of students it serves. Some emphasize technical greening of operations. Some are more philosophical in orientation, debating the meaning of sustainability and, at times, redesigning curricula. While the challenges and approach to sustainability vary widely from institution to institution, there are some fundamental commitments that must be achieved for an institution to be on the path to sustainability, however it is defined. First, a commitment to sustainability is central to the academic functions of the university, thus featured in statements of the mission and purpose, in all academic disciplines as well as in general and professional education requirements in research, and in the hiring, tenure and promotion of faculty. Second, the institution seeks to reduce its "ecological footprint" through sustainable practices and policies. Third, sustainability committees, audits, and celebrations are visibly present. Fourth, the institution is engaged in outreach and forming partnerships locally and globally to enhance sustainability.

There has been considerable progress in higher education institutions and in many disciplines, such as engineering and architecture, to introduce green design and ecoefficiency – reducing costs and environmental damage through energy conservation, recycling and other green practices. But reorienting general and specialized education toward sustainability has proven more difficult.

The major problem for higher education is that it is almost impossible to create a sustainable university in an unsustainable society. David Orr describes our plight as walking north on a southbound train. The train of economic globalization is barreling south. We, the advocates of sustainability in higher education, are taking significant steps to create a more humane, just and sustainable path for globalization. But as we walk north, we are still passengers of this accelerating train moving in the opposite direction (Orr, 2003). We urgently need to educate and motivate professionals, citizens, present and future leaders to change course toward a more sustainable future.

A major opportunity to strengthen our capacity and motivation to do this is provided by the United Nations' Decade of Education for Sustainable Development (DESD), 2005-2014. Education for sustainable development was a priority of the Earth Summit in Rio in 1992. Recognizing that too little had been done to implement this priority, the World Summit for Sustainable Development recommended, at the urging of the Japanese and other governments, that a DESD be adopted by the General Assembly. It was adopted in December of 2002.

UNESCO, in its role as task manager for this Decade, states in its *Framework for a Draft International Implementation Scheme*:

> Education for sustainable development has come to be seen as a process of learning how to make decisions that consider the long-term future of the economy, ecology and equity of all communities. …This represents a new vision of education, a vision that helps people of all ages better understand the world in which they live, addressing the complexity and interconnectedness of problems such as poverty, wasteful consumption, environmental degradation, urban decay, population grown, health, conflict and the violation of human rights that threaten our future. The vision of education emphasizes a holistic, interdisciplinary approach to developing the knowledge and skills needed for a sustainable future as well as changes in values, behavior, and lifestyles (UNESCO, 2003).

Many international partnerships and alliances have formed to strengthen the contribution of higher education to this new vision of education and to this Decade. The World Summit on Sustainable Development formally recognized a range of partnerships that were critical to articulating the agenda for sustainable development and promoting it through education at all levels. The Earth Charter was recognized for providing an integrated ethical agenda for framing the central principles of sustainable development. The Global Higher Education for Sustainability Partnership (GHESP) was launched at the UNESCO conference in Johannesburg. The Ubuntu Declaration brought together, for the first time, science, technology and education for sustainable development. These and other initiatives are collaborating to develop resources to assist colleges and universities in making sustainability central to their mission and functioning. This involves deep reflection on the nature of the educational and social transformation that is required to create a sustainable future, and practical examples of how institutions in diverse cultural settings have successfully reoriented their teaching and research, outreach, and operations to embody their own forms of sustainability.

The chapters in this book provide many critical resources for this task, helping us explore what sustainability is and is not, and learn from living examples of institutional practice. Its unique contribution is the depth with which it explores the problematics, the promise, and the practice of higher education for sustainability. Clearly we need a fuller vision of what development is for, and how to educate all, to create just, equitable, and ecologically sound societies. This book provides both the inspiration and the insights to help us accomplish this great task.

Richard M. Clugston
Executive Director
Association of University Leaders for a Sustainable Future

REFERENCE

Orr, D.W. (2003). Walking north on a southbound train. *Conservation Biology, 17*(2), pp. 348-351.

UNESCO (2003). *United Nations Decade of Education for Sustainable Development* (January 2005 - December 2014): Framework for a Draft International Implementation Scheme (p. 4). Paris, France: UNESCO.

PREFACE

Peter Blaze Corcoran and Arjen E.J. Wals (Editors)

Sustainability is becoming an integral part of university life. Universities around the world are re-thinking their missions and are looking to restructure their courses, their research programs, and the way life on campus is organized. Over one thousand university presidents, provosts, and deans have signed one or more international declarations which seek to promote sustainability in higher education (i.e. the Talloires Declaration, The Kyoto Declaration of the International Association of Universities, The Swansea Declaration, and the COPERNICUS Charter of the European Association of Universities). Universities increasingly realize that their environmental impact is tremendous, not only in terms of the energy they use and the waste they generate, but perhaps first and foremost in the way they equip their graduates in dealing with sustainability issues in both their personal and professional lives. University administrators, curriculum developers, researchers, teachers, and students are looking for concrete ways to integrate sustainability that are consistent with the responsibility of higher education in today's society.

In this book, sustainability is critically explored as an outcome and a process of learning, but also as a catalyst for educational change and institutional innovation. Questions are raised about both the meaning of sustainability and institutional resistance to change.

Scholars from a variety of established and emerging fields of education present their views on potential contributions to the development of sustainability in higher education. The chapter authors are leading practitioners, critics, and researchers who span several generations in their fields of education. They represent a variety of cultures, academic fields, and perspectives of sustainability. Taken together, the authors provide historical, philosophical, and pedagogical expertise related to environmental learning and organizational change in tertiary education. These multiple perspectives are integrated and complemented by a number of cases of colleges and universities from around the world that have re-designed critical dimensions of their institutions to deal with the challenges presented by the introduction of sustainability. Special attention is given to the issue of assessing progress toward integrating sustainability in higher education.

The book has three parts. *Part One: Problematics* provides a rationale for the book, a history of sustainability initiatives in higher education, and raises critical issues with regard to the meaning of both "education" and "sustainability" within the context of diverging interests, norms, values, and epistemologies.

In *Part Two: Promise*, scholars from a variety of "educations" present their views on learning about sustainability in a university setting. These include environmental education, environmental justice, deep ecology, ecofeminism, transformative education, natural resource management, whole systems thinking, and enriched disciplinary education.

In *Part Three: Practice*, these multiple perspectives are integrated and complemented by a number of institutional cases using a variety of methods. The cases are selected on the quality of the methodology, diversity of the institutions, and potential transferability towards other contexts. Part Three is introduced by a review and critique of case study literature including an analysis of the efficacy of the various methodologies and a synthesis of the issues raised by the various cases. The book concludes with a web-based resource section that includes analytic profiles of initiatives that speak to the challenge of sustainability in higher education and contains resources for further learning, including web-site addresses, a bibliography, and profiles of key organizations worldwide.

ACKNOWLEDGEMENTS

This book is dedicated to our Dutch colleagues who are showing the way toward sustainable development – to the many students, academicians, government officials, and NGO activists who are contributing to this significant cultural shift. They deserve great credit for their national and international efforts. The programs, partners, and individuals are too many to mention individually; taken together their work is leading to a sustainable path in higher education and in society.

We are enormously grateful to Wageningen University for its support during the long process of organizing and editing this book. We thank our patient and supportive colleagues of both the "Education and Competence Studies" and "Communication and Innovation Studies" groups.

We appreciate the help of University Leaders for a Sustainable Future, secretariat for the signatories of the Talloires Declaration. Rick Clugston, Executive Director, has been supportive of our concept for the book and of our travel to work together on it.

We thank Florida Gulf Coast University for its support of the new Rachel Carson Center for Environmental and Sustainability Education through which much of this work was accomplished. In particular, we thank Carolyn Gray, Dean of the College of Arts and Sciences.

We thank Corrie Pieterson of Florida Gulf Coast University and E. Rogier van Mansvelt of the Dutch Foundation for Sustainable Higher Education (DHO) for volunteering to organize the resource link part of the book.

We deeply value the commitment of our colleague Marja Boerrigter who so thoughtfully prepared the manuscript.

We are endebted to Tamara Welschot, our editor at Kluwer, for her openness and enthusiasm for bringing education for sustainable development to new audiences.

Most of all, we are indebted to our chapter authors for their willingness to contribute, for their patience with us, and for their accomplishments in higher education for sustainability. We also wish to acknowledge the work of many others in this critical field of study, especially those from cultures and languages not represented in this book.

PART ONE

PROBLEMATICS

CHAPTER 1

THE PROBLEMATICS OF SUSTAINABILITY IN HIGHER EDUCATION: AN INTRODUCTION

Peter Blaze Corcoran & Arjen E.J. Wals

The higher education community is called to respond to times of disastrous anthropogenic environmental crises, failing political systems, religious intolerance, and unsustainable and inequitable economic development. The scope and range of the negative impacts of university-educated people on the natural systems that sustain Earth are unprecedented.

Characterizing this crisis, leading environmental academic David Orr has written "the crisis of the biosphere is symptomatic of a prior crisis of mind, perception, and heart. It is not so much a problem in education, but a problem of education (Orr, 1994)." Orr goes on to say:

> Education is not widely regarded as a problem, although the lack of it is. The conventional wisdom holds that all education is good, and the more of it one has, the better.... The truth is that without significant precautions, education can equip people merely to be more effective vandals of the Earth. (Orr, 1994, p. 5)

Society has privileged institutions of higher education. We expect much of those on whom modernism has vouchsafed such a franchise. In an essay entitled "The Role of Higher Education in Achieving a Sustainable Society", Tony Cortese writes:

> Higher education institutions bear a profound moral responsibility to increase the awareness, knowledge, skills and values needed to create a just and sustainable future. These institutions have the mandate and potential to develop the intellectual and conceptual framework for achieving this goal. They must play a strong role in education, research, policy development, information exchange and community outreach and support.... They have the unique freedom to develop new ideas, comment on society, and engage in bold experimentation, as well as contribute to the creation of new knowledge (Cortese, 1992, p. 5).

Surely one of the aims of education must be to sustain the possibility of a good society of right living. Never has the opportunity to create the foundation for a sustainable future been greater. Higher education can play a pivotal role in turning society toward sustainability. We must rediscover and teach indigenous and ancient truths, generate new concepts and ways of thinking, and we must inspire students with a hopeful vision. The Secretary General of the United Nations, Kofi Annan, has argued that "our biggest challenge in this new century is to take an idea that sounds abstract—sustainable development—and turn it into reality for all the world's people" (UN, 2002). Certainly the principle of intergenerational responsibility is at

Peter Blaze Corcoran & Arjen E.J. Wals (Eds.), Higher Education and the Challenge of Sustainability: Problematics, Promise and Practice, 3-6.

the heart of formal education. But how might the concept of sustainability be defined in terms of formal higher education? What *is* the challenge of sustainability for higher education? What is demanded of us by moral responsibility to rising generations?

The assumption of human culture has been that the beauty and bounty of Earth would be transferred across generations, that the process of education would transfer the values, skills, and knowledge to survive and thrive in the cultural and natural systems of which we are a part. Universities have had, in the modern world, a pivotal position in defining education for this task. Yet certain core ideas embedded in disciplinary thinking and the practice of those ideas, are increasingly problematic. Hence, a challenge to higher education is to reconsider its disciplines, its institutional practices, and, indeed, its mission to account for economic and human development that is sustainable.

To accept the notion of the importance of the concept of sustainability for higher education is to accept something that constitutes a problem. In Part One, the authors, taken together, articulate the problematics of sustainability as they relate to the field of higher education.

The history of the concept, going back to its roots at the first United Nations meeting that concerned itself with the relationship between people and their social and natural environments, the Stockholm Conference on the Human Environment (1972) is outlined by Tarah Wright in Chapter 2. She relates sustainability declarations to the international development of environmental education through the Belgrade Charter (1975) and the Tbilisi Declaration (1977) and to the evolving development of sustainability in education through Chapter 36 of Agenda 21 from the United Nations Conference on Environment and Development (1992). The emergence of the view that higher education has a moral obligation to both teach and model environmental sustainability and that universities are also obligated to the communities in which they reside is described. Her analysis of the evolution of sustainability declarations, she argues, helps us understand key priorities and paths. She writes:

> The identification of these themes and patterns furthers the understanding of what universities believe are the key priorities to become sustainable institutions, and what paths universities believe they should take on the journey to sustainability.

Indeed the intellectual history of sustainability in higher education as articulated in these declarations provides the starting point discussing the problematics. Richard Bawden writes:

> The introduction of education for sustainability within the academy is not without its dilemmas however. Not the least of these is its role in furthering the divide between the understandings and professional discourse of the 'expert' and that of the 'lay' public with respect to differing perceptions about the nature of the problematique....
>
> In exploring this second path, a persuasive argument can be mounted in support of the need for there to be far greater 'engagement' between those in the academy and those in the citizenry, with the development of a systemic discourse appropriate to this 'interface domain.' An ethos of "sustainabilism' will be an essential characteristic of this new inclusive discourse that will be focused on the search for democratic public judgment

and responsible communicative actions with respect to what it is that we *should do* next!

In Chapter 3, he analyzes sustainability as emergence, as "sustainabilism," and argues that through a scholarship of engagement and through critical scholarship of a praxis of collaboration between experts and citizens we can learn what ought to be done next in the quest for sustainability.

In Chapter 4, John Huckle provides a critical focus on the act of education. He reminds us that however we might accommodate the many ways of knowing about and the many discourses on sustainability, we must use critical pedagogy that accommodates the terrible postmodern complexities faced by the young. He offers critical realism, a dialectical and materialist philosophy, and a framework to understand the structures and processes of the sciences of nature, and of politics. He applies critical realism to education and to mainstream environmentalism. This work helps us analyze the interests shaping different kinds of education for sustainability. In a larger sense by rejecting postmodern relativism it creates a possibility for defining critically "appropriate and morally" correct sustainability. Indeed, Huckle argues it "allows the construction of a new grand narrative of sustainability" that can lead us to such sustainability.

Stephen Sterling reminds us in Chapter 5 that the nature of sustainability requires a fundamental change of epistemology, and therefore, of education. He writes:

> Sustainability is not just another issue to be added to an overcrowded curriculum, but a gateway to a different view of curriculum, of pedagogy, of organizational change, of policy and particularly of ethos. At the same time, the effect of patterns of *unsustainability* on our current and future prospects is so pressing that the response of higher education should not be predicated only on the 'integration of sustainability' into higher education, because this invites a limited, adaptive, response.... We need to see the relationship the other way around—that is, the necessary transformation of higher education towards the integrative and more whole state implied by a systemic view of sustainability in education and society.

Using a systems perspective, he helps us to see the vast complexity of higher education and sustainability. He also helps us see how difficult a shift to the emergent ecological paradigm *is* in a failed system of higher education. A challenge for all of us in a higher education system that is part of the unsustainability problem is how we can address the problem from within by analyzing learning levels and learning responses. Sterling offers possibilities for deeper and transformative learning.

Sterling helps us see how sustainable education might emerge since "the process of sustainable development or sustainable living is essentially one of learning, while the context of learning is essentially that of sustainability." This is especially challenging at institutions; he uses Hawkesbury College and Schumacher College as examples of learning systems. These examples and Sterling's elucidation of the dimensions of the change to the ecological education paradigm in higher education help us see the possibility of mutual coevolutionary transformation of both education and society toward sustainability.

Finally in Part One, we look at how sustainability might be analyzed, assessed, and, even, measured across institutions. Michael Shriberg assumes that we lack clear criteria for assessment tools. In Chapter 6, he says it is desirable and possible to address process, motivations, and outcomes. He analyzes ten existing assessment tools by what they include and by what they exclude. He derives new parameters for decreased consumption, centrality of sustainable education, cross functional integration, cross institutional integration, and incremental and systemic progress. Shriberg suggests various strengths and weaknesses of the assessment tools and suggests deeper dimensions that ought to be considered.

As we enter the problematic terrain of higher education and sustainability, a context of diverging interests, norms, values, and epistemologies, emerges. The authors in Part One help us orient this terrain and provide a basis for further exploration.

REFERENCES

Cortese, Anthony D. (1992). Educational for an Environmentally Sustainable Future. *Environmental Science and Technology, 26*(6), 1108-1114.
Orr, David W. (1994). *Earth in Mind: On Education, Environment, and the Human Prospect.* Washington, DC: Island Press.
United Nations press release (2002). SC/SM/7739. *Secretary General Calls for Break in Political Stalemate over Environmental Issues.*

CHAPTER 2

THE EVOLUTION OF SUSTAINABILITY DECLARATIONS IN HIGHER EDUCATION

Tarah Wright

INTRODUCTION

The notion of sustainability in higher education (SHE) was first introduced at an international level by the United Nations UNESCO-UNEP International Environmental Education Programme in 1978. Since then, a number of national and international declarations directly relating to environmental sustainability in higher education have been developed. These declarations have gained acceptance in the higher education community and have subsequently been endorsed and signed by numerous universities. The Talloires Declaration of 1990, for example, has over 275 signatories (University Leaders for a Sustainable Future, 2002) and over 291 educational institutions have endorsed the COPERNICUS Charter (CRE-COPERNICUS, 2002). How has sustainability been defined in these international declarations? How has the concept of sustainability in higher education evolved over the past 30 years? Such an understanding is essential to contextualizing present practices and beliefs in higher education. If we are to fully comprehend the current state of sustainability in higher education and how we might proceed in the future, we must first understand the evolution of sustainability declarations and how such declarations have helped higher education frame their commitment to sustainability in the past.

This chapter will describe the evolution of environmental sustainability declarations in higher education from the 1970s to present, and examine the patterns and themes that emerge from these documents. The focus will be on major international declarations such as the Tbilisi, Talloires, Halifax, and Kyoto Declarations, the COPERNICUS Charter, and the most recently created Lüneburg Declaration. The chapter will conclude with a discussion of the implications the emergent themes in sustainability declarations have for the future.

A CHRONOLOGY OF SUSTAINABILITY DECLARATIONSFORMATTING INSTRUCTIONS

Environmental sustainability declarations specifically developed for higher education are relatively new, emerging in the early 1990s. There are some key

Peter Blaze Corcoran & Arjen E.J. Wals (Eds.), Higher Education and the Challenge of Sustainability: Problematics, Promise and Practice, 7-19.

international conferences, guidelines and directives, however, that paved the way for these declarations to come into being. The Stockholm Conference on the Human Environment in 1972, for example, discussed international sustainable development issues which had specific relevance to higher education. The Stockholm Declaration adopted at the conference discussed the interdependency between humans and the environment, the distribution of wealth, and the notion of intergenerational equity. Specifically related to educational institutions, the Stockholm Declaration called for environmental education for all people from grade school through adulthood so to "broaden the basis for enlightened opinions and responsible conduct by individuals, enterprises and communities in protecting and improving the environment in its full human dimension" (UNESCO, 1972, Principle 19). Also related to SHE declarations is the development of environmental education conferences and declarations. The Belgrade Charter (1975) and the Tbilisi Declaration (1977) for example, were both influential in the development of international environmental education and sustainability initiatives. The Tbilisi Declaration was the result of the UNESCO/UNEP Intergovernmental Conference on Environmental Education in 1977. It stated that in order for people to develop a better understanding of the human-environment relationship, formal and non-formal environmental education opportunities should be made available to people of all ages and level of academic aptitude. In a statement regarding the role higher education could play in achieving environmental sustainability, the Declaration asked colleges and universities to consider environmental concerns within the framework of the general university:

> Universities, as centres for research, teaching and training of qualified personnel for the nation, must be increasingly available to undertake research concerning environmental education and to train experts in formal and non-formal education. Environmental education…is necessary for students in all fields, not only natural and technical sciences, but also social sciences and arts, because the relationship between nature, technology and society mark and determine the development of a society (UNESCO-UNEP, 1977, p. 33).

Additionally, the Tbilisi Declaration asked universities to consider the development of environmental curricula, engage faculty and staff in the development of environmental awareness, provide specialist training, engage in international and regional co-operative projects, and inform and educate the public regarding environmental issues. As we shall see, all of these initiatives were echoed in the SHE declarations that began to emerge more than a decade later.

The United Nations Conference on Environment and Development of 1992 also had a profound influence on the development of environmental sustainability declarations. This conference focused on issues of environmental sustainability and their application to various disciplines and fields. The publication of Agenda 21 was a direct result of the conference. Chapter 36 of Agenda 21 – Education, Awareness and Training, specifically addressed issues related to sustainability within educational institutions. Offering similar sentiments to the Tbilisi Declaration, Chapter 36 identified a worldwide lack of environmental literacy, and posited that formal and informal education was the solution to environmentally unsustainable behavior amongst humans. It called for reorienting education towards sustainable development, increasing public awareness of environmental issues and promoting the training of educators in environmental issues. Section 36.1 offered tremendous support for the development of SHE declarations, stating that countries should assist universities and colleges in the development of plans to promote research and common teaching approaches to sustainable development.

Evidence of a global understanding of environmental issues can be found in the increase of conferences and declarations related to the environment in the 1970s and 1980s. Specific declarations related to sustainability within higher education, however, do not emerge until the early 1990 and continue through to the next millenium (Table 1). What provided the impetus for such declarations? Universities in the 1990s found themselves in a world of environmental concerns. Universities were looked upon by society as institutions that could seek knowledge and truth, and apply such knowledge in order to solve the complex problems of society (Brubacher, 1982). At the same time universities were being criticized for their inability to be models of sustainability both in greening their physical operations and in developing environmentally friendly curriculum (Bowers, 1997 Clugston, 1999; Orr, 1995). David Orr chastised universities stating that environmental degradation is not the work of ignorant people, "rather, it is largely the result of work by people with BA's, B.Sc.'s, LLB's, MBA's and PhD's" (Orr, 1992, p. 7). One reaction of colleges and universities to these criticisms was to create and sign international agreements and declarations related to sustainability in higher education (SHE).

Table 1. International Sustainability in Higher Education Declarations.

Year	Declaration	Country of Signatory Universities	Number of Signatories (as of June 2002)
1990	Talloires Declaration	International	275
1991	Halifax Declaration	Canada	20
1993	Kyoto Declaration	International	n/a [1]
1993	Swansea Declaration	International	n/a [2]
1994	CRE COPERNICUS Charter	Europe	291
1997	Declaration of Thessaloniki	International	n/a [3]
2000	Lüneburg Declaration	International	n/a [4]

1. Adopted at the Ninth International Association of Universities Round Table, however there are no individual signatories.
2. Adopted at the Association of Commonwealth Universities Conference, however there are no individual signatories.
3. Adopted at the UNESCO Conference on Environment and Society: Education and Public Awareness For Sustainability, however there are no individual signatories.
4. Adopted at the Higher Education for Sustainability - Towards the World Summit on Sustainable Development Conference, but no individual signatories.

The Talloires Declaration. The Talloires Declaration was the result of a conference held at the Tufts University European Centre in France in which twenty-two university presidents, vice-chancellors, and rectors met to discuss how higher education could contribute to an environmentally sustainable future. The conference asked participants to contemplate the role universities could play in working toward an environmentally sustainable future, and what their individual universities could accomplish in working toward this goal. Participants agreed that "by practicing what it preaches, the university can both engage students in understanding the institutional metabolism of materials and activities, and have them actively participate to minimize pollution and waste" (ULSF, 1990).

The result of the meeting was the development of the Talloires Declaration. This declaration was the first international document that focused specifically on sustainability in higher education and the first official statement given by university administrators of a commitment to environmental sustainability in academe. The declaration was signed by all participants at the meeting in France, with a promise that they would encourage their colleagues at other universities to sign the declaration as well. Since 1990, signatories to the Talloires Declaration have increased from 20 to over 275 signatories around the world (University Leaders for a Sustainable Future, 2002).

The Halifax Declaration. In December 1991, the Conference on University Action for Sustainable Development was held in Halifax, Canada. Conference participants came from a wide range of university sectors, including university presidents, administrators, faculty, students, and representatives from all levels of government in Canada, non-governmental organizations, and the business community. The principal goal of the conference was to consider the role universities could play in improving the capacity of countries to address environment and development issues, and to discus the implications the Talloires Declaration had for Canadian Universities. The result was the Halifax Declaration. This Declaration echoed the sentiments of the Talloires Declaration, emphasizing the moral obligation of universities to environmental sustainability:

> Universities are entrusted with a major responsibility to help societies shape their present and future development policies and actions into the sustainable and equitable forms necessary for an environmentally secure and civilized world (Lester Pearson Institute for International Development, 1992).

The Halifax Declaration gave a new dimension to SHE declarations in that it offered an Action Plan for signatory universities to follow. The Action Plan outlined short and long-term goals for universities to work toward and specific frameworks for action in order to become more sustainable institutions. The Action Plan emphasizes education and training, research and policy, increased recognition of inter-disciplinary work, and a pro-active approach by universities toward sustainable development. While the Action Plan was written to help the implementation of the Halifax Declaration, a recent implementation analysis found that many of the initiatives listed in the Action Plan were considered inappropriate or irrelevant by specific institutions, and therefore did little to help signatory universities (Wright, 2002).

The Kyoto Declaration. The Kyoto Declaration of 1993 differs from the Talloires and Halifax Declarations in that there are no formal signatory institutions. The Declaration was the result of discussions at the Ninth International Association of Universities Round Table in 1993 and was adopted by the 90 international university leaders assembled there. The Declaration was also formally endorsed by the International Association of Universities in South Africa, August 2000.

The Declaration challenged universities to promote environmental sustainability through both environmental education, and physical operations. The moral obligation of higher education to contribute to environmental sustainability that had been heard in other declarations, was reiterated in this document:

> Global sustainable development implies changes of existing value systems, a task which universities have an essential mission in, in order to create the necessary international consciousness and global sense of responsibility and solidarity (International Association of Universities, 1993, p. 4).

Like the Halifax Declaration, the Kyoto Declaration offered an Action Plan for individual universities to follow. This Action Plan recognized that while the suggested initiatives might not be appropriate for each university, there were certain actions that universities should be encouraged to take including the development of ecological literacy programs, developing partnerships amongst universities as well as industry and government, engaging in public outreach initiatives, encouraging sustainable research, and the development of more sustainable physical operations within the university.

The Swansea Declaration. At the same time that the Kyoto Declaration was being considered by the International Association of Universities, representatives from over 400 universities in 47 countries were discussing similar SHE issues at the Association of Commonwealth Universities' fifteenth Quinquennial Conference at the University of Wales. The theme of the conference was "People and the Environment - Preserving the Balance". Inspired by the development of the Talloires and Halifax Declarations, and disappointed by the lack of university presence at the United Nations Conference on Environment in 1992, conference participants created the Swansea Declaration to add their voice "to those many others worldwide that are deeply concerned about the widespread degradation of the Earth's environment, about the pervasive influence of poverty on the process and the urgent need for sustainable practices" (UNESCO, 1993, p. 1).

The Swansea Declaration repeated many of the tenets of past university sustainability declarations. However, it added an interesting dimension to the discussion of SHE in that it stressed equality amongst countries as an important factor in achieving sustainability worldwide. Recognizing that less developed countries may have more pressing priorities than environmental sustainability, the Swansea Declaration implored universities in richer countries to provide support for the evolution of SHE initiatives at less fortunate universities around the world.

The CRE COPERNICUS Charter. The CRE COPERNICUS Charter for Sustainable Development was created by Co-operation Programme in Europe for Research on Nature and Industry through Coordinated University Studies (COPERNICUS) which was established by the Association of European Universities (CRE). The Charter was created as an effort to further the efforts of the Magna Charta of European Universities, Talloires Declaration, Agenda 21, and the Halifax Declaration, and to mobilize European institutions of higher education to further develop their understanding of sustainability within their institutions. The

Charter was presented to over 500 universities within 36 nations at the CRE biannual conference in Barcelona in 1993. By 1994, over 213 European rectors had personally signed the Charter.

The Charter reiterated the desire for universities to become leaders in creating sustainable societies, and stressed the need for a new set of environmental values within the higher education community. The Charter highlights technology transfer, public outreach, environmental literacy programmes, developing environmental ethics amongst members of the university community, and encouraging partnerships as key elements to achieving SHE.

The Thessaloniki Declaration. The Thessaloniki Declaration was a direct result of the UNESCO Conference on Environment and Society: Education and Public Awareness for Sustainability, in Greece 1997. This event was considered a follow-up to the Tbilisi conference 20 years earlier. Unlike the Tbilisi Declaration which focused on environmental education alone, the Thessaloniki broadened its scope to encompass all issues of sustainability within higher education.

The declaration argued that the concept of environmental sustainability must be clearly linked with poverty, population, food security, democracy, human rights, peace and health and a respect for traditional cultural and ecological knowledge. In terms of institutions of higher education, the declaration affirmed that universities and colleges must address issues related to the environment and sustainable development and that universities must be reoriented towards a holistic approach to education. The Thessaloniki Declaration is similar to the Kyoto and Swansea Declarations, in that there are no formal individual signatory institutions. However, the declaration called for governments and leaders in education to honour the commitments they had already made in signing past declarations of environmental sustainability.

Lüneburg Declaration. The most recent SHE declaration was the result of the Higher Education for Sustainability - Towards the World Summit on Sustainable Development (Rio+10) Conference held in Lüneburg, Germany in October 2001. This conference was considered a preparatory event in the higher education sector for the Rio+10 Summit in Johannesburg 2002. The conference focused on developing a clear SHE statement to present at the Rio+10 Conference that was representative of higher education stakeholders from both the global North and South. The Lüneburg Declaration was drafted by members of the Global Higher Education for Sustainability Partnership (GHESP) before the conference, and finalized and adopted by the participants at the conference. The objective of the Declaration was to ensure that higher education was given priority in the international work programme to follow the Rio + 10 Summit.

The Lüneburg Declaration synthesizes the majority of declarations related to sustainability in higher education. It stresses the need to understand the interconnectedness of globalization, poverty alleviation, social justice, democracy, human rights, peace and environmental protection issues in relation to SHE. It is a unique declaration in that it recognizes the problems encountered with the implementation of sustainability declarations in the past and calls for the development of a "toolkit" for universities to use in order to translate their written commitment to sustainability to action. Further, it lists priorities for working toward

SHE in education institutions, NGOs, governments, and the United Nations. The declaration also calls for the empowerment of all people to work towards sustainability. The declaration does not ask for signatories, but promotes the endorsement and implementation of previous declarations.

EMERGING THEMES IN SUSTAINABILITY DECLARATIONS

An analysis of SHE declarations from the Talloires to Lüneburg, reveals how declaration authors and declaration signatories frame the central task of becoming sustainable institutions. While each declaration is different depending on the context in which it was written, Table 2 shows key themes that emerge from the various sustainability declarations. There are two themes that are common to all declarations. First, each SHE declaration discusses the moral obligation of universities to become sustainable institutions. Second, all of the declarations discuss the need for public outreach activities. The development of ecologically literate staff, faculty and students is a popular theme, as is the development of partnerships with all levels of government, non-governmental organizations (NGOs) and various industries. Surprisingly, the notion of developing more sustainable physical operations on the university campus does not seem to be a priority for the majority of the declarations. A closer look at these themes gives us a deeper understanding of the degree to which it is discussed in the SHE declarations.

Table 2. Common principles of Sustainability in Higher Education Declarations*.

Declaration	Moral obligation	Public outreach	Sustainable physical operations	Ecological literacy	Develop interdisciplinary curriculum	Encourage sustainable research	Partnership with government, NGOs and industry	Interuniversity cooperation
Tbilisi	x	x		x		x	x	
Talloires	x	x	x	x	x	x	x	x
Halifax	x	x		x			x	x
Kyoto	x	x	x	x		x	x	x
Swansea	x	x	x	x		x		x
CRE-COPERNICUS	x	x		x		x	x	
Thessaloniki	x	x		x	x		x	
Lüneburg	x	x			½(x)	x	x	x

*modified from Wright, 2002b

Moral Obligation. Universities have been a significant part of society for many years and have served various purposes over time. Brubacher (1982) suggests two philosophies underlying the functions of the modern university. The first is epistemological in nature, and states that the university's purpose is to answer the great questions of human existence. According to this philosophy, universities seek only knowledge and truth. Alternatively, the political philosophy of education states that universities not only seek knowledge, but also apply knowledge in order to solve the complex problems of society. The university educates the citizenry, and prepares students for an active life and social responsibility in the world. Wille (1997) relates the political philosophy of education to SHE stating that educational institutions are being challenged to take more responsibility for preparing graduates to deal with the environmental problems humanity faces:

> Colleges are being asked to prepare graduates with analytical and critical thinking skills, strong communication and technological skills, while at the same time preparing them for active participation in a rapidly changing environment with a commitment to maintaining the integrity of our global ecosystem (Wille, 1997, p. 331).

The idea that universities are morally obligated to teach and become models of environmental sustainability is echoed in all of the SHE declarations. Perhaps the best illustration of this is from the CRE-COPERNICUS Charter:

> Universities and equivalent institutions of higher education train the coming generations of citizens and have expertise in all fields of research, both in technology as well as in the natural, human and social sciences. It is consequently their duty to propagate environmental literacy and to promote the practice of environmental ethics in society (CRE-COPERNICUS, 1994).

Without a doubt, the wording in all SHE declarations is value laden. The language reflects the idea that the university has a special role in society and is morally bound to create change. This is common to all of the SHE declarations.

Public Outreach. The second theme common to all SHE declarations is the need for universities to engage in public outreach. World university participation rates have increased from 13 million students in 1960 to 65 million in 1991, however universities remain elitist in that the majority of earth's citizens do not attend college or university (UNESCO, 1998). Practitioners in the sustainability movement recognize the need for an environmentally literate citizenry in order to work toward a more environmentally sustainable future. As universities are considered answerable not only to their students, but also to the communities and regions in which they dwell, SHE declarations suggest that universities need to help in the education of the general population. The Swansea Declaration makes this explicit in calling for an increased awareness of sustainable development. Universities are encouraged to "utilize resources of the university to encourage a better understanding on the part of governments and the public at large of the inter-related physical, biological and social dangers facing the planet Earth, and to recognize the significant interdependence and international dimensions of sustainable development" (UNESCO, 1993). The need for public outreach points to a belief that universities have a responsibility to both its students, and the communities in which they reside.

Sustainable Physical Operations. An analysis of emerging themes in the international declarations show that sustainable physical operations is not a priority in the majority of SHE declarations. While the physical side of "greening" of campus is considered a key component to becoming more sustainable, it is not surprising that it is not featured in the declarations. Wright (2002b) found that institutional policies, rather than international declarations, focus more on physical operations. This is perhaps because of the recognition that physical operations are specific to the institution. An analysis of the implementation of the Halifax Declaration, for example, found that many of the specific guidelines given to signatory universities for action regarding physical operations were not implemented within each institution because the directives given in the Halifax Declaration Action Plan were seen as either irrelevant or inappropriate for the institution (Wright, 2002). While sustainable physical operations is not forgotten, the SHE declarations do not emphasize this theme.

Ecological Literacy. While there are many definitions of ecological literacy (Disinger & Roth, 1992; Golley, 1998; Hutchinson, 1998; Orr, 1992; Smith-Sebasto, 1997), the essence of the various definitions suggest that ecological literacy is the ability of an individual to comprehend the functions of the world with a realization that all human activities have consequences for the biosphere, and the translation of this understanding into action for the health of the earth This notion of ecological literacy is referred to frequently in the majority of SHE declarations. In some cases, the development of ecological literacy focuses solely on students within the university. The Talloires Declaration, for example, states that universities must "create programs to develop the capability of university faculty to teach environmental literacy to all undergraduate, graduate, and professional school students" (University Leaders For A Sustainable Future, 1990). The CRE-COPERNICUS charter expands the scope of ecological literacy initiatives to include faculty and staff at the university:

> Universities and equivalent institutions of higher education train the coming generations of citizens and have expertise in all fields of research, both in technology as well as in the natural, human and social sciences. It is consequently their duty to propagate environmental literacy and to promote the practice of environmental ethics in society (CRE-COPERNICUS, 1994).

As mentioned above, ecological literacy is also directly linked with the theme of public outreach. Many declarations, therefore, discuss the need for universities to aid the development of an environmentally literate public world citizenry. The Halifax Declaration offers a good example of this stating that universities must "enhance the capacity of the university to teach and practice sustainable development principles, to increase environmental literacy, and to enhance the understanding of environmental ethics among faculty, students, and the public at large" (Lester Pearson Institute For International Development, 1992).

Develop Interdisciplinary Curriculum. Developing interdisciplinary curriculum is closely related to the theme of ecological literacy. This theme is based upon the belief that if environmental literacy is going to occur, it will not happen by having students taking a mandatory course in environmental studies. Rather, students will become more ecologically literate if they see the connections between each subject they study and the environment. Principle 7 of the Talloires Declaration demonstrates this theme, directing deans and university environmental practitioners to develop interdisciplinary curricula for an environmentally sustainable future.

Encourage Sustainable Research. Many of the SHE declarations call for individual universities to encourage and promote faculty to conduct research that contributes to local, regional and global sustainability. Principle 4 of the Kyoto Declaration, for example, advocates universities undertaking research and action in sustainable development. While this theme is ideal in the minds of those already engaging in such research, it is riddled with difficulties in post secondary institutions that are built on academic freedom. To reward a faculty member who engages in research that contributes to sustainability may be considered by some academic favouritism. Clearly universities who choose to pursue this area will have to proceed with caution.

Partnerships. With the exception of the Swansea Declaration, SHE declarations are unanimous in their call for the development of partnerships between universities and individuals and institutions beyond the university in order to become more sustainable. This is well illustrated in the text of the Thessaloniki Declaration:

> In order to achieve sustainability, an enormous co-ordination and interrogation of efforts is required in a number of crucial sectors and rapid and radical change of behaviours and lifestyles, including changing consumption and production patterns. For this, appropriate education and public awareness should be recognized as one of the pillars of sustainability together with legislation, economy and technology (UNESCO, 1997).

Statements such as these reflect the emerging notion that the university cannot create societal change on its own. While universities are indeed agents of social change, these declarations recognize the need for cooperation at many levels including partnerships with government, non-governmental organizations, and industry. The types and degree of partnerships vary from declaration to declaration, but the majority emphasize global cooperation. The Halifax Declaration, for example, calls for increased interaction between the university community and all organizations concerned with sustainable development on local, regional and international levels. The Talloires Declaration also suggests that universities must work with national and international organizations to promote a worldwide effort toward a sustainable future.

The Lüneburg Declaration also highlights the need for developing partnerships on many levels. In the text of the Lüneburg Declaration, universities are asked to increase attention to international environmental sustainability and provide more opportunities for inter-cultural exchange in the learning environment; increase a focus on capacity development and intensified networking among institutions of education; and, promote stronger integration of training and research and closer interaction with stakeholders in the development process (Global Higher Education Partnership for Sustainability, 2001).

Inter-university Cooperation. Not only should universities engage in partnerships with the outside community, SHE declarations encourage universities to cooperate with each other. One example is the Swansea Declaration that states that signatory universities must "co-operate with one another and with all segments of society in the pursuit of practical and policy measures to achieve sustainable development and thereby safeguard the interests of future generations" (UNESCO, 1993). The CRE-COPERNICUS Charter also encourages cooperation in its call for sustainability networks. Further, the Action Plan in the Halifax Declaration calls for "establishing a network among universities in order to share information about the greening of the universities" (Lester Pearson Institute For International Development, 1992).

SHE declarations offer broad statements of intent for the role higher education will play in the future. Within these statements, we find that themes have emerged. The rise of themes common among the declarations suggest that there are certain priorities for sustainability in higher education. By identifying the themes, we are able to get a better understanding of how institutions have framed their commitment to sustainability and where they may go in the future.

HOW HAVE SHE DECLARATIONS EVOLVED?

The term evolution implies a gradual process in which something changes into a different, and usually more complex or better form. When we look at the evolution of SHE declarations, we must question if this has indeed happened. The identification of common themes gives an indication that there is some continuity between the SHE declarations. As Table 2.2 illustrates, however, there is no evolution of themes over time. In fact, the themes have remained fairly constant from the early 1990s to present. Yet this does not mean that the declarations have not evolved. While the evolution of the declarations cannot be found in the key SHE themes, there is evidence in the wording of the declarations that an evolution has taken place.

Many of the declarations state that they build on the work, and often the wording of previous SHE declarations. The Halifax Declaration, for example, specifically mentions the Talloires Declaration in its preamble, and the Swansea Declaration mentions both the Talloires and the Halifax Declarations. The most recent Lüneburg Declaration also discusses how it builds upon past SHE declarations. This implies both adaptation and evolution.

There is also evidence that SHE declarations are evolving in their understanding of the role a declaration can play in achieving sustainability. Those involved in the sustainability in higher education movement may have been naïve in the beginning to assume that signing a declaration also meant that the institution would implement it. Universities have been accused of attempting to "greenwash" their institutions by endorsing SHE declarations without taking any subsequent action. The signing becomes a public relations exercise to promote their university rather than an actual statement of intent towards becoming sustainable. SHE practitioners are realizing that monitoring implementation is essential to the success of a declaration (Walton, 2000; Wright, in press). CRE-COPERNICUS is currently assessing the potential for systematic monitoring of signatory universities to their Charter. The Lüneburg Declaration also makes specific reference to the implementation of the declarations:

> Furthermore, the EUA-COPERNICUS, the International Association of Universities (IAU), and the Association of University Leaders for a Sustainable Future (ULSF) commit to achieving the following targets within next five years: Promote expanded

endorsement and full implementation of the Talloires, Kyoto and COPERNICUS declarations (Global Higher Education Partnership for Sustainability, 2001).

Accountability systems are suggested to ensure that declarations are both meaningful and effective. This may have financial, political, and social ramifications for universities who sign declarations in the future, but will undoubtedly strengthen the impact these documents have on endorsing universities in the end. In terms of evolution, the Lüneburg Declaration represents a new model for SHE declarations. It clearly expresses the sentiment that it is no longer adequate for universities to create and sign statements of intent towards environmental sustainability. It asks for universal cooperation, and claims that environmental change will only occur when rhetoric is turned into reality.

CONCLUSION

This chapter has explored where SHE declarations have been in the past, where they are now, and provides glimpses for SHE declarations in the future. It has highlighted key themes that have emerged in SHE declarations since the early 1990s, including the ethical and moral responsibility of the university to contribute to local, regional and global sustainability; the need for public outreach and universities to become models of sustainability in their own communities; encouraging sustainable physical operations; fostering ecological literacy; the development of interdisciplinary curriculum; encouraging research related to sustainability; forging partnerships with government, non-governmental organizations and industry; and, cooperation amongst universities. The identification of these themes and patterns furthers the understanding of what universities believe are the key priorities to becoming sustainable institutions, and what paths universities believe they should take on the journey to sustainability. This provides a starting point for an exploration of the challenges to sustainability in higher education.

Further, the chapter discussed the evolution of SHE declarations from the 1990s to present. While it is anticipated that SHE declarations will continue to evolve, it is likely that the themes will remain the same. These themes represent a utopian vision of SHE and confirm research on indicators of sustainability in higher education. The themes are a constant reminder of what sustainability in higher education should look like, and a vision for SHE practitioners to keep in mind while struggling with the practicalities of adaptation and evolution in the quest to realize the dream of sustainability in higher education.

REFERENCES

Bowers, C. (1997). *Education For an Ecologically Sustainable Culture.* New York: State University of New York Press.

Brubacher, J. (1982). *On the Philosophy of Higher Education.* San Francisco: Jossey-Bass.

Clugston, R. (1999). Introduction. W. Leal Filho (ed.), *Sustainability and University Life: Environmental Education, Communication and Sustainability* (pp. 9-11). Berlin: Peter Lang.

CRE-COPERNICUS (1994). *CRE-COPERNICUS Declaration.*Geneva: CRE-COPERNICUS Secretariat.

CRE-COPERNICUS. (*CRE-COPERNICUS Homepage* [Web Page]. URL http://www.copernicus-campus.org [2002, July].

Disinger, J. &. R.C. Roth (1992). Environmental Literacy.
    Http://Www.Ed.Gov/Databases/ERIC_Digests/Ed351201.Html.
Global Higher Education Partnership for Sustainability. (2001). Lüneburg Declaration. *Proceedings of the Higher Education for Sustainability - Towards the World Summit on Sustainable Development (Rio+10) Conference.* Germany: Cre-COPERNICUS.
Golley, F. (1998). *A Primer for Environmental Literacy.* Connecticut: Yale University Press.
Hutchinson. (1998). *Growing Up Green: Education For Ecological Renewal.* New York: Teachers College Press.
International Association of Universities (1993). *The Kyoto Declaration.* Kyoto, Japan: International Association of Universities.
Lester Pearson Institute For International Development (1992). *Creating A Common Future: Proceedings of the Conference On University Action For Sustainable Development.* Halifax: Atlantic Nova Print.
Orr, D. (1992). *Ecological Literacy: Education and Transition to a Postmodern World.* Albany: State University of New York Press.
Smith-Sebasto, N.J. (1997). Education For Ecological Literacy. Patricia Thompson (Ed.), *Environmental Education For The 21st Century* (pp. 279-289). New York: Peter Lang.
UNESCO (1972). *The Stockholm Declaration.* Stockholm: UNESCO.
UNESCO (1993). *The Swansea Declaration.* Gland: UNESCO.
UNESCO (1998). Higher Education in the Twenty-first Century Vision and Action. *Proceedings of the World Congress on Higher Education.* Paris: UNESCO.
UNESCO-UNEP (1977). Mockba: UNESCO-UNEP Press.
University Leaders For A Sustainable Future (1990). *The Talloires Declaration.* Washington: ULSF.
University Leaders for a Sustainable Future (2002). *ULSF Homepage* [Web Page]. URL http://www.ulsf.org/ [2002, April 20].
Walton, J. (2000). Should Monitoring be Compulsory within Voluntary Environmental Agreements? *Sustainable Development, 8*(3), 146-154.
Wille, R. (1997). P. Thompson (Ed.), *Environmental Education for the 21st Century* (pp. 331-337). New York: Peter Lang.
Wright, T. (in press). Examining The Implementation of the Halifax Declaration. *Canadian Journal of Environmental Education.*
Wright, T.S.A. (2002a). *Environmental Sustainability, Policy, and the University.* Unpublished doctoral dissertation, Edmonton Alberta: University of Alberta.
Wright, T.S.A. (2002b). A Review of Definitions and Frameworks for Sustainability in Higher Education. *International Journal for Sustainability in Higher Education, 3*(3).

BIOGRAPHY

Dr. Tarah Wright is Assistant Professor and Director of Environmental Programmes for the Faculty of Science at Dalhousie University, Canada. Tarah's PhD, undertaken at the University of Alberta, examined the implementation of the Halifax Declaration in signatory universities and investigated the various challenges and barriers to sustainability initiatives in Canadian and international post secondary institutions. Since then she has continued her research in the area of sustainability policy implementation analysis, collaborative development of institutional sustainability policies, and has worked on a participatory action research project focused on the development of indicators of sustainability for Canadian universities.

CHAPTER 3

SUSTAINABILITY AS EMERGENCE: THE NEED FOR ENGAGED DISCOURSE

Richard Bawden

INTRODUCTION

Every generation seemingly feels a need to establish what it is that it might contribute to civilization writ large: To state in bold and imaginative terms that which it hopes will be its heritage. This one is no exception. "Let ours be a time remembered for the awakening of a new reverence for life, the firm resolve to achieve sustainability, the quickening of the struggle for justice and peace, and the joyful celebration of life" (Earth Charter Initiative, 2000). Little to argue with in that rhetoric, save perhaps to note that each of these noble aims are not without some contestation of specified ends, nor downright disagreements about means by which such ends might be - or more poignantly, ought to be - met.

The matter of the "firm resolve to achieve sustainability" is a fine case in point – a classic example, it might be argued, of post-modern ambiguity of both means and ends. The very concept of *sustainability* is replete with imprecision, while the notion of having a *firm resolve* to achieve anything gives little evidence of how it might actually be achieved in practice even if we could agree on what it was we were hoping to achieve! Davison (2001) frames the dilemma well when he asks: "What are we to sustain above all else? Why? And how may we do so?" Yet as that writer, among others posit, it is this very same ambiguity that, somewhat paradoxically, provides the focus for discourse about a variety of issues within the contemporary problematique, while undeniably also providing fuel for those who would procrastinate in their resolve to seek sustainability!

"The ideal of sustainability gives rise to an agenda of good questions, practical questions that bear directly on our forms of life, drawing out and giving practical substance to our disquiets and to our hopes....[t]hese questions are valuable to us because they command our attention in an age of ecological crisis while simultaneously defying resolution and closure: they demand that we hold open for questioning our assumptions about what a resolution of this crises might involve." (Davison, 2001, p. 213).

And there is certainly no shortage of evidence to support the claim here that if not a crisis, then at least an ecological dilemma (Hajer, 1996) does indeed characterize this age of ours - where the word *ecological* is taken to embrace both

Peter Blaze Corcoran & Arjen E.J. Wals (Eds.), Higher Education and the Challenge of Sustainability: Problematics, Promise and Practice, 21-32.

the bio-physical and socio-cultural aspects of the global environment in which we human beings are firmly embedded.

It is the matter of *agenda of good questions* that provides the focus, in all of this, for the academy, for questioning is (or at least ought to be) at the very heart of its enterprise. Few formal institutions furthermore, can claim to have such a record of sustainability as the academy: There must be lessons to be learned from such persistence that would add considerably to further profound understandings of the concept of, and actions to achieve, sustainability in its broadest socio-ecological context. Ironically and tragically however, it is at this very same moment of extreme relevance that universities are showing their own vulnerability and possible irrelevance – suffering their own crisis of identity and purpose.

THE ACADEMY, SCIENCE, AND SOCIETY

Over recent years, calls for the academy to become "...a more vigorous partner in the search for our most pressing special, civic, economic problems..." (Boyer, 1996) – for academics to become more directly and openly *engaged* with the issues of the day - have become increasingly strident. Wengers (1998) captures the essence of such intended engagement well when he describes it as "active involvement in mutual processes of negotiation of meaning". And the matter of 'sustainability', invoked in reaction to widespread evidence of environmental deterioration at all levels of the biosphere, is clearly recognized as being as pressing as any, both within the academy itself, and beyond, in civil society at large, and thus worthy of engagement. At least in most parts of the Western capitalist world, "[a] well-founded ecological consciousness has asserted itself down into the last recesses of everyday life" (Berking, 1996). Overtures for a fuller engagement are an indication that, even within the University itself, there is a growing feeling that it is failing to fulfill its part of the social contract to help citizens to learn how to live lives in manners that can be sustained across future generations. Yet while the academy can justifiably claim considerable credit for helping to raise such public consciousness of issues to do with sustainability, through the wide promulgation of the results of its scientific inquiries, it is seemingly reluctant to engage with those in civil society in genuinely collaborative efforts to do something about the situation. Institutions of higher learning have become disconnected from the context for such learning; the 'issues of the day' with which those in civil society have to grapple, are no longer the key focus of the attention of either professors or their students. The reasons for this apparent disconnect or disengagement, are complex.

In the first place, "[i]t is no longer clear" as Readings (1996) posited "what the place of the University is within society nor what the exact nature of that society is". In an era of globalization, the nexus that had long persisted between the University and the nation-state, no longer holds. "The University thus shifts from being an ideological apparatus of the nation-state to being a relatively independent bureaucratic system": A system in which, incidentally, the process of scientific research and the associated scholarship of discovery, are increasingly privileged, albeit in a very restricted manner that is dictated both by paradigm and by political economy. To this ambivalence of social purpose must be added some 'guilt by association': A growing insecurity among some, born of the recognition that globalization is not confined to the transnational flow of 'goods' but also to the planetary flow of 'bads' to which science and technology, more or less unwittingly, has contributed. In the process of techno-scientific modernization, so many different kinds of hazards and threats have been released, that we can be said to living in a 'risk society' on a global scale (Beck, 1992). And as science, scientific rationality, and scientific research, are all central aspects of the modernization paradigm that prevails within academia, universities must accept that they are now as much 'part of the problem' as they were once an almost unrivalled 'source of the solution'.

We can certainly marvel at the myriad ways by which science-based technologies have transformed the way we live; have produced the machines that have relieved us from the drudgery of 'mechanical work', the medicines that have saved us such pain and suffering, the agriculture that now can feed so many billions of us, and the instruments of communication through which we can converse and share knowledge. But we need to concern ourselves much more than we have so far, with the unintended consequences of the application of such technologies; the chemical pollutions, the soil degradations, the water depletions, the species extinctions and biodiversity reductions, the global warming, and so on. There are moral reasons for such concerns, and indeed they extend beyond the mere consequences of our technological exuberances into the realms of rights and privileges and constraints – and the very concept of what it is that we *should be* seeking to make sustainable.

Paradoxically while technoscience, as a powerful mode of knowing, has identified these risks for us, it is epistemologically ill-equipped to help us deal with issues that are the province of philosophy. This raises the imperative for those within the academy to think about and to do things differently: to be concerned "no longer exclusively with making nature useful, or with releasing mankind from traditional constraints, but also and essentially with problems of techno-development itself. Modernization is itself becoming reflexive; it is becoming its own theme" (Beck, 1992). A fundamental question arises as a key theme to such critical review: Can the use of technology ever be truly compatible with the pursuit of the good life? Albert Borgmann, who is among those philosophers of technology who have taken this matter extremely seriously, argues that "the peril of technology lies not in this or that manifestation, but in the pervasiveness and consistency of its pattern" (Borgmann 1984). He makes the strong distinction between technologies as 'focal things', which engage mind and body, are embedded within a complicated web of relations, and demand a focal practice, with 'devices', which to the contrary, are isolated entities which allow (encourage) mind, body and world to be dissociated from each other.

The formulation of Borgmann's 'device paradigm' is based on the argument that the steady replacement of the former by the latter, essentially in parallel with the 'progress' of modernization, has led to cultural and psychological impoverishment. This leads to the need for technology, and indeed of science itself, to be critically reflective. However, by definition, the objectivist technoscientific paradigm is ill-equipped to critically explore its own nature in other than purely objectivist terms: And that leaves a lot to be desired. "The way we make scientific facts and build them into coherent theories and descriptions sets limits to the kinds of things we can come to understand about nature" argues Hubbard (1990), and that, it can be argued, includes the nature of knowing about knowing and knowledge (epistemology). Scientists usually do not acknowledge these limits, nor do most other people. "And the overestimation of science as a way to know, hence of the extent of the knowledge we can gain through science has led us to undervalue other kinds of knowledge" (Hubbard, 1990). It has also led to peculiarly distorted views of reality where knowledge has become divorced from values – technical decisions separated from normative judgments – and the essence of 'the whole' has become lost through an exclusive focus on the 'fragmented parts'. The forest is invisible to those who see only separate trees!

David Orr (1992) has further observed that "modern science has fundamentally misconceived the world by fragmenting reality, separating observer from observed, portraying the world as a mechanism, and dismissing non-objective factors all in the service of the domination of nature". The result, he contends, has been a radical miscarriage of human purposes where the domination of nature has led to "the domination of other persons". He cites C.S. Lewis (1947) in support of this contention: "At the moment then of man's victory over nature, we find the whole human race subjected to some individual men, and individuals subject to that in themselves which is purely 'natural' – to their irrational impulses."

This emphasis on 'man', while probably primarily reflecting the (nevertheless inexcusable) conventions at the time of writing, has a special poignancy. For here is yet another great divide which, in this instance, is obscenely asymmetric. It is not just a lack of integration between differentiated genders, but a very distorted emphasis on one sex over the other. "Knowledge", as Hubbard (1990) reminds us, "has become gendered", with science, that seemingly most legitimate form of knowing, being colonised by objectivity, which, in turn, is identified as masculine. As Mary Belenky and her colleagues argue, "conceptions of knowledge and truth that are accepted and articulated today have been shaped throughout history by the male dominated culture. ...[d]rawing on their own perspectives and visions, men have constructed the prevailing theories, written history and set values that have become the guiding principles for men and women alike" (Belenky et al., 1986). Yet this identification is of course not inherent in nature, nor the nature of science, but results from "the ways scientific works, facts, and theories are constructed and from the ways we construct sex and gender" (Hubbard, 1990).

All of these matters that are of profound importance to the matter of sustainability for judgments about what we *should do* next in confronting the perils of a risk society, take us way beyond the realm of decisions about what we *could do* next, in the relatively simple circumstances demarcated by reductionist technoscience. As already hinted, 'Oughts' and 'Shoulds' are the focus not of science but of moral philosophy, and in a 'reflexive modernity' we must learn how to deal not just with the ethics of consequences as well as the efficiencies of production, but also with the rights and responsibilities of all living beings: A focus that is about as complex as the human mind can envision, yet one replete with the 'glorious distinctions of difference' that can lead, when nurtured appropriately, to systemic emergence.

Our focus therefore needs to shift from a 'techocentricity' with 'egocentric' overtones, to a holocentric systemic one. We need to seek not just the self-gratification of the good life, but what Prozesky (1999) has referred to as "inclusive well-being", and in order to do that effectively, he argues, we need nothing less than "an ethical renaissance". Or indeed to an entire paradigmatic renaissance, as we learn to embrace complexity and learn to explore it systemically. This will force us to move beyond the instrumentality of technoscientific rationality and embrace ways of knowing that free us from what Yankelovich (1991) refers to as the "tyranny of objectivism" that insists that as there is only one form of genuine knowledge: "the claims made by religious truths, the insights of art and literature, the truths of history, the judgment of the public, and the truths of psychological insight and intuitive understanding get lumped together with the claims of astrology people, people who see flying saucers with extraterrestrial beings on them, those convinced that they have knowledge of previous incarnations, and all are held suspect as knowledge claims" (p. 221).

Just as we are beginning to appreciate the fact that the global problematique must concern our ethical and aesthetic values and moral judgments as much as they concern scientific reason, we are beginning to recognize and rue the epistemological fact that "moral judgement has been eliminated from our concepts of rationality as far as they are actually built into existent scientific and systems paradigms" (Ulrich, 1993).

The reference to systems paradigms here is of considerable significance, for it is the 'systems logic' of the 'whole being different (if not greater) than the sum of its parts', that gives power to the very notion of 'emergence'. To the systemist, unpredictable and novel properties emerge whenever different sub-systems are allowed to mutually associate, both within systems, and between different 'levels' within the nested hierarchies in which they are presumed to exist. And this is as relevant to 'learning systems', that rely for their existence on a range of different differences inherent in collective discourse, as it is for any other system in 'nature' (Bawden, 2000). If you want emergence, then you have to allow 'differences' like normative discourse with empirical discourse, to 'interact', and conversely, if you concentrate attention solely on the fragmented aspects of discourse, emergence will be forever denied. As it happens, Ulrich is one among few who are taking these systemic distinctions seriously, with his quest for the development and promulgation of an inclusive discourse that engages experts with the citizenry in a context in which he promotes the use of systems heuristics, "as if people mattered" (Ulrich, 1998).

MORAL DISCOURSE AND PUBLIC JUDGMENT

There are a number of reasons why moral dimensions should have become disassociated from our conventional rationalities and thence from the discourse around such matters as 'sustainability'. Not the least of these is the issue of understanding of the nature of morals and morality in the first place. "Is it [morality] a matter (as modern philosophers have assumed) of a certain kind of *judgement* – maybe even of certain 'evaluative' *words*? Is it a matter, as Kant argued, of action and practical reason? Is it a matter of emotion, sympathy, motive, as Hume supposed? Or is it, as Aristotle suggested, a matter of character and moral education"? (Scruton, 1994). As Singer (1994) observes, the distinctions between the positions held by Kant and Hume take us to the fundamental question of whether ethics is objective or subjective. The debate becomes between those who hold the primacy of reason as the source of ethics, and those who, in contrast, promote intuition. "Different terms have been used to frame this question, but behind it always lies the division between, on the one hand, those that hold that there is somehow a true, correct or best-justified answer to the questions 'What ought I to do?' no matter who asks the question, and, on the other hand, those who hold that when different individuals or different societies disagree on ethical issues then there is no standard by which one could possibly judge one answer to be better than another" (Singer, 1994, p. 7).

A further impediment to the re-inclusion of moral dimensions into contemporary rationalities is sheer lack of practice at engagement with issues from an ethical perspective. Busch (2000) has presented the notion that we have spent several centuries now, secure in the abdication of our individual moral responsibilities to the care of one Leviathan or another. Whether it has been acceptance of Bacon's scientism, or Hobbe's statism, or Smith's economism, we have been seemingly content to place our trust respectively in scientists, authoritarian monarchs, or the market (or God for that matter) to tell us what is good or bad. Confronted now by the need to recapture our ethical competencies, we need to re-engage with what E.P. Thompson termed the 'moral economy' (Thompson, 1971) - the structure of rights, privileges and constraints that endows any particular social order with its identity and culture – such that we can use it as the ethical framework for our choices.

Paul Thompson, who has written extensively and with enviable clarity about ethical perspectives on sustainability within an agricultural context, has made an invaluable contribution to the role that moral discourse plays in the evolution of a moral economy. Building on ideas developed by E.P. Thompson and James Scott, he argues that the constant reproduction, testing and revision of rights, privileges and constraints (processes often accompanied by affronts of conflict) "constitute the practical moral discourse that aims to reform moral economy" (Thompson, 1998). Moral discourse, he thus envisages as a level of 'normativity' that lies between the structures of those rights on the one hand, and "political and ethical theory or systematic idealizations of practical discourse" on the other. Moral discourse involves both explicit and tacit, formal and informal, linguistic and non-linguistic modes of communication between individuals within groups of people, as networks of communication, engaged in the reproduction and challenge of the rules that they themselves agree to live by.

Reinforcing the significance of relationships to the moral enterprise, Busch (2000) argues that "moral responsibility lies not in individuals or society, but in the social relations that we create both through our own volition and through choices that society gives us". In this manner it is from within (democratic) social relationships that the 'moral economy' emerges.

What such networks represent is the opportunity to present the key problematic issues of the day "in a manner in which they may be addressed by informed citizens. They can help citizens become informed through participation in the networks. They can ensure that all citizens have an opportunity (without being forced) to engage in deliberation, discussion, and debates about the issues affecting their lives" (Busch, 2000). Such networks are closely akin to the sorts of 'communities of practice' envisioned by Wengers (1998) where: "[o]ur engagement in practice may have patterns, but it is the production of such patterns anew that gives rise to the experience of meaning". Meaning emerges through our engagement in communities of practice. As Scruton (1994) sees it, communities are not formed through agreements between rational individuals rather it is rational individuals who are formed through communities. It is through moral discourse *as* such communities that we learn together to bring forth the world that we believe that we *should* bring forth, through communication with each other.

A further way to view such communities of practice is to link them specifically with the process of collective critical learning. Such learning involves groups of individuals learning with and from each other in a manner where critique is applied not just to the situation at hand but also to the processes of learning themselves, and of the epistemic foundations of such processes. Such learning groups can be presented as critical learning systems (Bawden, 2000) that of have all the attributes of any self-organizing system including the ability to co-adapt with the environments in which they must operate.

This sense of 'meaning' emerging through social communication and collective learning, is pervasive. Even critics of modernity and the legacy of the Enlightenment, such as Alasdair MacIntyre hold to the belief that "what matters at this stage is the construction of local forms of community within which civility and the intellectual and moral life can be sustained" (MacIntyre, 1981). Bernstein (1983) is perhaps even more even more passionate in stating his support for communal discourse in the face of the very real perils of modernity: "... at a time when the threat of total annihilation no longer seems to be an abstract possibility but the most imminent and real potentiality, it becomes all the more imperative to try again and again to foster and nurture those forms of communal life in which dialogue, conversation, *phronēsis*, practical discourse and judgment are concretely embodied in our everyday practices" (p. 229).

There are strong resonances here with the nature of so-called "communicative ethics" where ethics " is not simply a matter of individual consciousness but rather a concern indissolubly connected with language and communication" (Dallmayr, 1990). Inspired in particular by the work of Karl-Otto Apel and Jurgens Habermas, communicative ethics is a cognitivist approach to ethics that is grounded in a belief of the importance of rational argumentation as the basis for moral judgement. As an approach, it lies in stark contrast to other philosophical positions "of the metaphysical type and intuitionist value ethics on the one hand, and noncognitivist theories like emotivism and decisionism on the other" (Habermas, 1990). Communicative ethics draws many critics including those who would argue that moral discourse is more than reasoned argumentation.

SUSTAINABILITY AND THE SCHOLARSHIP OF ENGAGEMENT

These two matters, (i) of meaning clarification of morality, and (ii) of the practice of moral discourse, indicate at least two vital roles that the academy can assume with respect to re-engaging with civil society in the context of 'sustainability' which can be construed as one of Boyer's *pressing problems* of the day. In this regard it might prove useful to explore aspects of what Boyer (1996) referred to as the Scholarship of Engagement in reference to his proposition that the (American) academy must become a much more rigorous partner in the search for answers to such problems.

In the first place, in what I might refer to as first order of engagement, there is the need to find ways of reintegrating the ethical with the scientific into paradigms or systems of inquiry that permit the expression of a true synergy of ways of knowing and appreciation: A conceptual engagement with a practical issue, as it were. Charles Muscatine (1990) is among those who see this as an imperative, at least as far as the sustainability of the academy itself is concerned: "Either the university of the future will take hold of the connection between knowledge and human values, or it will sink quietly and indiscriminately into the non-committal moral stupor of the rest of the knowledge industry." Knowing becomes valuing becomes knowing!

Secondly, there is a need to develop strategies for collective inquiry that embrace both instrumental and practical rationalities into a *praxis* of collaboration between 'experts' and 'lay people' that allows the development of a democratic discourse appropriate to addressing the question of what it is that *ought to be done next* in the quest for 'sustainability'. From this perspective, what constitutes 'sustainability' is *emergent* from a contextual discourse that is ethically defensible in terms of its constitution and its processes – the 'means' as it were – while leading to 'ends' that reflect moral judgment as well as scientific rationality. Here, I would argue, we are talking of a second order of engagement - a practical and 'fulsome' engagement with the task itself.

The latter in particular, has the essence of one of Borgmann's 'focal things' that demands a 'focal practice' reflecting strong embeddedness and practical reason, and Davison (2001) expands on this theme with his insistence that "the search for moral orientation in the technological world is nothing less than a practical craft of skilled judgment about, and experimentation with, the ways technology constitutes our self-defining and world-defining relationships".

Both of these orders of engagement demand critical scholarship of praxis – critical application of theory, concept, and principle to practice itself. In the former, the discourse, while informed by practical action, is about inquiry into the nature of action (hands off engagement); in the latter, the discourse *is* the action that informs both the nature of that action and the actual action to be taken (hands on engagement). 'Emergence' will be most powerful in the latter case.

It is not enough to argue that the "essential purpose of the university is not to carry out research, nor even, in the conventional understanding of the term, to teach, but to furnish a critical commentary upon the assumptions, beliefs, values, knowledge, and technologies that inform and support the social order" (Downey, 1983). This is not engagement in either of the senses developed above. There is no 'on the ground mix it with them' discourse here, nor any hint of concern for 'moral economy'. There is little sense of democratic participation, little appreciation of difference, little respect for 'lay' knowledge or ways of knowing, and very little, if any, humility. There is accordingly, little chance that any shared resolution will be reached about what might be the 'best thing to do next' in the name of 'sustainability'.

Not that the alternative of communal discourse is easy. Bryan Wynne is among those who have written tellingly about the difficulties associated with the expert-lay knowledge divide, and how it might best be treated. There is for instance the very matter of reactions to the legitimacy of 'lay' knowledge. "Although prevailing treatments do recognize reflexivity among lay people, this is inadequately restricted to the intimate and interpersonal. Thus alternative, more culturally rooted and legitimate forms of collective public knowledge – and of corresponding public order – which could arise from the informal non-expert public domain are inadvertently but still systematically suppressed" (Wynne, 1996, p. 46). This again reflects a lack of 'engagement' of 'experts' with 'lay-persons' in the systemic sense that I have been trying to establish as the basis for emergence.

CONCLUSION

The matter of 'sustainability' is arguably the most pressing issue of our times. As 'sustainabilism', it provides the ethos, and thus the ethical context, for so much contemporary debate about what constitutes 'better' – for society, for the earth, for both together. Little by little it is replacing the blatant productivism that has preceded it – at least in the rhetoric. Yet can anything ever be designed to be sustainable, and even if it could, would it necessarily be best for the common good? Can we sensibly ever address, and/or act towards, a sustainable state of inclusive well-being? Should the real center for our commitment as human beings be on the sustainability of our concerns for 'inclusive betterment', and for the learning processes that are fundamental to such a commitment? And what role the academy, under such circumstances?

The only way we are going to find out, I have argued, is to get out there and act within its context: To engage in a deliberative democratic discourse that is characterized by practical and emancipatory rationalities that allow a synergy to develop between 'expert' and 'lay' knowledge and ways of knowing, between scientific reasoning and moral argumentation, between the empirical and the normative, between the academy and civil society.

This is - or at least *ought to be* - the 'new' social contract for universities.

There are those who argue for the academy to wholeheartedly embrace the scholarship and praxis of engagement – for universities to become *engaged institutions* (cf. Simpson, 2000, for instance). Indeed a seven-part test has even been devised to detect the 'truly engaged university' (Kellogg Commission, 1999) where engagement is interpreted as a slightly transmuted form of the process of extension/outreach that has long constituted one of the three fundamental missions part of the Land-Grant universities in the United States of America. Yet as Fear et al. (2001) argue, that is to miss the nature of the essential collaborative discourse at what they term the engagement interface – that 'space' where academicians and citizens come together with mutual respect for each other's ways of knowing and doing, and beliefs and values, and feelings and appreciations, in the quest for 'communicative action'. This, they submit, calls indeed for a new scholarship – a new mode of scholarly expression that embraces far more than 'outreaching' from an academy still constrained by its cognitive foundations in technoscientific or instrumental rationality, and its structural foundations in discipline- distinguished academic 'silos'. This is a scholarship of and by moral people – not abstract institutions - and it is from a broad, civil concern for such scholarship and the practical and emancipatory discourse that it promotes, that notions of what it is that we should attempt to sustain are most likely emerge.

All of this notwithstanding, it would be naïve of us to believe that there is an easy 'row to hoe' here. As Dresner (2002) has recently reminded us, "we can be sure that any attempt to bring about sustainability will meet enormous resistance from many people and vested interests". And that includes resistance even to the very idea of encouraging (or in some circumstances, even allowing) learning systems to engage with sustainabilism as a topic of discourse, with the risk that new insights-for-action that might disturb the ambitions of the powerful, could indeed emerge.

But then as Dresner also concludes, there is little choice about the matter, for "the alternative to the pursuit of sustainability is to continue along the present path of unsustainability, leading to disaster" (Dresner, 2002).

REFERENCES

Bawden, R.J. (2000). Valuing the Episteme in the Search for Betterment: The Nature and Role of Critical Learning Systems. *Cybernetics and Human Knowing (7)*, 5-25.

Beck, U. (1992). *Risk Society: Towards a New Society*. London: Sage.

Belenky, M.F., Clinchy, B.V., Goldberger, N.R, and Tarule, J.M. (1986). *Women's Ways of Knowing: The development of self, voice and mind*. New York: Basic Books.

Berking, H. (1996). *Solidary Individualism: The moral impact of cultural moderniation in late modernity*. Paul Knowlton (tr). In: S. Lash, B. Szernzynski, and B. Wynne (Eds.) "Risk, Environment, and Modernity: Towards a new ecology". London: Sage Publications.

Bernstein, R.J. (1983). *Beyond Objectivism and relativism: Science, hermeneutics and praxis*. Philadelphia: University of Pennsylvania Press.

Borgmann, A. (1984). *Technology and the Character of Contemporary Life: A philosophical inquiry*. Chicago: University of Chicago Press.

Boyer, E.L. (1996). The Scholarship of Engagement. *Journal of Public Service and Outreach (1)*, 11-20.

Busch, L. (2000). *The Eclipse of Morality*. New York: Aldine de Gruyter.
Dallmayr, F. (1990). Introduction. In: S. Benhabib and F. Dallmayr (Eds.), *The Communicative Ethics Controversy*. Cambridge Mass: MIT Press.
Davison, A. (2001). *Technology and the Contested Meaning of Sustainability*. New York: State University of New York Press.
Downey, J. (1983). The University as Court Jester (May Issue), *University Affairs*. Montreal.
Dresner, S. (2002). *The Principles of Sustainability*. London: Earthscan Publications Limited.
Earth Charter Initiative (2000). http://www.earthcharter.org.draft/charter.htm.
Fear, F., Rosaen, C., Foster-Fishman, P.G., and Bawden, R. J. (2001). Outreach as Scholarly Expression: A Faculty Perspective. *Journal of Higher Education Outreach and Engagement*, *(6)*, 21-34.
Habermas, J (1990). *Discourse Ethics: Notes on a Program of Philosophical Justification*. In: S. Benhabib and F. Dallmayr (Eds.). *The Communicative Ethics Controversy*. Cambridge, Mass: MIT Press.
Hajer, M. (1996). *Ecological Modernisation as Cultural Politics*. In: S. Lash, B. Szernzynski, and B. Wynne (Eds.) "Risk, Environment, and Modernity: Towards a new ecology". London: Sage Publications.
Hubbard, R. (1990). *The Politics of Women's Biology*. New Jersey: Rutgers University Press.
Kellogg Commission of the Future of State and Land-Grant Universities (1999). *Returning to Our Roots: The engaged institution*. Washington DC: National Association of Universities and Land Grant Colleges.
Lewis, C.S. (1947). *The Abolition of Man*. New York: Macmillan.
MacIntyre, A. (1981). *After Virtue: A study in moral theory*. Notre Dame, Indiana: University of Notre Dame Press.
Muscatine, C. (1990). Cited in Page Smith: *Killing the Spirit: Higher education in America*. New York: Viking Press.
Orr, D.W. (1992). *Ecological Literacy: Education and the transition to a postmodern world*. New York: State University of New York Press.
Prozesky, M. (1999). *The Quest for Inclusive Well-being: Ground work for an ethical renaissance*. Inaugural lecture. Pietermaritzberg, South Africa: University of Natal.
Readings, B. (1996). *The University in Ruins*. Cambridge, Mass:Harvard University Press.
Scruton, R. (1994). *Modern Philosophy*. London: Sinclair-Stevenson.
Singer, P. (1994). Introduction. In: Singer, P. (Ed) *Ethics*. Oxford: Oxford University Press.
Simpson, R.D. (2000). Towards A Scholarship of Outreach and Engagement in Higher Education *Journal Education, Outreach and Engagement (*6), 7-21.
Thompson, E.P. (1971). *The Moral Economy of the English Crowd in the Eighteenth Century*. Reprinted in: Thompson, E.P. 1993 *"Customs in Common: Studies in Traditional and Popular Culture*. New York: The New Press.
Thompson, P.B (1998). *Agricultural Ethics: Research, teaching and public policy*. Ames: Iowa State University Press.
Ulrich, W. (1993). "Some Difficulties of Ecological Thinking Considered from a Critical Systems Perspective: A plea for critical holism". *Systems Practice* (6), 584-609.
Ulrich, W. (1998). *Systems Thinking as if People Mattered: Critical systems thinking for citizens and managers*. Lincoln School of Management, Working Paper No 23, University of Lincolnshire and Humberside.
Wengers, E. (1998). *Communities of Practice: Learning, meaning and identity*. Cambridge: Cambridge University Press.
Wynne, B. (1996). *May the Sheep Safely Graze? A reflexive view of the expert-lay knowledge divide*. In: S. Lash, B. Szernzynski, and B. Wynne (Eds.) "Risk, Environment, and Modernity: Towards a new ecology". London: Sage Publications.
Yankelovich, D. (1991). *Coming to Public Judgment: Making democracy work in a complex world*. Syracuse, NY: Syracuse University Press.

BIOGRAPHY

Richard Bawden has been a Visiting Distinguished University Professor at Michigan State University since his retirement, in 1999, from the University of Western Sydney Hawkesbury, where, for more than 15 years, he had been the Dean of Agriculture and Rural Development. It was during his tenure as Dean, that the Hawkesbury agriculture faculty first established and subsequently nurtured the "systemic learning turn" that identified that institution internationally, as a centre committed to an approach to sustainable development that had both 'experiential learning' and 'systems theories and practices' as its foundations. As the author of many articles and book chapters on "The Hawkesbury Experiences", as well as a consultant to a number of international development agencies, Profesor Bawden has long promulgated the need to integrate ecological and ethical concerns and responsibilities with economic dimensions into practical rural development strategies.

CHAPTER 4

CRITICAL REALISM: A PHILOSOPHICAL FRAMEWORK FOR HIGHER EDUCATION FOR SUSTAINABILITY

John Huckle

... environmental education must weave an analysis of power, politics and the state into an ecology's sense of sustainability, survival and the environment. This kind of interdisciplinary effort could develop a deeply contextual understanding of nature and society as holistic cluster of interdependent relations (Luke, 2001, p. 200).

INTRODUCTION

In addressing an appropriate philosophy of knowledge and education to enable the kind of interdisciplinary effort that Timothy Luke recommends, this chapter begins by considering a textbook series that is widely used in the UK.

In the year 2000 Routledge listed forty nine published and forthcoming titles in the series *Introductions to the Environment*. Fifteen of these were environmental science texts in such areas as environmental biology and soil systems; eight were environment and society texts in such areas as environment and economics or environment and planning; the remaining twenty six were environmental topics texts, including one on environmental sustainability. At that time the list did not include a text on the environment and education.

In their preface to the environment and society titles the editors, David Pepper and Phil O'Keefe (2000), write of the mushrooming of research and scholarship on the relationships between the social sciences and humanities on the one hand and the processes of environmental change on the other. This has been reflected in the proliferation of associated courses at undergraduate level, while at the same time changes in higher education mean that an increasing number of such courses are being taught and studied within modular frameworks offering maximum choice or flexibility. Finding more traditional textbooks inadequate the authors and editors had responded to these new challenges by writing texts based on their own course materials.

While seeking 'the right mix of flexibility, depth and breadth' and 'maximum accessibility to readers from a variety of backgrounds' as it sketches 'basic concepts and map(s) out the ground in a stimulating way', the series as a whole, like other attempts to classify and present environmental knowledge, raises important issues of

Peter Blaze Corcoran & Arjen E.J. Wals (Eds.), Higher Education and the Challenge of Sustainability: Problematics, Promise and Practice, 33-47.
© 2004 *Kluwer Academic Publishers. Printed in the Netherlands.*

interdisciplinarity in higher education for sustainability (HEfS). Are there texts (aspects of environmental science, environment and society, environmental topics) that are more important than others? Are there key concepts, ideas, and values that link the texts together and provide for a common focus on sustainability? How do the texts deal with issues of philosophy, ethics and politics? Do they accommodate local, non-academic, environmental knowledge? Do they tell a grand story or narrative of the transition to sustainability or many small stories? Above all, do they empower students as ecological citizens who are capable of playing an informed and active role in this transition? Do they embody a critical pedagogy that fosters such citizenship? Do they reflect the kind of interdisciplinary effort to which Luke refers?

In addressing an appropriate philosophy to underpin such interdisciplinarity, this chapter argues that the key requirement of institutions and courses that seek to educate for sustainability is a philosophy of knowledge that integrates the natural and social sciences and the humanities, accommodates local knowledge, supports critical pedagogy, and continues to regard education as a form of enlightenment linked to a vision of more sustainable futures. It suggests that critical realism provides such a philosophy. This can resolve the tensions between mainstream, Marxist and postmodern environmentalisms in progressive ways, and underpin an HEfS based on a constructive postmodern cosmology, science and grand narrative (Gare, 1995).

DIVISIONS OF ACADEMIC KNOWLEDGE

Peter Dickens (1996) reminds us that the crisis of sustainability is both a crisis of the ways in which modern capitalist societies combine with nature and a crisis of understanding whereby the citizens of those societies fail to understand their relations with nature. The rise of modernity and new forms of industrial production separated people from nature with new kinds of knowledge contributing to this alienation. People were separated from the land, from the products of their labour, from one another, and from their own inner nature, by new social, technical and spatial divisions of labour that also separated them from knowledge that enabled them to make sense of the world. New forms of generalised and abstract knowledge, that could be applied to the control and management of nature and society, gained power and displaced the local knowledge that people used to monitor, understand and control the consequences of their actions.

The modern university became an institution that reflected modern reductionism and dualism. Academic divisions of labour separated knowledge into discrete compartments with separate natural and social sciences largely talking past one another. Students failed to understand how knowledge connects, how processes in the social world might combine with those in the biophysical world to produce sustainable development, and how people's local knowledge can combine with academic knowledge to foster such development.

Dickens (1996) argues that HEfS requires a unified science that can explain how social processes as understood by the social sciences combine with ecological and biophysical processes as understood by the physical and natural sciences. Critical realism (Archer et al., 1998; Collier, 1994) provides an appropriate philosophical foundation for such a science which is socialist in that it predicts the need for greater self management and new kinds of ecological (Barry, 1996; Dryzek, 1997) or cosmopolitan democracy (Held, 1995) if development is to realise the social, cultural and personal dimensions of sustainability alongside the ecological and economic.

INTERDISCIPLINARITY

While the rise of the new social movements and the impact of radical politics on universities in the late 1960s and 1970s led to interdisciplinary courses, including those in environmental and development studies, Jones and Merritt (1999) draw on reports from the Higher Education Funding Council to suggest a dearth of interdisciplinarity in contemporary British environmental higher education. Like *Introductions to the Environment* most courses are multidisciplinary rather than interdisciplinary, juxtaposing knowledge in often unrelated parts rather than realising a genuine integration of disciplines. Interdisciplinarity challenges academics to reconcile ideas about the nature of reality, how that reality can be known, and what procedures should guide enquiry (ontology, epistemology, and methodology) and we will see that critical realism offers a philosophical framework for accommodating different knowledge claims. It is particularly relevant for HEfS which focuses on an ambiguous and contested concept (Bourke & Meppem, 2000; Sachs, 1997) and where knowledge (in such areas as climate change or the impact of genetically modified organisms) is often uncertain and provisional in nature.

Geography is a particularly significant for HEfS since it has long concerned itself with the relations between the biophysical and social worlds. Advances in the subject that draw on ideas reviewed in this chapter now allow an intradisciplinary approach that may be superior to some interdisciplinary and multidisciplinary approaches (Huckle & Martin, 2001). Readers can assess the relevance of UK university geography by visiting the Geography Discipline Network (http://www.chelt.ac.uk/gdn/) and the Learning and the Teaching Support Network for Geography, Earth and Environmental Sciences (http://www.gees.ac.uk/planet/). In the latter's resource database they will find Judy Chance's paper on curriculum integration at Oxford Brookes University, a leader in HEfS (Pepper, 1996). In the remainder of the chapter an asterisk indicates a geographer (e.g. Pepper*, 1996) in order to highlight the subject's potential.

LOCAL KNOWLEDGE AND CITIZEN SCIENCE

Before moving to a consideration of critical realism it should be noted there is currently much evidence of individuals and workers' and citizens' movements attempting to re-embed themselves in nature by discovering new ways of working, living, and knowing. Those who reject science are clearly not progressive but others do seek to engage private corporations and the state in new forms of consultation and participation aimed at creating forms of knowledge or citizen science (Irwin,

1995; Eden*, 1998) that have greater relevance to their lives. As citizen science, a unified science for sustainability should combine relevant aspects of academic or abstract knowledge with relevant elements of the local (tacit and lay) knowledge that people develop in their everyday lives. Tacit knowledge is that which cannot be easily described or encoded in the form of words, written documents or other impersonal means (e.g. the farmers' knowledge of soil, children's knowledge of their playground), while lay knowledge is popular, commonsense knowledge that may enable people to live sustainably with one another and the rest of local nature. New information technologies such as the internet allow people to link abstract and local knowledge in new ways and so provide for a critical postmodern pedagogy (Castells et al., 1999). Universities can clearly assist in developing citizen science, use postmodern pedagogy, and so help to empower their students and the wider community with new ideas and outlooks.

DIALECTICAL MATERIALISM

Critical realism is a materialist and dialectical philosophy. Materialism maintains that the world should be understood primarily in terms of matter and material causes rather than spirit, mind or ideas (idealism), while dialectics suggests that such matter is best viewed as a system of processes, flows and relations, rather than a complex of ready-made things (mechanical materialism, positivism) (Cornforth, 1987; Harvey*, 1996). People and other organisms do not exist outside of or prior to the processes, flows and relations that create, sustain or undermine them. They are constituted by flows of energy, material and information in ecosystems, made possible by the relations between things in the biophysical and social worlds. The new biology and the life sciences support dialectics, seeing a constant two way exchange between organisms and their environment such that the one shapes the other with no sharp dividing line between them. The idea that people constantly change nature, and develop in relation to a nature that they modify or socially construct, accords with our practical or commonsense view of the world and offers a starting point for collapsing the dualism between the natural and social sciences.

In addition to seeing everything in nature as related and in a constant state of transformation, dialectics also regards nature as undergoing an evolutionary process towards higher states or self-organisation and complexity (Lewin, 1997; Manson*, 2001). Organisms contain latent structures and potentialities which are realised in different ways in different environments and some organisms are more successful in changing environments and adapting to them than others. Humans have been particularly successful but there is accumulating evidence that they are modifying or constructing nature in unsustainable ways and that it is taking its revenge. Contradictions between the promise and reality of modern development now challenge them to reshape the processes, flows and relations in ecosystems (by developing new technologies and forms of social organisation) in order to put development on a more sustainable path. Sustainable development requires the co-evolution of society and the rest of nature, and dialectics suggests that the prospects of such development are linked to the struggles of opposing forces that are inherent in all things. Particularly significant for education is the struggle of ideas (Sneddon*, 2000).

Hartmann (1998) reminds us that ideas about sustainable development are inevitably contradictory since its advocates have different values and interests and wish to sustain different sets of ecological, environmental and social relations. Attention to all three sets of relations leads him to suggest that maintaining the metabolism between bio-physical and social systems in ethically and politically acceptable ways, involves sustaining:

1. Relations among humans (social relations) based on mutual respect and tolerance. Just relations allow equitable access to food, clothing, health care, shelter and meaningful work, provide for freedom of thought and mental development, and promote democratically determined political and economic decisions.
2. Relations among humans and other species (environmental relations) that minimize human domination of and impact on other species and their environments or habitats.
3. Relations among organisms and their environment (ecological relations) which have created the climate, hydrological cycle, radioactive levels, and other environmental conditions (ecological processes) that we have experienced throughout most of human history.

Creating and maintaining these relations requires us to care for the welfare of other human beings, future generations and other species, and requires us to translate this concern into appropriate forms of governance and citizenship (Christie & Warburton, 2001). Appropriate forms of education can guide such development but dialectical materialism suggests that education should be a form of praxis that is process rather than product based (Gadotti, 1996). Since all knowledge starts from activity in the material world and returns to it dialectically, theory becomes a guide to practice and practice a test of theory. Critical pedagogy is developed around concepts of structure, power, ideology, emancipation and critique (Janse van Rensburg et al., 2000,) and claims that knowledge and truth should not be products to be transmitted to students, but practical questions to be addressed as students and teachers reflect and act on significant events and issues that affect their everyday lives. Efforts to realise sustainability on the campus, and in the surrounding community, can provide opportunities for praxis and for evaluating academic ideas alongside lay and tacit knowledge.

CRITICAL REALISM

Critical realism is a development of dialectical materialism. It acknowledges that the mind only knows the world by means of perception, thought and language, but clings to the ontological assumption that there is a real objective knowable material world. This real world displays three levels of abstraction at which mechanisms can be examined and knowledge generated. At the deepest or more abstract level are the real objective powers of objects, the processes made possible by relations between things. At an intermediate level are more contingent factors, specific to given historical and social circumstances, which determine whether or not objective powers are realised (whether processes cause events). At the surface level are experienced phenomena which arise out of the combination of objective powers with contingent factors and can be observed at a given place and time. Realist explanation consists of connecting experience in the empirical domain (e.g. warmer summers, more frequent storms) to structures and processes in the real domain (e.g. the workings of the atmosphere and global energy economy) through contingent factors in the actual domain (e.g. increased use of fossil fuels, failure of politicians to control carbon emissions).

Critical realism offers a unified approach to the natural and social sciences while recognising real but different structures and processes with the physical, biological and social worlds. The biological world is emergent from the physical world and the social world emergent from the physical and biological worlds. The causal mechanisms and properties of inorganic and organic nature combine with human nature in dialectical ways allowing each to grow and develop in ways that are more or less sustainable. The new physical and life sciences enable us to understand the dialectical and systemic nature of the physical and biological worlds and the processes of emergence that underpin the principle of qualitative change. The critical social sciences enable us to understand the ways in which social institutions (e.g. markets, systems of production, governments, universities) facilitate or undermine the interactions between human and non-human nature that foster sustainability.

Social science needs to be combined with natural science to understand how society is embedded in nature, while natural science needs to be combined with social science to understand the forms that nature takes in specific social (historical and geographical) circumstances. Critical realism offers a unified science with the methods of the natural and social sciences sharing common principles but adopting different procedures due to their different subject matters. The social sciences can be sciences in the same sense as the natural sciences but not in the same way. This is because:

–   The subject matter of the social sciences cannot be reduced to that of the natural sciences (e.g. human behaviour cannot be reduced to biochemical reactions), there are qualitative differences;

–   Social reality is pre-interpreted. Society is both produced and reproduced by its members and is therefore both a condition and an outcome of their activity (social relations and structures). The social sciences have a subject-subject relationship with their subject matter, rather than a subject-object one of the kind that characterises the natural sciences;

–   Social structures, unlike biophysical structures, are usually only 'relatively enduring'. The processes they enable are not universal or unchanging over time and space.

Critical realism is anti-positivist since it claims that to explain a phenomena it is not sufficient to show that it is an instance of well established regularities or connections, but necessary to discover its connections with other phenomena via knowledge of the underlying structures and mechanisms that work to produce these connections. It accepts a weak social constructivism (Dickens, 1996) by recognising that social reality is pre-interpreted and that language, discourse and ideology shape its production and reproduction. At the same time it rejects a strong social constructivism that denies the material reality of nature.

Critical realism regards nature as socially constructed or produced in two senses: it is materially shaped by social practices and it is existentially produced as cultural meanings, discourses and representations. Nowhere on the surface of the earth is there a 'first nature' untouched by human influence but when speaking of the social construction of nature we should not imply that such natures as the countryside, food, our bodies, and landscape, are wholly artefacts of society or culture. To do so would deny a realist concept of nature that *refers to the structures, processes and causal powers that are constantly operative within the physical world, that provide the objects of study of the natural sciences and condition the possible form of intervention in biology or interaction with the environment. The nature whose laws we are always subject, even as we harness them to human purposes, and who processes we can neither escape nor destroy* (Soper, 1995, p. 155/6).

Nature in the realist sense sets elastic limits on how people can live in the world, but for the critical realist nature is a theoretical, explanatory concept, not a source of value. It tells us the facts about our predicament but we ourselves must decide what forms of ethics, politics and governance should regulate our relations with the rest of human and non-human nature. Democratic socialist politics appear rational to critical realists (Collier, 1994, p. 200) because:

– Socialism suggests that change comes about by changing social structures and mechanisms not by changing the way we view the world (idealism);

– There is a correspondence between critical realism's world view and certain models of socialist cosmopolitan or ecological democracy. Just as the world is one of stratified mechanisms, with wholes not reducible to parts nor parts to wholes, so a genuine democracy should embrace all sites of power (the body, social welfare, the economy, culture, civil society, coercive relations and organised violence, regulatory and legal relations) at all levels from the local to the global;

– A socialist political philosophy should be partly based on knowledge of those constraints which prevent human nature (and the rest of nature) from realising its potential. In that critical realism, linked to critical theory and Marxist political ecology, reveals the contradictions of capitalism, and the associated causes and possible solutions to the crisis of sustainability, it is an appropriate foundation for revolutionary praxis and HEfS.

CRITICAL REALISM AND HEfS

Critical realism's approach to education seeks to overcome the epistemic fallacy that suggests that reality is simply what experience or experiment tells us it is. It claims that the world cannot be changed rationally unless it is interpreted adequately. Such interpretation requires teachers to engage dialectically with students to:

- probe experience;
- liberate knowledge of deeper realities (structures, processes and events);
- reveal those structures and processes that produce and reproduce powerful interests that prevent people from realising their potential;
- expose knowledge or ideology that sustains such interests; and
- reflect and act on alternative structures, processes and knowledge which allow a greater degree of self determination and democracy.

Malcolm Plant illustrates such critical pedagogy by providing accounts of dialectical encounters with students in Chapter 22. For the moment let us consider the example of teachers and students using the internet to interpret corporate social responsibility (CSR), a strategy whereby business claims to be encouraging more sustainable forms of development (Crowe, 2002). Applying Corson's outline of Bhaskar's conception of discovery (Corson, 1991), suggests that this would involve four stages of enquiry or praxis that would reveal the ideological nature of much current CSR and assess its potential given laws forcing environmental and social responsibility on business (Table 1). Such reflection and action is likely to suggest real limits to CSR under capitalism and prompt consideration of ecosocialist alternatives.

While the example in Box 4.1 draws on economics, business studies, politics, sociology and ethics, it does not require knowledge from the natural sciences. A similar four stage enquiry into global warming, the impact of genetically modified crops, or the conservation of fish stocks, would require such knowledge and readers may wish to consider such a topic and the manner in which the competing knowledge claims of environmental scientists should be handled and bio-physical knowledge combined with that from the social sciences and humanities.

Having examined the potential of critical realism to provide a philosophical framework for interdisciplinarity in HEfS, it remains to examine how the framework can combine elements of mainstream, Marxist and postmodern environmentalism.

Table 1. Critical Education and Corporate Social Responsibility.

1.	*An effect (result or regularity) is identified and described.* Students use the internet to obtain corporate social responsibility (CSR) reports from a number of firms including Tesco (www.tesco.com/everylittlehelps) and BPAmoco (www.bpamoco.com/alive). They use the Sustainable Development Commission site (http://www.sd-commission.gov.uk/) to link such reports to triple bottom line accounting and other approaches to sustainable development. They conduct interviews to assess fellow students awareness and understanding of CSR (their lay and tacit knowledge of corporations).
2.	*A creative model of the 'mechanism' involved is postulated, as a solution or explanation or response to the problem, which if it were to exist would explain the effect.* Students read extracts from Naomi Klein's *No Logo* (Klein, 2000) and consider her explanation of CSR: that by regulating (or appearing to regulate) themselves corporations avoid or delay regulation by government or democratically controlled law. Such action is an attempt to reconcile the demands of shareholders, consumers, governments, NGOs and anti-corporate campaigners; is encouraged by the decline of class politics and the retreat of the state from public interest issues; and reflects identity politics in risk society (Beck, 1992) that encourages corporations to compete for the public's trust (Swift, 1999). As a 'mechanism' CSR is essentially public relations designed to forestall the imposition of stronger regulation.
3.	*Research of two kinds is undertaken to demonstrate the existence and operation of the mechanism: the first kind, to isolate and in some instances observe the mechanism in action; the second kind to eliminate alternative plausible hypotheses.* Students relate Klein's explanation of CSR to Labour MP Linda Perham's private members' bill that calls for social, financial and environmental reporting to be made mandatory; requires companies to consult on big projects; and demands rights of redress for citizens negatively impacted by business activities. The bill would place specific duties and liabilities on directors and companies and proposes the establishment of a new regulatory body (Macalister, 2002). Does the background to the bill and its support by a coalition of NGOs, including Friends of the Earth (www.foe.co.uk/campaigns/corporates/), suggest CSR is working as Klein suggests? Alternatively, is CSR 'an important part of modern business thinking (in which) British firms are leading the way by showing how they can make a difference on the ground' (Confederation of British Industry, www.cbi.org.uk) and where 'excessive intervention risks stifling innovation' (UK Government Report on CSR, www.ukonline.gov.uk/).
4.	*The postulated mechanism, once shown to be real, becomes available as evidence for interpreting the world (as it is or has recently been); action to replace unwanted with wanted forms of determination provides the critical concluding phase in this emancipatory process of discovery.* Students decide their own position on CSR and its implications for their behaviour as citizens and consumers. Some embark on a campaign to extend their fellow students' understanding of CSR and link this with an audit the university's suppliers and action to replace those with a poor record on CSR (e.g. Exxon www.stopesso.com) with those they consider to have a better record.

MAINSTREAM ENVIRONMENTALISM

Mainstream environmentalism and environmental education are technocratic, pervaded by positivism, and place the environment outside society, beyond the grasp of ordinary citizens, to be managed by experts in such areas as resource management, risk assessment, and curriculum planning (Luke, 2001). These experts have varying amounts of power to define environmental issues (generally as 'green' issues at the expense of 'brown' issues) and prescribe technical, behavioural and legislative 'fixes' that leave existing social relations relatively undisturbed. Environmental economists are particularly influential in shaping the mainstream discourses of sustainable development and ecological modernisation, but the mechanisms they advocate to put value on ecological capital, 'price the environment', and increase resource productivity, face real limits in an era of globalisation and deregulation. Evaluating elements of the environment solely or mainly in terms of their monetary or exchange value encourages people to regard them simply as commodities. Such commodity fetishism mystifies the relationship between people and the rest of nature and prevents them from understanding and controlling the system of which they are a part.

Clearly HEfS should not ignore mainstream environmentalism and approaches to sustainable development. Students should read mainstream texts, understand the substance, processes and tools of sustainability as advocated by mainstream reformers, and recognise that while the 'greening of capitalism' is to be encouraged it may not deliver social, cultural and personal sustainability (social justice, cultural diversity, physical and mental health) along with ecological and economic sustainability (Sachs, 1999). As we have seen such limits are revealed by giving due attention to the power relations that shape the fate of proposed and real reforms to environmental relations.

MARXIST ENVIRONMENTALISM

My accounts of dialectical materialism and critical realism have already sketched some elements of Marxian thought that explain the development of human nature alongside the rest of nature. Marxist environmentalism regards the contemporary capitalist world order as one that is unsustainable because the drive for capital accumulation results in inter-related economic and ecological crises. These crises prompt workers and citizens' movements to struggle for a more sustainable order assisted by transformative intellectuals who present a range of critical ideas for validation in praxis or critical action research.

Central among these ideas are those relating to the production of nature or the way in which nature and capital co-constitute one another in temporally and geographically varied and contingent ways (Castree*, 2000; Castree* & Braun*, 2001; Braun* & Castree*, 1998; Dickens, 1997; Smith*, 1984; 1996). Both capital and realist nature exert power or agency in such production but the diversity of capital/nature relations in time and space means that we should be cautious about making (teaching) universal statements about the causes of unsustainable development or the route to sustainability. Some productions of nature are more beneficial to humankind than others and questions of environmental ethics can similarly not be addressed in general terms. What is sustainable and beneficial in one time, place and culture may be unsustainable and destructive in another.

In developing a political ecology that integrates ecology and environmental issues into political economy (Keil et al., 1998) Marxist environmentalists seek to overcome the latent dualism and industrialism in Marx's thought; apply critical ideas about such topics as the state (Johnston*, 1989), globalisation (Held et al., 1999), and feminism (Dordoy & Mellor, 2000); and develop the theory and practice of a sustainable eco-socialism based on new economies of time and nature and new forms of welfare and citizenship (Pepper*, 1993; Little, 1998; Soper, 1999). The journal *Capitalism Nature Socialism* (*CNS*) provides a guide to these developments and its contributors include those who draw on critical theory and related theories of reflexive modernisation (Beck, Giddens & Lash, 1994; Blowers*, 1997). The articles in *CNS*'s Teaching Political Ecology series (http://gate.cruzio.com/~cns/syllabus/) are particularly relevant to the theme of this chapter (e.g. Walker, 1998).

Critical theory shifts the focus of Marxist environmentalism from the economy and capitalism to technocracy and modernity (Barry, 1999; Goldblatt, 1996). Instrumental reason (positivism) rather than capital accumulation is now the prime target of critique, and Habermas' theories of legitimation crisis, knowledge constitutive interests, communicative action, and colonisation of the lifeworld, provide insights into the crisis of sustainability, the interests shaping different kinds of EfS (technical, hermeneutic and critical), critical pedagogy, and the role of new social movements in the creation of an ecological democracy (Huckle, 1996). Sustainable development requires the erosion of instrumental reason and its control by communicative reason that can balance considerations of what is technically possible with considerations of what is culturally appropriate and morally and politically right.

Habermas' ideal speech situation provides a context for balancing such considerations free from social structures that systematically distort communication, so allowing people's common interest in sustainability to emerge. It is a model for organising and evaluating the knowledge claims advanced by different disciplines during an interdisciplinary enquiry and for cultivating the kind of critical thinking and values awareness sought by the Teaching and Learning at the Environment-Science-Society Interface (TALESSI) project (www.greenwich.ac.uk/~bj61/talessi/). Readers are encouraged to visit the project's website, consider its rationale in relation to the arguments advanced in this chapter, and evaluate some of its teaching and learning resources on sustainability after these have been downloaded and used in the classroom.

POSTMODERN ENVIRONMENTALISM

While the development of postmodern science (the new physical and life sciences including quantum theory, complexity theory, and postmodern ecology) confirms dialectical materialism and critical realism in their assertion of a dynamic world of structures and processes, postmodern popular and academic culture raises contradictions for HEfS. On the negative side it fosters ironic detachment and a nihilistic indifference to the world that undermines any prospect of co-ordinated political action for sustainability. On the positive side it brings a new sensitivity to marginalized voices and knowledge claims concerning science, environmental issues and sustainability.

Postmodern approaches to the social sciences and humanities reject critical realism's ontology, claiming that there are no universal foundations for knowledge in realist nature or that there is no reality outside language and discourse. Since there is no single reality, there can be no grand theories to explain how reality works and hence no prospect of progress or utopia based on related grand narratives (e.g. Marxism, ecological modernisation). Since all 'truths' including scientific ones are particular to the groups or societies that believe them and have no universal validity, the grounds for common agreement (communicative action) together with the emancipatory power of social criticism and critical pedagogy are undermined.

While critical realism rejects postmodernism's ontological relativism or strong social constructivism (Gandy*, 1996; Proctor*, 1998) it can, as noted above, accept a weak social constructivism that accommodates epistemological pluralism (there are many ways of knowing and many perspectives and discourses on the environment and sustainability). The related challenge for HEfS is to ensure that its critical pedagogy is also a constructivist pedagogy (Janse van Rensburg et al., 2000) that builds upon student's existing knowledge and interests, accommodates lay and tacit knowledge, and acknowledges how power is wielded through language and discourse. By engaging with cultural politics, marginalized voices, and texts of all kinds, such pedagogy can reinvigorate the modern vision of education as enlightenment (Parker, 1997).

Risk society produces a new generation of youth between the borders of a modern world of certainty and order informed by the culture of the West and its technology of print, and a postmodern world of hybridised identities, electronic technologies, local cultural practices, and pluralized public spaces. Consequently many students experience programmed instability and transitoriness, and are condemned to 'wander across, within or between multiple borders and spaces marked by excess, otherness, difference, and a dislocating notion of meaning and attention' (Giroux, 1999, p. 103). Critical pedagogy should address their shifting attitudes, representations and desires by fostering aesthetic and cognitive reflexivity (Beck, Giddens & Lash, 1994); enabling them to reflect and act on the structural roots of their own subjectivities; and so developing a shared language of resistance that points to possibility and hope. HEfS has a key role to play in this regard.

CONCLUSION

Dialectics and critical realism do then offer a philosophy that transcends the limits of mainstream, Marxist and postmodern environmentalisms, acknowledges their achievements, and allows the construction of a new grand narrative of sustainability that both relegitimates and radicalises modernity (Gare, 1995; Jencks 1996; Myerson, 2001). This is a narrative about the co-evolution of society and the rest of nature, their generative qualities, and the need for improved rationality and governance so that tendencies towards self organisation and complexity can lead towards greater sustainability. It allows students and teachers of HEfS to see themselves as part of an unfinished story and provides an overarching rationale for curriculum development and interdisciplinary.

REFERENCES

Archer, M., Bhaskar, R., Collier, A., Lawson, T., & Norrie, A. (Eds.) (1998). *Critical realism: essential readings*. London: Routledge.

Barry, J. (1996). Sustainability, Political Judgement and Citizenship: Connecting Green Politics and Democracy. In: B. Doherty & M. de Geus (Eds.), *Democracy and Green Political Thought*. London: Routledge.

Barry, J. (1999). *Environment and Social Theory*. London: Routledge.

Beck, U. (1992). *Risk Society*. London: Sage.

Beck, U., Giddens, A. & Lash, S. (1994). *Reflexive Modernisation: Politics, Tradition and Aesthetics in the Modern Social Order*. Cambridge: Polity.

Blowers, A. (1997). Environmental Policy: Ecological Modernisation and Risk Society, *Urban Studies*, 54/5-6, 845-71.

Bourke, S. & Meppem, T. (2000). Privileged Narratives and Fictions of Consent in Environmental Discourse, *Local Environment*, 5/3, 299-310.

Braun, B. & Castree, N. (Eds.) (1998). *Remaking Reality: nature at the millennium*. London: Routledge.

Castells, M., Flecha, R., Freire, P., Giroux, H.A., Macedo, D. & Willis, P. (1999). *Critical Education in the New Information Age*. Oxford: Rowman & Littlefield.

Castree, N. (2000). Marxism and the Production of Nature, *Capital and Class*, 72, 5-36.

Castree, N. & Braun, B. (Eds.) (2001). *Social Nature: Theory, Practice, and Politics*. Oxford: Blackwell.

Christie, I. & Warburton, D. (Eds.) (2001). *From Here to Sustainability: Politics in the Real World*. London: Earthscan.

Collier, A. (1994). *Critical Realism, an introduction to Roy Bhaskar's philosophy*. London: Verso.

Corson, D. (1991). Bhaskar's Critical Realism and Educational Knowledge, *British Journal of Sociology of Education*, 12/2, 223-241.

Cornforth, M. (1987). *Materialism and the Dialectical Method*. London: Lawrence & Wishart.

Crowe, R. (2002). *No Scruples?* London: Spiro Press.

Eden, S. (1998). Environmental Knowledge, Uncertainty and the Environment, *Progress in Human Geography*, 22/3, 425-432.

Dickens, P. (1996). *Reconstructing Nature, Alienation, Emancipation and the Division of Labour*. London: Routledge.

Dickens, P. (1997). Beyond sociology: Marxism and the environment. In: M. Redclift & G. Woodgate (Eds.), op. cit., pp. 179-194.

Dordoy, A. & Mellor, M. (2000). Ecosocialism and Feminism: Deep Materialism and the Contradictions of Capitalism, *Capitalism Nature Socialism*, 11/3, 41-61.

Dryzek, J. (1997). *The Politics of the Earth*. Oxford: Oxford University Press.

Gare, A. (1995). *Postmodernism and the Environmental Crisis*. London: Routledge.

Gadotti, M. (1996). *Pedagogy of Praxis: a dialectical philosophy of education*. New York: SUNY.

Gandy, M. (1996). Crumbling Land: the postmodernity debate and the analysis of environmental problems, *Progress in Human Geography*, 20/1, 23-40.

Giroux, H. (1999). Border Youth, Difference and Postmodern Education. In: M. Castells, R. Flecha, P. Freire, H. A. Giroux, D. Macedo, & P. Willis, (1999) op. cit. pp. 93-116.

Goldblatt, D. (1996). *Social Theory and the Environment*. Cambridge: Polity.

Hartmann, F. (1998). Towards a Social Ecological Politics of Sustainability. In: R. Keil, D. Bell, P. Pentz & L. Fawcett (Eds.) op. cit., pp. 336-352.

Harvey, D. (1996). *Justice, Nature and the Geography of Difference*. Oxford: Blackwell.

Held, D. (1995). *Democracy and the Global Order, from the modern state to cosmopolitan governance*. Cambridge: Polity.

Held, D., McGrew, A., Goldblatt, D. & Perraton, J. (1999). *Global Transformations: Politics, Economy and Culture*. Cambridge: Polity.

Huckle, J. (1996). Teacher Education. In: J. Huckle & S. Sterling (Eds.) *Education for Sustainability*, London, Earthscan, pp. 105-119.

Huckle, J. & Martin, A. (2001). *Environments in a Changing World*, Harlow, Prentice Hall.

Irwin, A. (1995). *Citizen Science*. London: Routledge.

Janse van Rensburg, E., Lotz, H., Du Toit, D., Mhoney, K. & Oliver C. (2000). *Learning for Sustainability: an environmental education professional development case study informing education policy and practice*. Johannesburg: Learning for Sustainability Project.

Jencks, C. (1996). *What is Post-Modernism?* London: Academy Editions.

Johnston, R. (1989). *Environmental Problems, nature, economy and state*. London: Bellhaven.

Jones, P.C. & Merritt, J.Q. (1999). The TALESSI Project: promoting active learning for interdisciplinarity, values awareness and critical thinking in environmental higher education, *Journal of Geography in Higher Education*, 23/3, 335-348.

Keil, R., Bell, D., Pentz, P. & Fawcett, L. (Eds.) (1998). *Political Ecology: global and local*. London: Routledge.

Klein, N. (2000). *No Logo*. London: Flamingo.

Lewin, R. (1997). *Complexity: life on the edge of chaos*. London: Phoenix.

Little, A. (1998). *Post-industrial Socialism: towards a new politics of welfare*. London: Routledge.

Luke, T. W. (2001). Education, Environment and Sustainability: what are the issues, where to intervene, what must be done?, *Journal of Philosophy of Education*, 33/2, 187-202.

Macalister, T. (2002). MP leads call for corporate ethics bill, *The Guardian*, 13.6.02, p. 24.

Manson, S. (2001). Simplifying complexity: a review of complexity theory, *Geoforum*, 32/3, 405-414.

Myerson, G. (2001). *Ecology and the End of Postmodernity*. Cambridge: Icon Books.

Parker, S. (1997). *Reflective Teaching in the Postmodern World: a manifesto for education in postmodernity*. Buckingham: Open University Press.

Pepper, D. (1993). *Eco-socialism, from Deep Ecology to Social Justice*. London: Routledge.

Pepper, D. (1996). *Modern Environmentalism: an introduction*. London: Routledge.

Pepper, D. & O'Keefe (2000). Series editors' preface. In J. Barry (2000) op. cit.

Proctor, J.D. (1998). The Social Construction of Nature: Relativist Accusations, Pragmatist and Critical Realist Responses, *Annals of the Association of American Geographers*, 88/3, 352-376.

Redclift, M. (1987). *Sustainable Development: exploring the contradictions*. London: Routledge.

Redclift, M. & Woodgate, G. (Eds.) (1997). *The International Handbook of Environmental Sociology*, Cheltenham: Edward Elgar.

Sachs, W. (1997). Sustainable Development. In M. Redclift & G. Woodgate (Eds.), op.cit., pp. 71-82.

Sachs, W. (1999). *Planet Dialectics; explorations in environment and development*. London: Zed Books.

Smith, N. (1984). *Uneven Development; nature, capital and the production of space*. Oxford: Blackwell.

Smith, N. (1996). The Production of Nature. In: G. Robertson, M. Mash, L. Tickner, J. Bird, B. Curtis & T. Putnam (Eds.), *Future Natural: nature, science, culture*. London: Routledge, pp. 35-54.

Sneddon, C.S. (2000). Sustainability in ecological economics, ecology and livelihoods: a review, *Progress in Human Geography*, 24/4, 521-549.

Soper, K. (1995). *What is Nature? Culture, politics and the non-human*. Oxford: Blackwell.

Soper, K. (1999). The Politics of Nature: reflections on hedonism, progress and ecology, *Capitalism Nature Socialism*, 10/2, 47-70.

Swift, R. (Ed.) (1999). Mind Games, the rise of corporate propaganda, *The New Internationalist*, July.

Walker, P. (1998). Politics of Nature: An Overview of Political Ecology, *Capitalism Nature Socialism*, 9/1, 126-131.

BIOGRAPHY

John Huckle is a geographical and environmental educator who formerly taught at De Montfort and South Bank Universities. He is the principal author of *Reaching Out*, WWF-UK's programme of teacher education, and a co-editor of *Education for Sustainability* with Stephen Stirling. John can be contacted via his website at www.john.huckle.org.uk.

CHAPTER 5

HIGHER EDUCATION, SUSTAINABILITY, AND THE ROLE OF SYSTEMIC LEARNING

Stephen Sterling

The development of ecological understanding is not simply another subject to be learnt but a fundamental change in the way we view the world. (John Lyle, 1994)

INTRODUCTION

This chapter argues that sustainability implies a double learning challenge to higher education, concerning both 'paradigm' and 'provision'. The possibility of reorientation of higher education in the context of sustainability depends on widespread and deep learning within the higher education community and by policy makers - and this has to both precede and accompany matching change in learning provision and practice. Whilst discussion often centres on this latter aspect, which may be called 'education for change', sufficient attention also needs to be given to the first aspect, which concerns 'change in education' particularly as regards ethos, purpose and policy. A systems-based staged model of learning is offered as a tool for thinking about the difficulty and possibility of such deep change, and the idea of co-evolution as learning process between institutions and their communities is briefly outlined as a promising way forward.

In January 2002, in his inaugural professorial lecture at the University of Bath, UK, Peter Reason – an authority on cooperative inquiry and action research - laid down a challenge to his institution, to place the 'twin crises' of justice and sustainability at the centre of its educational and research efforts (Reason, 2002). Sitting in the auditorium, I sensed that members of his audience greeted this radical notion with differing reactions varying from enthusiasm to incredulity. I will argue below that the probability of his or any other higher education (HE) institution responding fully to such a challenge depends on a deep appreciation of three fundamental areas of concern, which can be summarised metaphorically as: the nature of the territory now occupied as regards both paradigm and provision, the nature of territory that sustainability implies, and the journey that is required to shift from one grounding to another.

To help map some of this ground, I will use ideas and tools drawn from systemic thinking, which offers some clarity and overview in a complex and difficult terrain. Systems thinking argues that 'valid knowledge and meaningful understanding comes

Peter Blaze Corcoran & Arjen E.J. Wals (Eds.), Higher Education and the Challenge of Sustainability: Problematics, Promise and Practice, 49-70.
© 2004 *Kluwer Academic Publishers. Printed in the Netherlands.*

from building up whole pictures of phenomenon, not by breaking them into parts' (Flood, 2001, p. 133). Given the complexity of this subject area - involving worldviews, the nature of sustainability, policy and practice in higher education, organisational learning, and transformative change - a systems perspective which seeks to illumine the relationships involved is both necessary and helpful. Systems thinking addresses any problematic nexus such as this by increasing the level of abstraction or overview, rather than the conventional reductionist route of examining detail and dividing the issues into smaller parts.

This chapter argues that sustainability does not simply require an 'add-on' to existing structures and curricula, but implies a change of fundamental epistemology in our culture and hence also in our educational thinking and practice. Seen in this light, sustainability is not just another issue to be added to an overcrowded curriculum, but a gateway to a different view of curriculum, of pedagogy, of organisational change, of policy and particularly of ethos. At the same time, the effect of patterns of *unsustainability* on our current and future prospects is so pressing that the response of higher education should not be predicated only on the 'integration of sustainability' into higher education, because this invites a limited, adaptive, response. Rather, I will argue, we need to see the relationship the other way round - that is, the necessary transformation of higher education towards the integrative and more whole state implied by a systemic view of sustainability in education and society, however difficult this may be to realise. In sum, this is an argument for what I have termed 'sustainable education' (Sterling, 2001). My emphasis here, therefore, is less the detail of curriculum, pedagogy and management that a changed educational paradigm implies (which, though critically important, may be found in other chapters in this volume), but rather the deeper issue of why and how sustainability requires a changed paradigm, and how such change through deep learning may or may not occur. A distinction is thus made between 'learning *through* higher education' (relating to *provision*) which is the usual subject of discourse, and 'learning *within* higher education' (relating to the guiding *paradigm*).

The key issue is one of *'response-ability'*: how far institutions and higher education as a whole are able to respond sufficiently to the wider context of the crisis of unsustainability and the opportunities of sustainability. The common perception is often that little more than a change in teaching or curriculum is necessary – that is, an adaptive adjustment in learning provision. A full response, however, commensurate with the size of the challenge, implies a change of educational *paradigm* – because sustainability indicates a change of cultural paradigm which is both emergent and imperative. Many commentators maintain the fundamental issue at stake is a 'crisis of perception' which most of us are part of, and that a change of cultural worldview based on some form of systems thinking is both necessary and emerging, if still fragile (Harman, 1988; Clark, 1989; Bohm, 1992; Wilber, 1996; Capra, 1996). This appears to entail a shift of emphasis from relationships based on fragmentation, control and manipulation towards those based on participation, appreciation and self-organisation. Increasing numbers of writers are pointing to the emergence and nature of this ecological worldview, predicated on the notion of a co-created or participative reality. Thus this worldview is variously called 'participative' (Heron, 1996; Reason & Bradbury, 2001) 'co evolutionary'

(Norgaard, 1994), or 'living systems' (Elgin, 1997). Evidence of this emergent paradigm can be seen in aspects of ecological and integrative thinking, particularly ecophilosophy, social ecology, eco-psychology and creation spirituality, as well as more practical expressions in major areas of human endeavour such as holistic science, ecological economics, sustainable agriculture, holistic health, adaptive management, ecological design and architecture, and efforts to develop sustainable communities.

At the level of root metaphor, this change involves a shift from the influence of mechanism towards the promise of a living systems or ecological metaphor. The emergent postmodern ecological paradigm suggests a change of epistemology, from reductionism towards holism, from objectivism towards critical subjectivity, and from relativism to relationalism. Without the deep learning that this implies, on the part of policymakers, administrators, curriculum developers, lecturers and all the actors in higher education, the response of HE to sustainability is always likely to be partial and accommodatory rather than full and transformative. Yet, as the incontrovertible educational maxim states, 'learning needs to start where people are'. Herein lies the paradox and profound challenge of change in higher education – how do we work towards transformative learning in a system that itself is intended to be a prime agency of learning? There is a double problem here: first, higher education institutions are not primarily reflexive learning systems (learning organisations) but teaching and research systems. Second, higher education is not primarily engaged in the provision of deep learning to students, but in first-order learning: the transmission of information and the development of instrumental skills aligned (increasingly) to the perceived needs of the economy.

To return to the metaphor above, it is possible to map out in some detail (as far as anybody knows) the new educational territory, but unless we perceive where we are - and are aware that our current territory is increasingly untenable - we are unlikely to shift from that which we know and are familiar with. So we might begin by recognising our current position: that within the last fifteen or so years, the influence of neo-liberal and neo-conservative thinking and language has dominated educational thinking and practice, bringing in a narrowly-cast vocationalism, instrumentalism and managerialism at the cost of more liberal and humanistic interpretations of the role and nature of education (Smyth & Shacklock, 1998). Further, that this managerial education paradigm is underpinned by a deep-rooted cultural paradigm of modernism and mechanism, despite the intellectual currency of postmodern deconstructionism.

To help with the argument I will present a number of models which are based on systems approaches. These act as maps or tools for thinking about the territories mentioned above.

SUSTAINABILITY, NESTING SYSTEMS AND SYSTEMS FAILURE

Let's start with a systems view of sustainability. This is seen as a qualitative condition or emergent property arising from the relationships involved in any system whether considered at local level or global level, and demonstrating the *survival*, the *security*, and the *well-being* of 'the whole system'. So if we regard a set of relationships as a 'system' - such as a family, a community, a farm, a local economy, a school, a university, an education system or an ecosystem – then the health of any such system depends on the health of its subsystems, and they on their subsystems and so on. *Sustainability is the ability of a system to sustain itself in relation to its environment*, given that all systems are made up of subsystems and parts of larger meta-systems. A system that either undermines the health of its own subsystems or of its meta-system is unsustainable. Systems thinkers use 'nesting systems' models to illustrate such relationships between subsystems and metasystems thus:

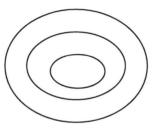

Figure 1. Nesting systems.

This simple model provides a basis to consider an ecology of 'human activity systems' at any systemic level. The fundamental problem revolves around how to achieve, or at least work towards - what systems thinkers term - 'goodness of fit' or coherence between the nesting contextual levels of ecosphere, society/economy, and education (and their subsystems): so that increasingly, each becomes - and together become - a 'viable' or healthy system. According to Bossel (1998, p. 75) a viable system is one which is 'able to survive, be healthy, and develop in its particular environment'. Whilst necessarily imprecise, this notion of the healthy, sustainable system is a guiding idea in the discussion below, and applies at any and every system level.

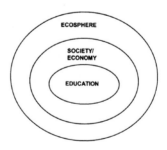

Figure 2. Education, society/economy, and ecosphere as nesting systems.

Starting at the centre, it is possible to regard any educational system (a system of related components including policies, institutions, curricula, actors *etc.*) as a subsystem of wider society: it is organised by, financed by, and mandated by this society. It is shaped and oriented by the needs, policies, values and norms of the social context which it serves. However, there is a co-evolutionary relationship which may be seen as a key to change in both system levels, and I will consider this further, below.

At this point, we can introduce the idea of 'systems failure' (Chapman, 2002). This can refer to 'objectives not met' or 'inappropriate objectives', or the 'undesirable side effects' of a system. Criticism of higher education, particularly in political debate, often centres on the first meaning, but in terms of the ecology of systems outlined above, I will argue that higher education largely 'fails' in terms of the latter two aspects of failure: the purposes or objectives of higher education largely fail to take into account sustainability, while undesirable side-effects include widespread ecological illiteracy and its consequences (Orr, 1994; Jucker, 2002).

This failure of educational systems reflects a more fundamental failure at the higher level of the nesting systems model. The fundamental 'system failure' is our continuing inability to sufficiently adapt our social and economic systems to their ecological context – the limits, 'laws' and systemic nature of the ecosphere (the outer ring in Figure 2 above). According to Brown of the Worldwatch Institute - which has been charting progress relating to sustainability for some two decades - the 'bottom line' is how far our global economy *fits* into the global ecology. The biggest issue, he says, is whether we can see the *economy as part of the environment*, rather than the *environment as part of the economy*. In other words, which is the larger context. As Meadows (1992), Daly (1996), Clayton and Radcliffe (1996), Brown (2001) and The Natural Step programme point out, socio-economic systems should be regarded as subsystems of the encompassing biophysical or ecospheric system, upon which they are entirely dependent. But as Brown (2001, p. 3) notes:

> Economists see the environment as a subset of the economy. Ecologists see the economy as part of the environment.

From this simple but profound difference of perception, it is clear that at a deeper level still, the root of this system failure is our shared worldview or social paradigm. The next step in the argument is that our dominant worldview and epistemology gives rise to an unsustainable relationship with the ecosphere, and that this same epistemology is dominant in Western educational systems. Further, as education is a subsystem of society, then by an inexorable logic, education is largely part of the overall system failure in the relationship between society and the ecosphere.

If this argument is valid, then clearly the answer to the crisis of unsustainability cannot be a simple tweaking of educational policy and practice, nor the current rush to 'improvement'. This is an issue that E.F. Schumacher, the radical economist, pondered nearly thirty years ago:

> the volume of education has increased and continues to increase, yet so do pollution, exhaustion of resources, and the dangers of ecological catastrophe. If still more

> education is to save us, it would have to be education of a different kind: an education
> that takes us into the depth of things. (Schumacher, 1997, p. 208).

Similarly, David Orr states that much of what is wrong with the world is not a result of a deficit of education but is the continuing legacy of a kind of education which:

> alienates us from life in the name of human domination, fragments instead of unifies, overemphasizes success and careers, separates feeling from intellect and the practical from the theoretical, and unleashes on the world minds ignorant of their own ignorance. (Orr, 1994, p. 17).

This is not so much - to use Orr's distinction - a crisis *in* education of the sort that occupies politicians and editorial writers, as a crisis *of* education, which is far less noticed. This bigger crisis begs the most central of questions which concerns the *purpose* of education, and by association, the purpose of any institution and learning programme.

Of course, there have been considerable efforts to 'reorient' education to take more account of sustainability, the international movement beginning with the Stockholm United Nations Human Conference on the Environment of 1972 which first identified the critical role of education in addressing environmental issues. Yet, as a UNESCO report on progress since the 1992 Rio Summit, prepared for the 2002 World Summit on Sustainable Development notes, "much of current education falls far short of what is required", and it calls for a "new vision" and "a deeper, more ambitious way of thinking about education" (UNESCO, 2002).

All this leaves us with a profound paradox: the agency that is charged with the provision of education and learning - that is, the education system and its component parts including higher education - is largely part of the unsustainability problem it needs to address. The fundamental challenge then, is how to achieve significant rather than superficial orientation of higher education, and this calls for a theory of learning which can help clarify the nature and possibility of the required change.

Again, systems thinking provides some insight here, with its distinction between learning levels based on Bateson's ideas (1972). While paradigm change is essentially about learning - if there is no learning, there can be no paradigm change – it is clear that most learning that goes on within and outside learning institutions makes no difference at all to individuals' or society's overall paradigm. This is because, applying Bateson's theory of learning levels, it is *first order learning* or basic learning. Bateson distinguished three orders of learning and change, corresponding with increases in learning capacity, and these have been adopted by learning and change theorists, particularly in the field of systemic learning and organisational change.

LEARNING LEVELS

Using systems terms, learning (by an individual, a group or an organisation) can be seen as having two aspects – *self-correction* and *meaning-making* in response to a change in the system's environment. Such learning can serve either to *keep a system stable*, or enable it to *change to a new state* in relation to its environment. The second response is a much deeper change which requires new meaning-making and examination of existing assumptions.

These two types of learning are variously described as 'single-loop' and 'double-loop learning' (Argyris & Schon, 1996), 'adaptive' and 'generative' learning (O'Connor & McDermott, 1997), 'basic learning' and 'meta-learning' (Bawden, 1997a) or 'first order' and 'second order' change (Ison & Russell, 2000). These categories are often used to describe organisational change, but they can also be applied to change in worldview or belief systems amongst individuals and groups. The first level of learning is a limited response to change in the system's environment (which in this case is the whole sustainability imperative). It keeps the system and its 'theory-in-use' stable, whether we are considering the dominant educational paradigm or our belief system. This *can* be an appropriate response, except where the challenge from the system's environment is so great that second-order learning is required.

This distinction between learning levels helps us perceive the nature of the learning challenge and crisis we now face. The new postmodern conditions of unsustainability, complexity, and uncertainty require higher-order learning not just by students, but by the whole education community, and indeed, society as a whole. Let's look at this important distinction further.

Single-loop learning does not normally impinge on or change the values of the learner, the educator, the educational institution, or indeed society. It is an essentially non-critically reflective, adaptive, response (to the concerns of sustainability in this case) based on the values and *modus operandi* of instrumental rationality. Such learning which serves stability tends to be characterised by negative feedback loops, which dampen change. Double-loop learning/second-order change, by contrast, is deeper learning where change tends to be characterised by positive feedback loops between the system and its environment, whereby both attain a new state (Banathy, 1992).

> Second-order change is change that is so fundamental that the system itself is changed. In order to achieve (this) it is necessary to step outside the usual frame of reference and take a meta-perspective. First-order change is change within the system, or more of the same. . (Ison & Russell, 2000, p. 229)

Thus, first-order learning and change is akin to what Clark (1989, p. 236) calls 'change within changelessness', and is often geared towards effectiveness and efficiency – 'doing things better', rather than 'doing better things'. Much of the movement towards 'raising standards' in Western education systems is clearly of the first kind.

Beyond first and second order change, systems thinkers – again, drawing on Bateson - recognise a *third* learning level which is described as transformative learning or epistemic learning. Arguably, such a quality of learning is key to the realisation of a more sustainable cultural paradigm – in individuals, in education systems, and in society as whole. The logic of this is that learning *within* paradigm does not change the paradigm, whereas learning that facilitates a fundamental recognition of paradigm and enables paradigmatic reconstruction is by definition transformative. According to Wenger (1998, p. 226) 'learning – whatever form it takes – changes who we are by changing our ability to participate, to belong, to negotiate meaning'. Transformative learning does this to an unusual degree. It engages and involves the whole person, and affects change in deep levels of values and belief through a process of re-perception and re-cognition. It is not then just a matter of intellectual or conceptual learning, but engages our emotional and intuitive selves as well. In learning theory terms, it signifies a move from first-order learning to second-order learning where values, beliefs and paradigm are critically realised and examined, and a further stage where a new paradigm emerges. Systemic thinking in this epistemic sense then, is not simple familiarity with some systems ideas, but 'a way of thinking that is independent of the content of systemic concepts' (Brown & Packham, 1999, p. 11).

According to the Center for Transformative Learning at OISE at the University of Toronto, transformative learning involves experiencing:

> ...a deep structural shift in the basic premises of thought, feelings and actions. It is a shift of consciousness that dramatically and permanently alters our way of being in the world. Such a shift involves our understanding of ourselves and our self-location: our relationships with other humans and with the natural world (Morrell & O'Connor, 2002, p. xvii).

The three learning levels are summarised in Table 1.

Table 1. Three learning levels summarised.

Learning I: basic learning	learning	thinking	knowing
Learning II: meta-learning	learning *about* learning	thinking *about* thinking	knowing *about* knowing
Learning III: epistemic learning	learning *about* learning *about* learning	thinking *about* thinking *about* thinking	knowing *about* knowing *about* knowing

The common saying that one 'can't see the wood for the trees' perhaps provides a useful analogy: Learning I might be only 'seeing the trees', or working within the paradigmatic 'wood'; Learning II might be stepping out and recognising the wood as a whole; Learning III might be the 'helicopter view', seeing that a number of alternative woods or paradigms exist. Another way of putting it is:

I   –'doing things better'
II –'doing better things'
III –'seeing things differently'

What these models clearly suggest is that 'lower levels' of learning are less difficult and more everyday in nature. Indeed, theorists make a distinction between basic learning and 'higher order' learning levels. Arguably, sustainability requires higher order learning, that is, epistemic or transformative learning, which can – in turn – offer alternative operative paradigms and sets of practice at 'lower' levels of learning and knowing. We need to be clear then what level we have in mind when discussing learning.

This model of learning levels not only clarifies orders of learning, but also provides a 'map' of the staged learning responses journey towards deeper learning that any higher institution and its members might experience over time.

LEARNING RESPONSES IN EDUCATION AND IN WIDER SOCIETY

This idea of progressive responses parallels O'Riordan and Voisey's helpful notion of the 'sustainability transition'. In a major study, *The Politics of Agenda 21 in Europe* (1998), these authors suggest that a four-stage shift in the transition to sustainability is necessary, from 'very weak sustainability' to 'very strong sustainability', characterised by changes in environmental and economic policies, and in degrees and types of public awareness, with the last phase involving:
- much closer integration between environmental and economic policy;
- a cultural shift in public awareness; and a
- renewal of emphasis on local democracy and activity.

However, the initial learning response (equating with 'very weak sustainability', which is at least a step beyond ignorance or outright denial) is to adapt just sufficiently to *accommodate* this disturbance, without fundamentally changing the whole system. Disturbingly, O'Riordan and Voisey (1998, p. 2) suggest that many of the institutions 'that need to be readjusted in order to embrace the sustainability transition' actually thrive in a non-sustainable world:

> The innate logic of these institutions encourages them to vary marginally the status quo, though never more than is suboptimally tolerable.....No wonder sustainable development is taking time to be credibly articulated in policy and day-to-day behaviour.

We can suggest a parallel and link between this *social* learning response and the *educational* response to sustainability, bearing in mind (as discussed above) the notion of education as a subsystem of society.

A model of possible learning responses by both education and wider society follows:

Table 2. Staged learning responses to the challenge of sustainability.

	Type of response	Resultant change	Type of learning	
1	No response	No change	Denial/ignorance (no learing	
2	Accommodation	Green gloss	Adaptive	
3	Reformation	Serious reform	Critically reflective adaptation	
4	Transformation	Whole system redesign	Transformative	

This range of learning responses is linked to a range of action responses:

Table 3. Comparing staged social and educational responses to sustainability.

Sustainability transition	Response	State of sustainability	State of education
1 Very weak	Denial, rejection or minimum	No change (or token)	No change (or token)
2 Weak	'Bolt-on'	Cosmetic reform	Education *about* sustainability
3 Strong	'Build-in'	Serious greening	Education *for* sustainability
4 Very strong	Rebuild or redesign	Wholly integrative	Sustainable education

These models beg some explanation.

The first level 'response' is no response (or if there is *some* awareness, minimum response). This may be through ignorance of the challenge of sustainability or denial.

The second level is accommodation: a 'bolt-on' of sustainability ideas to existing system, which itself remains largely unchanged. This is an adaptive, first order change or learning. Through this response, the dominant paradigm maintains its stability.

The third level is reformation: this is a 'build-in' of sustainability ideas to the existing system, through which the system itself experiences significant change. This is critically reflective, adaptive response, or second-order change, where paradigmatic assumptions are called into question.

The fourth level is transformation: this is a deep, conscious reordering of assumptions which leads to paradigm change.

A number of points should be made about this model of change, that:
- these responses can be seen as *consecutive stages* that learners in the sustainability transition (that is, all of us) need to move through;
- however, this is not a simple linear progression of discrete stages but is better seen as reflecting the nesting (and therefore subsuming) levels of simple learning, meta-learning, and epistemic cognition;
- movement beyond the accommodatory response (second level above) involves a good deal of learning by all actors – and particularly policymakers, managers, practitioners who shape institutions and organisations – and such learning is difficult;
- learning is more likely to stop at or become stuck 'in' level 3 above because of the difficulty of paradigm change and the resistance of any belief system to such profound change;
- 'education as a whole' – as a subsystem of society – cannot shift through the transition faster than the shift in 'society as a whole' allows without education becoming 'reined in'.
- Thus, there needs to be both correspondence and recursion between these parallel shifts. However, this recursive relationship indicates that as well as

constraint, co-evolution through interaction between progressive elements in both education and society (or institutions and their community) is possible.

Some researchers are critical of staged learning models, arguing that they reflect the normative orientation of those that wish to equate higher education with deep learning (see e.g. Haggis, 2003), yet given the pressing context of the sustainability transition, it seems that some model of deep change is necessary. My experience of teaching on the MSc 'Education for Sustainability' programme at South Bank University (SBU), London, including witnessing change in students as they grapple with real-life contexts, has demonstrated the value and validity of such a model.

RESPONSE LEVELS IN EDUCATION

Let's now look in more detail at the fourth column of Table 3 above. Whilst teaching on the SBU programme, I developed a model of progressive engagement and change which follows the same logic as the models above, which applies to the education system as a whole, and to institutions and actors within the system (including policymakers, theorists, researchers and practitioners). A first stage, as noted above, is no response. This is common enough, and may be due to ignorance of sustainability, denial or sheer difficulty. Beyond this, a first actual response is often:

- *Accommodation*: a *bolt-on* of sustainability ideas to existing system, which itself remains largely unchanged. This is an adaptive response to the concerns of sustainability based on the values and *modus operandi* of instrumental rationality. There is minimal effect on the institution, and the values and behaviour of teachers and students. This is often a content-oriented response, but it is characterised by incoherence and conflict between reflected educational values. For example, sustainability concepts such as biodiversity or carrying capacity may be added into some parts of the curriculum and some subjects, which in other respects carries messages supporting unsustainability. The idea of sustainability and of sustainable development are interpreted in ways which are consistent with the prevailing worldview. The descriptive term here is 'education *about* sustainability', or 'learning about change'. Whilst a long way from leading us to sustainable living, it is much better than nothing, and can open the door to deeper change.

At a deeper level, the response is:

- *Reformation*: a *building in* of sustainability ideas into existing systems. More coherent coverage of content, an attempt to teach values and skills perceived to be associated with sustainability, and attempts to 'green' the operation of the institution. There is some critical recognition of the dominant educational paradigm, its inadequacies and contradictions. The paradigm is modified and this is expressed in some change in policy and practice. The descriptive terms here are 'education *for* sustainability', and 'learning for change'.

At a deeper level still, the response may be transformative:

- *Transformation*: a *re-design* on sustainability principles, based on a realisation of the need for paradigm change. This response emphasises process and the quality of learning, which is seen as an essentially creative, reflexive and participative process. Knowing is seen as approximate, relational and often provisional, and learning is continual exploration through practice. The shift here is towards 'learning *as* change' which engages the whole person and the whole learning institution, whereby the meaning of sustainable living is continually explored and negotiated. There is a keen sense of emergence and ability to work with ambiguity and uncertainty. Space and time are valued, to allow creativity, imagination, and cooperative learning to flourish. Inter- and transdisciplinarity are common, there is an emphasis on real-life issues, and the boundaries between institution and community are fluid. In this dynamic state, the process of sustainable development or sustainable living is essentially one of learning, while the context of learning is essentially that of sustainability. In this way, sustainability becomes an emergent property of the sets of relationships that evolve. This response is the most difficult to achieve, particularly at institutional level, as it is most in conflict with existing structures, values and methodologies, and cannot be imposed. The descriptive term here is 'education *as* sustainability' or 'sustainable education'.

TOWARDS TRANSFORMATIVE LEARNING

This journey through higher orders of learning involves experience of:
- greater challenge/threat to existing beliefs/ideas – and so more resistance;
- greater 'perturbation' required to stimulate learning;
- greater reconstruction of meaning;
- greater engagement and breadth of response in the learner;
- achievement of greater flexibility and less rigidity of thought;
- higher order of consciousness or mindfulness;
- more emergence as a result of learning;
- the difference between 'unwitting self-reference' and knowing self-reference and therefore the possibility of transcendence.

It is clear from these descriptive models of change that achievement of the higher orders of learning is difficult, although easier for the individual than entire institutions involving many actors. As Ison and Stowell suggest, drawing on Prigogine's theory of dissipative structures:

> ...each learner goes through a period of chaos, confusion and being overwhelmed by complexity before new conceptual information brings about a spontaneous restructuring of mental models at a higher level of complexity thereby allowing a learner to understand concepts that were formally opaque (2000, p. 6).

The alternative response, for individuals or institutions not ready for change, is shut-down or denial –through which the existing paradigm is maintained against perceived threat. All this raises questions of methodology – 'how is transformative learning facilitated?' and of possibility, 'is there evidence that deep institutional change in HE institutions can occur?'

On the first question, evidence suggests that in formal education, there has to be an intention on the part of the designers/teachers born of their own experience, to construct a learning situation through which they can encourage others to explore epistemic learning as a shared experience of inquiry. This is not simple:

> To understand and deliver a pedagogy which enables and provokes students to move across levels of epistemic competence is in itself challenging. To do so requires an awareness on the part of the curriculum designer and personal tutor so that they can facilitate the emergence of these changes (Ison & Stowell, 2000, p. 6).

However – and in response to the second question above – there is evidence of such change, although unsurprisingly, less at the level of entire institutions than that of micro-situations. One notable institutional example is that of Hawkesbury College, and agriculture and land use institute in Australia which, for some twenty years starting in the late seventies, explored the possibility and problems of systemic change in education and learning. Philosophically, the Hawkesbury story was founded on disillusion with the 'the inadequacy' of reductionist science in agriculture and other areas of human endeavour, and a determination to explore the nature, implications of a new paradigm:

> The language of reductionism and positivism does not entertain the very complex and dynamic phenomenon associated with sustainable practices…it is clearly time to argue loudly for a shift in thinking from the Age of Productivity to the Age of Persistence…a new research paradigm in the tradition of what has been called the science and praxis of complexity (Bawden, 1991, p. 2363).

Thus the mission at Hawkesbury became to help 'people in rural communities … learn their way forward to better futures, in the face of immensely complex, dynamic and slowly degrading environments, socio-economic, politico-cultural, and biophysical, in which they increasingly recognised they were deeply embedded' (Bawden, 1997b, p. 1). It was recognised that this would necessitate the provision of experience that would encourage a shift in perception. For the Hawkesbury team, this meant learning to perceive and think more systemically. It led, over time, to the evolution of what Bawden called 'a self-organising critical learning system'. Such a system, he suggests, is able to:

– 'connect with the environments about it, and learn about and from them;
– create meaning both experientially and inspirationally;
– design 'meaning informed' strategies for desirable and feasible changes;
– deal with inherent tensions of difference both within and without;
– deal with conflicts, paradoxes, complexity and chaos;
– have requisite variety;
– have requisite redundancy; and
– be self-referential and critically self-reflexive'
(Bawden, 1997a, p. 30).

This suggests a basic recipe that might be applied, re-learnt and adapted according to context and participant mix, in any learning situation where the existing structures and ethos allowed such experimentation. Clearly, such characteristics are more appropriate or applicable to non-formal adult learning situations – see for example the methodologies that go under the title of Participatory Rural Appraisal (PRA) (Chambers, 1977). Yet many formal educators can recall specific learning experiences which in some way significantly engaged or moved the learners on.

In 2002, I conducted an evaluation at Schumacher College (Sterling & Baines, 2002) - a small, privately-run 'international centre for ecological studies' at Dartington, UK – where there is significant evidence of, and an unusually high incidence of – transformative learning compared to the quality of the learning experience in most formal mainstream institutions. Whilst the College is in many respects unique, there is a good deal of interest in how far aspects of its learning experiences might be emulated in more mainstream situations. Whilst the latter are often characterised by *systematic* management and organisation including top-down control, explicit rules, defined structures and areas of responsibility, and a degree of rigidity, Schumacher College demonstrates a high degree of *systemicity:* that is, internal connection, relatedness and coherence which is in many ways the key to understanding its operation and distinctiveness. The aims and objectives of the education programme and of the individual courses are less tightly defined, subject to change and evolution depending on how a course evolves, not spelt out at the detailed level of course aims and learning outcomes, and where expressed, relate to levels of personal change and long-term change in the wider world. There is a fine and dynamic balance between explicitness and implicitness, autonomy of the part and integration within the whole, structure and spontaneity, and healthy emergence and synergy is inherent to the mode of operation. In brief, Schumacher College has evolved a 'learning system' which can, and often does, facilitate transformative learning. This learning environment is characterised by its fluidity, integration, multidimensionality, intensity, ethical integrity, caring and synergy.

WHOLE SYSTEMS CHANGE

Obviously, Schumacher College, and the Hawkesbury experiment are highly unusual exemplars, and the standard higher education institution with thousands of students might wonder what it could possibly learn from them. At root level, the *paradigmatic shift* that informs these experiments is the key, and the *whole systems change* that this implies. Fundamentally, this means a shift from the machine metaphor that informs prevailing views of educational management and the learning process towards a view of the institution as a living system and learning organisation (Senge in De Geus, 1997). So far, such thinking is far more in evidence in business management discourse associated with the implications of complexity theory (see e.g. Stacey, 1996; De Geus, 1997) than it is in education.

However, Banathy (1991; 1992; 1999) who has been a leading writer on systemic change in education, argues that we need to move beyond the traditional paradigm, through which - he says - our inquiry is still dominated by reductionism, 'objectivity' and determinism. This approach cannot, 'possibly cope with the complexity, mutual causality, purpose, intention, uncertainty, ambiguity, and ever accelerating dynamic changes that characterise our systems and larger society environment' (1991, p. 10). His work makes a series of useful distinctions that first, clarify the nature of what he terms the 'design journey' towards a more systemic conception of education, and second, give intellectual credence to the idea that indeed, there is a necessary difference of paradigm at issue here, rather than a tinkering within existing boundaries. These distinctions include the difference between:

Table 4. Difference between first order change and second order change.

Improving/reforming educational systems AND	Transforming educational systems
Making adjustments in existing system	Redesigning education systems
Piecemeal change	Whole system/systemic change
Planning process	Design process
Designing for the future	Designing the future
Adaptive learning	Transformative learning

This is the difference between first order change and second (and possibility of third) order change, reviewed above. The difference between these applied paradigms is the collective intelligence we bring to bear. Thus, Banathy suggests that the nature of the questions surrounding education and learning undergo a qualitative shift, from such 'in paradigm' questions as:

> How can we improve the system to make it more efficient/effective? How can we improve student and teacher performance? How can we establish better standards, and how can we test for those better standards? *Et cetera.*

to:

> What is the nature and what are the characteristics of the current post-industrial information age? What should be the role and function of education in this new era? *Et cetera.* (Banathy, 1991, p. 17)

Similarly, Ison makes a critical distinction between the dominant top-down, expert-led 'teaching paradigm' giving rise to a 'teaching system' which stifles creativity, initiative and critical thought and ignores the multidimensionality of complex problems: on the other, the need for a 'learning paradigm' and consequent 'learning systems' which encourage and allow such qualities to emerge and encompasses multiple perspectives. There is a need he says, 'to re-establish universities as communities of learners, [lecturers] must become involved in learning about learning, facilitating the development of learners, and in exploring new ways of understanding their own and others' realities' (Ison, 1990, p. 9). The conventional teaching paradigm, says Ison, is so different from the learning paradigm which sustainability requires, that the possibility of sustainable agriculture (the subject of Ison's paper) is threatened. Here again, is an argument not for just for an add-on change of method or content, but for a profound change of epistemology.

Elsewhere, there is evidence of growing recognition that sustainability necessarily requires a change of ethos, epistemology and praxis in higher education, and that it is for each institution to grapple with the difficult transition this implies. One example is the EU Socrates Thematic Network for Agriculture, Forestry, Aquaculture and the Environment (AFANet), which between 1997 and 2000 explored in some detail - and confirmed that sustainability necessarily implies - a shift from transmissive methodology towards transformative methodology and a fundamental rethink of the academic missions of institutions.

These examples bear out the need for and possibility of a 'whole system shift' which can be summarised simply as four 'P's:

Paradigm	instead of higher education reflecting a paradigm founded on a mechanistic root metaphor and embracing reductionism, positivism, and objectivism, *it begins* to reflect a paradigm founded on a living systems or ecological metaphor and view of the world, embracing holism, systemisism and critical subjectivity. This gives rise to a change of ethos and *purpose*...
Purpose	instead of higher education being mostly or only as preparation for economic life, *it becomes*: a broader education for a sustainable society/communities; sustainable economy; sustainable ecology. This expanded sense of purpose gives rise to a shift in *policy*...
Policy	instead of higher education being viewed solely in terms of product (courses/materials/qualifications/educated people) *it becomes*: much more seen as a process of developing potential and capacity through life, at individual and community levels through continuous learning. This requires a change in methodology and *practice*...
Practice	instead of higher education being largely confined to instruction and transmission, *it becomes*: much more a participative, dynamic, active learning process based more on generating knowledge and meaning in context, and on real-world/situated problem solving.

These can be drawn using an iceberg metaphor – reflecting that the deeper levels of paradigm and purpose guiding policy and practice in higher education tend to be hidden from view and consequently also, most debate.

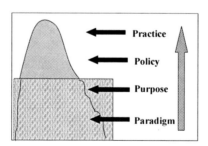

Figure 3. The four 'P's iceberg.

The four P's can also be seen and drawn as a nesting systems figure, suggesting that what an institution does (provision) is ultimately informed by its dominant view of reality and its epistemology (paradigm).

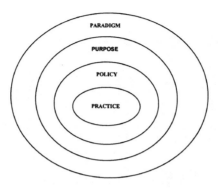

Figure 4. The four 'P's as nesting systems.

Clearly, these P's are relevant to any systemic level – from the national system to the institutional level, and even the departmental level within the institution.

At the level of the individual institution, a whole systems approach to realising a more integrative and ecological paradigm involves recognising the potential for *systemic coherence* and *healthy emergence* within and between the dimensions of its operation. Within any particular educational institution, we might identify at least seven dimensions of its operational life:

- ethos;
- curriculum;
- pedagogy, research, learning and inquiry;
- organisation/management style;
- resource management and use;
- physical structures/architecture;
- community links and relationships.

These can be represented in relationship as follows:

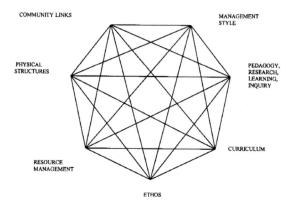

Figure 5. Seven operational dimensions of an educational institutions.

Hence, any dimension has at least six relational paths to be considered. The systemic view recognises that the existing relationships in the system may be characterised by dysfunctionality, lack of synergy or by negative and unintended emergent properties, conflict and contradiction, not least in terms of effects on people. To help move towards a more sustainable state, a whole systems view will pose such inclusive questions as:

- how far are these seven dimensions regarded as a systemic whole?
- how far are the relationships within and between these dimensions characterised by systemic coherence and healthy emergence, or by fragmentation and contradiction?
- how far is planning and change systemic and collaborative – keeping the effects on the whole system and emergence in mind - or piecemeal and imposed?

The general shifts sought might be summarized thus:

Table 5. The general shifts summarized.

Towards sustainable institutions	
From	*To*
Incoherence and fragmentation	Systemic coherence and positive synergy
Large scale, loss of connectivity	Human scale, high connectivity
Closed community	Open, 'permeable' community
Teaching organisation	Learning organisation
Microcosm of unsustainable society	Microcosm (as far as possible) of sustainable society

Again, using a systems perspective, what is actually possible in any institution will be partly conditional upon the wider context within which that institution operates. Looking again at Figure 2 above, any institution may be seen as occupying a subsystem level within the centre of the 'Education' system level. Thus, its response to sustainability will be influenced by the contextual level of what is happening in the education system as a whole, and beyond that, the context of its immediate community and society as a whole.

This leads to a critical question which radically reorients the prevailing notion of 'education for change', that is, education as an agent for social change. Thus, the question 'how can education change people's behaviour in respect to the environment and sustainability', which has underlain much environmental education discourse over the last 30 years is superseded. This linear view of the relationship between education and society is replaced by a systemic view, which generates a very different question: 'how can education and society change together in a *mutually affirming* way, towards more *sustainable patterns* for both?' Banathy (1991, p. 129) suggests this signals a change from education focusing on maintaining the existing state and operating as a rather closed system, towards helping shape society 'through co-evolutionary interactions, as a future-creating, innovative and open system'. This is a vision of on-going re-creation where both education and society, institution and community, are engaged in a relationship of mutual transformation, ongoing inquiry and reflexivity which can explore, develop and manifest sustainability values. This is itself a transformative, co-evolutionary relationship characterised by positive feedback loops which drive the metasystem (society) and subsystem (education) – or at micro level, community and institution - to a new state. This is what Henderson (1993) would call a 'breakthrough' scenario.

In this way, the meaning of the 'learning society' becomes much more than one which learns new skills, but one which is better able to understand itself. The initial driving forces in this process may be less to do with education (that is, the effects of 'education for change'), than increasing awareness in society – and therefore, amongst some actors in education - of deep systemic crisis in the ecological suprasystem and in our relationship with it. Hence, it is growing awareness of systems failure, including recognition of the inadequacy of current assumptions and values that is the current and potential spur to systemic change in higher education both in terms of paradigm and provision. As Chapman states, people:

> will not change their mode of thinking or operating within the world until their existing modes are proved beyond doubt, through direct experience, to be failing (Chapman, 2002, p. 14).

This perhaps gives equal grounds for pessimism and optimism as regards the probability of large-scale change in higher education. Meanwhile, and to conclude this section, I will summarise some of the key points that increase the possibility of deep change in higher education as a sufficient response to the challenge and opportunity of sustainability:
– the importance of conscious intent and leadership;
– the importance of second order learning as a precursor to epistemic change;

- the need for epistemological change towards a more participative or ecological paradigm;
- the importance of attention to context;
- the need for systemic rather than piecemeal change;
- the importance of a co-evolutionary rather than linear view of the relationship between education and society.

CONCLUSION – LEARNING BY DESIGN, OR BY DEFAULT?

This chapter has set out some 'tools for thought' and theoretical models of change in higher education as a response to the challenge of advancing the sustainability transition. Using insights from systems thinking particularly with regard to learning levels, some of the implications and dimensions of deep change towards a more participative and ecological education paradigm which in turn could reorient provision in HE have been mapped out at a general level. It is hoped that such mapping might make such change more possible and practicable.

In current postmodern conditions of complexity, instability, and unsustainability, the response of policy makers is too often to 'order the mess' by increasing central control and regulation, a first-order response that is likely to stifle rather than release the creativity and innovation that these conditions require. In short, the ethos of HE needs to move from 'systematic control to systemic inquiry'. Universities need to 'become knowledge generating, rigorous and transformative in ways and on terms that we can still barely envisage from our current positions' (Weil, 1999, p. 197). Otherwise says Weil, they are in danger of becoming both valueless and visionless and complicit in the market-driven vision of the role of higher education.

In sum, learning can either reinforce the prevailing worldview, or precipitate the 'movement of mind' (Senge, 1990, p. 13), the *metanoia* or re-perception of meaning that many commentators now advocate. In the end, transformative learning depends on the nature of the learning experience we have ourselves and can help to and hope to provide for others. As a society and in the higher education sector, we can choose either to strive towards deep learning and reorientation by conscious design, or have it thrust upon us by default, through the effect of mounting crisis.

REFERENCES

Argyris, C. & Schon, D. (1996). *Organisational Learning II*. New York: Addison Wesley.
Banathy, B. (1991). *Systems Design of Education*. New Yersey: Educational Technology Publications.
Banathy, B. (1992). *A Systems View of Education*. New Yersey: Educational Technology Publications.
Banathy, B. (1999). 'Systems Thinking in Higher Education' *Systems Research and Behavioral Science*, John Wiley, 16(2), 133-145.
Bawden, R. (1991). 'Systems Thinking and Practice in Agriculture', *Journal of Dairy Science*, 74(7).
Bawden, R. (1997a). 'Leadership for Systemic Development' in Centre for Systemic Development, *Resource Manual for Leadership and Change*. Hawkesbury: University of Western Sydney.
Bawden, R (1997b). 'The Community Challenge: The Learning Response', invited plenary paper, 29th Annual International Meeting of the Community Development Society, Athens, Georgia 27-30 July 1997.
Bateson, G. (1972). *Steps to an Ecology of Mind*. San Franscisco: Chandler.
Bohm, D. (1992). *Thought as a System*. London: Routledge.

Bossel, H. (1998). Earth at a Crossroads - Paths to a Sustainable Future, Cambridge University Press.

Brown, L. (2001). *Eco-Economy - Building an Economy for the Earth,* Earth Policy Institute. London: Earthscan.

Brown, M. and Packham, R. (1999). *Organisational Learning, Critical Systems Thinking, and Systemic Learning,* Research Memorandum no 20, Centre for Systems Studies, School of Management. Hull: University of Hull.

Capra, F. (1996). *The Web of Life.* London: HarperCollins.

Chambers, R. (1997). *Whose Reality Counts? Putting the first last.* London: Intermediate Technology Publications.

Chapman, J. (2002). *System Failure.* London: Demos.

Clark, M. (1989). *Ariadne's Thread - The Search for New Ways of Thinking.* Basingstoke: Macmillan.

Clayton, A., Radcliffe, N. (1996). *Sustainability - A Systems Approach.* London: Earthscan Publications.

Daly, H. (1996). *Beyond Growth – The Economics of Sustainable Development.* Boston: Beacon Press.

Elgin, D. (1997). *Global Consciousness Change: Indicators of an Emerging Paradigm.* California: Millenium Project.

Flood, R. (2001). 'The Relationship of "Systems Thinking" to Action Research'. In: Reason, P. and Bradbury, H. (Eds.) *Handbook of Action Research – Participative Practice and Enquiry.* London: Sage Publications.

Haggis, T. (2003). 'Constructing images of ourselves? A Critical Investigation into "Approaches to Learning" Research in Higher Education', *British Educational Research Journal,* Carfax Publishing, 29(1).

Harman, W. (1988). *Global Mind Change.* Indiapolis: Knowledge Systems.

Henderson, H. (1993). *Paradigms in Progress – Life Beyond Economics.* London: Adamantine Press.

Heron, J. (1996). *Cooperative Inquiry - Research into the Human Condition.* London: Sage.

Ison, R. (1990). *Teaching Threatens Sustainable Agriculture,* IIED Gatekeeper Series, no 21, IIED, London.

Ison, R. and Russell, D. (2000). *Agricultural Extension and Rural Development – Breaking out of Traditions, a second-order systems perspective.* Cambridge: Cambridge University Press.

Ison, R. and Stowell, F. (2000). 'Systems Practice for Managing Complexity', discussion paper, Open University.

Jucker, R. (2002). *Our Common Illiteracy.* Frankfurt am Main: Peter Lang.

Lyle, J. (1994). *Regenerative Design for Sustainable Development.* New York: John Wiley.

Meadows, D.H., Meadows, D.L. and Randers, J. (1992). *Beyond the Limits - Global Collapse or a Sustainable Future.* London: Earthscan.

Morrell, A. and O'Connor, M. (2002). 'Introduction'. In: O'Sullivan, E., Morrell, A., and O'Connor, M'. (2002), *Expanding the Boundaries of Transformative Learning'.* New York: Palgrave Macmillan.

Norgaard, R. (1994). *Development Betrayed - The end of progress and a co-evolutionary revisioning of the future.* London: Routledge.

O' Connor J., McDermott, I., (1997). *The Art of Systems Thinking.* London: Thorsons.

O'Riordan, T., Voisey, H. (1998). *The Politics of Agenda 21 in Europe.* London: Earthscan.

Orr, D. (1994). Earth in Mind – on education, environment and the human prospect. Washington: Island Press.

O'Sullivan, E., Morrell, A., and O'Connor, M. (2002). *Expanding the Boundaries of Transformative Learning.* New York: Palgrave Macmillan.

Reason, P. and Bradbury, H. (Eds.) (2001). *Handbook of Action Research – Participative Practice and Enquiry.* London: Sage Publications.

Reason, P. (2002). 'Justice, Sustainability, and Participation', Inaugural Professorial Lecture, Centre for Action Research in Professional Practice, University of Bath, www.bath.ac.uk/carpp/.

Schumacher, E.F. (1997). *'This I believe' and other essays.* Dartington: Green Books (essay first published in 1974).

Stacey, R. (1996). *Strategic Management and Organisational Dynamics.* London: Pitman.

UNESCO (2002). *Education for Sustainability – From Rio to Johannesburg: Lessons learnt from a decade of commitment.* Paris: UNESCO.

Senge, P. (1990). *The Fifth Discipline.* New York: Doubleday Currency.

Senge, P. (1997). 'Foreword'. In: De Geus (Ed.) *The Living Company.* London: Nicholas Brealey.

Smyth, J. and Shacklock, G. (1998). *Re-Making Teaching – Ideology, policy and practice.* London: Routledge.

Sterling, S. and Baines, J. (2002). *A Review of Learning at Schumacher College*. Dorchester: Bureau for Environmental Education and Training.

Sterling, S. (2001). *Sustainable Education – Re-Visioning Learning and Change*, Schumacher Society Briefing no 6. Dartington: Green Books.

Weil, S. (1999). 'Re-creating Universities for Beyond the Stable State', *Systems Research and Behavioral Science* (16) 2, 171-190.

Wenger, E. (1998). *Communities of Practice – Learning, Meaning and Identity*. Cambridge: Cambridge University Press.

Wilber, K. (1996). *A Brief History of Everything*. Dublin: Gill and Macmillan.

UNESCO (2002). *Education for Sustainability – From Rio to Johannesburg: Lessons learnt from a decade of commitment*. Paris: UNESCO.

BIOGRAPHY

Dr. Stephen Sterling is a founder member of the Bureau for Environmental Education and Training (BEET), and a consultant in environmental and sustainability working in the academic and NGO fields in the UK and internationally. He was a founder of the MSc in Environmental and Development Education at South Bank University (SBU), London, where he is an academic tutor. He coordinated the WWF-UK 'Reaching Out' programme of in-service training on education for sustainable development between 1997-2001. He has an extensive publications record, including *Good Earth-Keeping: Education, Training and Awareness for a Sustainable Future* (UNEP UK, 1992), *Education for Sustainability* (Earthscan 1996), *Education for Sustainable Development in the Schools Sector* (Sustainable Development Education Panel, 1988), and *Sustainable Education – Re-visioning Learning and Change*, (Green Books, 2001). His interest lies in the interface between systemic thinking, ecological thinking, learning and sustainability and this was the subject of his doctoral research. He is a member of the IUCN Commission on Education and Communication.

CHAPTER 6

ASSESSING SUSTAINABILITY: CRITERIA, TOOLS,
AND IMPLICATIONS

Michael Shriberg

INTRODUCTION

Efforts to assess campus' responses to the quest for sustainability have proliferated in recent years, driven in part by the desire to translate the idealist language of sustainability into concrete environmental and interrelated social goals and comparisons.[1] These efforts reveal much about the state of campus sustainability through their structure and content (in addition to their outcomes). They illuminate what experts believe are the essential attributes of a sustainable college or university as well as provide insight into the organizational processes involved in moving toward this fleeting goal. Constructed and implemented wisely, cross-institutional sustainability assessment tools can be a powerful force for organizational change. However, poorly constructed and implemented assessment tools provide misleading or irrelevant information, which can cause unnecessary alarm or complacency. To help avoid these potential problems and reach these potential benefits, this chapter identifies criteria for assessment tools, evaluates current efforts against these criteria, and generates conclusions about the state of sustainability in higher education and directions for future assessment research and practice. This focus reflects a bias toward process as opposed to outcomes, which is necessary because most tools are in the relatively early developmental stages, and have not yet been extensively used and thus cannot be evaluated in terms of effect.

THE NEED FOR ASSESSMENT

Defining and assessing sustainability across campuses has proven to be difficult, due in large part to the ambiguities involving in operationalizing and standardizing environmental and social principles. Therefore, many administrators as well as advocates question the wisdom of investing in a cross-institutional sustainability

---

[1] The initial research for this chapter was conducted for the March 2001 Consultation on "Assessing Progress Toward Sustainability in Higher Education" (Washington, D.C.), co-sponsored by University Leaders for a Sustainable Future (ULSF) and the National Wildlife Federation's Campus Ecology Program. The author wishes to acknowledge the organizers and attendees of this event generally, as well as Wynn Calder and Heather Tallent of ULSF specifically, for their assistance and feedback.

Peter Blaze Corcoran & Arjen E.J. Wals (Eds.), Higher Education and the Challenge of Sustainability: Problematics, Promise and Practice, 71-86.
© 2004 *Kluwer Academic Publishers. Printed in the Netherlands.*

assessment. The response to this relevant and potentially paralyzing concern depends heavily on the stage of campus sustainability efforts. For institutions with strong initiatives, the simplest response is that by participating in benchmarking and assessment activities, efforts can be advanced through internal evaluation of strengths and weaknesses, discovering "best practices", and promoting accomplishments to date. For institutions with weaker efforts, assessments can "jumpstart" a sustainability commitment by systematically identifying potential benefits and costs as well as strategies for success.

Simply put, campuses require methods of comparison to each other as well as to a vision of a "sustainable college or university" to ensure and affirm that they are moving in the right (or wrong) direction. The concept that Onisto (1999, p. 37) outlines for the economy as a whole applies to institutions of higher education: "Without a measure and value attached for the rates at which an economy consumes nature, there is no possibility for the market to act in any other interest than economic." In other words, to get to the "bottom line" of sustainability, institutions require a natural, social and economic capital balance sheet. Although circumstances vary considerably on each campus, cross-institutional assessment tools can minimize the effort involved in developing these balance sheets by sharing common experiences and goals.

Cross-institutional assessment tools can identify sources of support and resistance for sustainability initiatives, which helps lead to effective sustainability policies, objectives, and programs. In a theme echoed by campuses across the world, Monteith and Sabbatini (1997, p. 56-57) found that "people were supportive of the sustainability mantra, but when the implications became more clearly defined, disparities in approach and implementation became apparent." Therefore, assessment tools are important in operationalizing charters and policy statements about sustainability in higher education such as the Talloires Declaration (1990), Halifax Declaration (1991), Kyoto Declaration (1993) and Copernicus Charter (1993). "Although these documents contain important guidelines for education, none of them offers concrete prescriptions on an operational level for what Higher Education should do exactly in order to contribute maximally to sustainable development," claims Roorda (2000). Assessment tools can help alleviate this problem through focusing efforts on continual improvement. These tools can also facilitate communication of progress within and across institutions, which is key to mutual success in moving toward the ambitious and amorphous target of sustainability in higher education. These benefits are potentially far-reaching, but not guaranteed. Cross-institutional sustainability assessment can be an extremely frustrating exercise for all stakeholders if the assessment tool and process is not constructed and implemented prudently. Therefore, analysts need to carefully reflect upon what constitutes strong and useful sustainability assessment. My advice for this endeavor is outlined in the following section.

CRITERIA

Currently, there are no explicit guidelines for how to create cross-institutional assessment tools, although many criteria are implicit within current tools and theory. Perhaps the best way to begin the process of articulating criteria is by examining the broad principles underlying sustainability in higher education. Nobody has done this more clearly and thoughtfully than David Orr of Oberlin College. Orr, as quoted by the Penn State Green Destiny Council (2000, p. 4), proposes five criteria to rank campus sustainability: 1) What quantity of material goods does the college/university consume on a per capita basis? 2) What are the university/college management policies for materials, waste, recycling, purchasing, landscaping, energy use, and building? 3) Does the curriculum engender ecological literacy? 4) Do university/college finances help build sustainable regional economies? 5) What do the graduates do in the world? These questions, although difficult to quantify and answer, do not "tinker around the edges", as is the tendency of many environmental assessments; they deal with core issues of ecologically, socially and fiscally sustaining a society and campus. They move farther "upstream" and "downstream" than typical environmental criteria because they address physical and social impacts, intentions, and, of course, the key (ideal) "output" of higher education: educated citizens.

To paraphrase an often-used line of William McDonough: "Being less bad is not the same as being good." The most common pitfall of assessment tools is that they measure eco-efficiency (Fussler, 1996) (i.e. "being less bad") instead of true sustainability (i.e. "being good"). This distinction is crucial as eco-efficiency indicators stress material utilization, environmental performance and regulatory compliance, while sustainability indicators stress issues at the nexus of the environment, society and economy with the goal of no negative impacts (O'Connor, 1995). For example, an eco-efficiency energy indicator would measure energy conservation, while a sustainability indicator would measure total greenhouse gas emissions against a goal of zero. The difference is of mindset in promoting incremental (i.e. eco-efficient) or systemic (i.e. sustainable) change; eco-efficiency ends with the incremental while sustainability incorporates both approaches.

As Onisto (1999, p. 41) points out, the danger of relying solely on eco-efficiency indicators "comes from the appearance that something substantive is being done. It lulls people into feeling that the environment has been, and is adequately, considered." Since "sustainability is a process, not a destination" (Bandy II, 1998), the tools to measure sustainability must delve deep into decision-making by asking about mission, rewards, incentives and other process-oriented outcomes. In this way, analysts capture dynamic processes and motivations – including direction, strategy, intent and comprehensiveness – as well as present impacts. To identify levers for organizational change, assessment tools must ask "why" and "how" campuses pursue sustainability in addition to "what" they are currently doing. In other words, assessment tools must analyze processes and motivations in addition to outcomes.

One of the more difficult tasks involved in creating assessment tools is deciding which factors to measure. The best tools address contextually appropriate issues of major importance to campus environmental, social and economic efforts and effects. Since many facets of colleges and universities potentially fall under the rubric of sustainability, the problem here is of parsimony. The task of the creator and user of assessment tools is to identify issues with broad effects and influence, yet specific measurement possibilities. Moreover, the tools must provide mechanisms to prioritize sustainability-related issues.

Campuses need quick, yet penetrating ways to measure status, progress, priorities and direction. Therefore, the ability to calculate and compare progress toward sustainability is often a limiting factor in assessment. However, assessment tools need not be exclusively quantitative. In fact, quantitative tools in isolation have little chance of fully expressing progress toward sustainability since there is no well-defined "sustainable campus" upon which to base measures. On the other hand, qualitative data must be collected and analyzed in a manner that allows for cross-campus comparisons. The key is to find measurement methods that are flexible enough to capture organizational complexities and differences, yet specific enough to be calculable and comparable.

Perhaps most importantly, sustainability assessment tools must be comprehensible to a broad range of stakeholders. Without this accessibility and communicability, assessments will have little impact. Therefore, analysts must develop mechanisms for reporting that are verifiable and lucid. Given their potential importance as cross-campus communication tools in both process and outcome, comprehensibility should not be sacrificed for precision. However, this criterion does not preclude complicated methodology, as long as translation into understandable outcomes is possible (U.S. Interagency Working Group on Sustainable Development Indicators, 1998). The ecological footprint (Wackernagel & Rees, 1996) is a positive example of this principle, as complex calculations translate into an understandable and demonstrable geographic area.

The creators and users of cross-institutional sustainability assessment tools have a difficult task in measuring up to these criteria. They must not only portray the status of colleges or universities (as measured against the ever-evolving baseline of sustainability) but also integrate motivations, processes and outcomes into a comparable, understandable and calculable framework that moves far beyond eco-efficiency. These tools need to decipher directions and processes while stressing prioritized opportunities for change. Although no tool – and certainly no individual indicator – will capture all these attributes, the next section reviews efforts that excel at meeting different facets of these challenging criteria (Table 1).

Table 1. Evaluation of Campus Sustainability Assessment Tools.

Assessment Tool	Major Strengths	Major Weakness
State of the Campus Environment	– Comprehensive; Combines eco-efficiency & sustainability – Identifies barriers, drivers, incentives and motivations – Identifies processes and current status	– Little use of the term "sustainability" – Small sample within each college or university
Sustainability Assessment Questionnaire	– Emphasizes (cross-functional) sustainability as a process – Useful as a conversational and teaching tool – Probing questions that identify weaknesses and set goals	– No mechanisms for comparisons or benchmarking – Difficult for large universities to complete
Auditing Instrument for Sustainability in Higher Education	– Flexible framework for institutional comparisons – Process-orientation which helps prioritize and set goals through developmental stages	– Difficult to comprehend – Motivations are potentially excluded
Environmental Report and Workbook	– Useful in strategic planning and prioritizing – Collects baseline data and best practices	– Operational eco-efficiency and compliance focus – Difficult to aggregate and compare data – Motivations are largely ignored
Greening Campuses	– Comprehensive, action orientation incorporating processes – Explicitly and deeply addresses sustainability – User-friendly manual with case studies, recommendations	– Calculations and comparisons difficult – Focus on Canadian community colleges – Resources out-of-date
Campus Ecology	– Cross-functional, practical "guide" and framework – Baseline for current tools	– Environmentally focused (i.e. not sustainability) – No longer "state-of-the-art"
Environmental Performance Survey	– Process-oriented – Compatible with environmental management systems	– Operational eco-efficiency – Neglect of sustainability and cross-functional initiatives
Indicators Snapshot/Guide	– Quick, prioritized environmental "snapshot" – Opportunity for more depth on issues of concern	– Operational, eco-efficiency focus with little reference to processes, motivations, benchmarking and sustainability
Grey Pinstripes with Green Ties	– Model for data collection and reporting – Links programs and reputations	– Not sustainability-specific – Neglects decision-making processes and operations
EMS Self-Assessment	– Rapid self-assessment focused on processes	– Operational eco-efficiency focus

EVALUATION OF ASSESSMENT TOOLS

Perhaps because of the difficulties involved in implementing and reporting on cross-institutional assessment tools, the relatively new field of management for sustainability in higher education suffers from a lack of empirical data, as pointed out by Filho (2000) and others. Herremans and Allwright (2000, p. 169) wrote, "Even though the literature provides some excellent case studies of environmental initiatives that have been implemented throughout the world, most of the information available is in the form of examples of 'this is what we did on our campus.'" The major works in the field adhere to the trend of providing case studies and practical advice – mixed with some theory – but with little empirical crosscutting data (e.g. Cortese, 1999a; 1999b; 2001; Creighton, 1998; Eagan & Keniry, 1998; Eagan & Orr, 1992; Filho, 1999; Filho, 2002; Keniry, 1995; Smith & The Student Environmental Action Coalition, 1993). However, ten efforts – which vary greatly in scope, scale and phase – have emerged to alleviate this problem:[2]

"State of the Campus Environment" Survey and Report (U.S.)

The most comprehensive and ambitious assessment effort to date is the National Wildlife Federation (NWF) Campus Ecology Program's "State of the Campus Environment" project (McIntosh, Cacciola, Clermont, & Keniry, 2001). NWF's far-reaching goal is to provide a "national profile of environmental performance on America's colleges and universities (National Wildlife Federation, 2001)". To this end (and after an extensive review process), NWF developed the "first-ever large-scale (campus) environmental performance survey" – funded in part by the Educational Foundation of America, co-sponsored by 14 organizations, and administered by Princeton Survey Research Associates. The survey – which is web-based in order to reduce waste without sacrificing features such as the ability to pause and save data – was sent (in December 2000) to presidents, provosts and chief facilities officers at all 4,100 accredited two- and four-year colleges and universities in the U.S. The long-term goal is to conduct the survey every 2-3 years to assess national trends over time (K. Cacciola, personal communication, March, 2001).

The NWF survey effectively combines measures of incremental eco-efficiency (e.g. water conservation and recycling) with more systemic, sustainable processes (e.g. faculty training in sustainability, land stewardship practices, and use of life-cycle assessment). Moreover, the survey combines accountability for environmental performance and history of environmental initiatives with detailed issue-based questions. The survey also takes the unique step of explicitly identifying barriers, drivers, incentives and motivations for pursuing campus environmental change from a leadership perspective. The mixture of qualitative and quantitative measures ensures comparability, contextual richness and a comprehensible set of best practices. However, NWF emphasizes that the survey is not designed to rank

---

[2] The tools chosen for assessment are the most far-reaching and successful identified in the literature by the author and several other experts. However, this list is not comprehensive, as assessment tools have been omitted intentionally and unintentionally, due to lack of space and information.

individual campuses on sustainability, but rather to provide nationwide trends on managerial practices.

A weakness of NWF's assessment tool is the lack of explicit reference to sustainability, as the term only appears in the context of curriculum. NWF opted to use the term "management" or "environmental" instead of "sustainability" to ensure comprehension by administrators. However, since sustainability is widely-regarded as different from "environmental responsibility", campus leaders might attach different meanings to survey questions based on their interpretations, none of which might approach theorists' and practitioners' meaning of "sustainability". Without explicit reference to sustainability, social issues – and their interaction with environmental issues – tend to be neglected. An unavoidable weakness (given the broad scope of the survey) is that characterizing an entire campus with input from a maximum of the top three decision-makers (and, possibly, their staffs) is difficult and potentially misleading.

NWF received responses from 1,116 out of 12,300 individuals (9.1%) and 891 out of 4,100 institutions (21.7%) (McIntosh et al., 2001). While summarizing the results of the survey is beyond the scope of this chapter, NWF's Campus Environmental Scorecard represents a major step forward in our knowledge of campus environmental performance and decision-making processes. This process of "grading" the environmental status of U.S. campuses can and should become a foundation for future assessments by NWF and other individuals or institutions.

Sustainability Assessment Questionnaire (Global)

The Association of University Leaders for a Sustainable Future's (ULSF) Sustainability Assessment Questionnaire (SAQ) – which is currently being utilized at select campuses across the world – complements NWF's efforts. While NWF focuses on benchmarking, the SAQ is a largely qualitative "teaching tool" that stimulates "discussion and further assessment" (University Leaders for a Sustainable Future, 1999). The SAQ can be downloaded from ULSF's website (www.ulsf.org), and ULSF encourages institutions to use the SAQ as a "3-4 hour exercise on your campus with a group of approximately ten representatives including staff, students, faculty and administrators". The goals of the SAQ are to offer its users "a comprehensive definition of sustainability in higher education as well as to provide a snapshot of their institutions on the path to sustainability". The SAQ emphasizes decision-making mechanisms and processes, with responses on both a 5-point likert scale and in open-ended paragraphs.

The greatest strength of the SAQ is its clear focus on sustainability and sustainable processes. The major weakness of the SAQ is identified by ULSF in its cover letter for the tool (University Leaders for a Sustainable Future, 1999): "Since the questions are primarily qualitative and impressionistic, we cannot use the responses to rate or compare institutions." However, the results are helping to determine the perception of sustainability in higher education. An additional potential problem is that large institutions may not be able to answer many of the questions comprehensively, such as listing courses and research efforts related to sustainability.

SAQ users are asked to fill out a 7-question survey about their response to and usage of the SAQ. The results from this questionnaire (H. Tallent, Personal Communication, 2002) show that many users across the world find the SAQ useful as a way to frame campus sustainability as well as to design more specific assessments and strategies for their own campuses. Therefore, the SAQ seems poised to continue to be successful as a discussion-generating and progress-reporting tool for campus sustainability scholars and practitioners.

Sustainability is explicitly outlined in the cover letter and through a page of definitions placed before the survey. These definitions emphasize the social side of sustainability as well as the inherent ambiguities of moving toward and measuring sustainability as a campus. Another major strength of the SAQ is that it poses probing questions about sustainability and its integration into the campus in terms of strengths, weaknesses, goals and desires, such as "the institution's contribution to a sustainable economy and sustainable local communities". ULSF stresses sustainability, not eco-efficiency, in institutional operations by inquiring about source reduction, social responsibility in investing, and sustainable landscaping. In addition, the SAQ assesses crosscutting organizational structures and processes – such as integration of sustainability into incentives, rewards, staffing, and formal statements.

Auditing Instrument for Sustainability in Higher Education (Global)

The major goals of the Dutch working group which designed the Auditing Instrument for Sustainability in Higher Education (AISHE) include: providing criteria and a framework for internal and external sustainability audits; measuring the success in campus implementation of sustainability; and creating a mechanism to exchange experiences and motivations (Roorda, 2000; 2002). The goal is for AISHE to expand across Europe and the world, resulting in certificates, awards, and other forms of official recognition for users and the instrument itself (Roorda, 2000). As of this writing, AISHE had been tested at "a number of universities in the Netherlands and in Sweden" (Roorda, 2002).

AISHE consists of 24 "criteria" evaluated on five developmental "stages" (activity oriented, process oriented, system oriented, chain oriented, society oriented). For example, "staff development plan" is in the "society oriented stage" (the highest) if "the organisation policy on sustainability is based on societal and technological developments. There is systematic feedback to society (Roorda, 2002)." By assessing and prioritizing the stage of each item (in groups of 10-15 over a 4-6 hour span), a college or university forms a matrix (24 x 5) of status and goals complete with assistance tools for advancement. AISHE explicitly focuses on process over content, qualitative over quantitative measures, and descriptive over prescriptive measures (Roorda, 2002).[3] Thus, AISHE is both an auditing method and

---

[3] A useful description of how the creators of AISHE wrangled with these assessment tool construction issues is portrayed in Roorda (2002).

a policy instrument around which other sustainability tools, such as ISO 14001, can form. AISHE's process-orientation captures dynamic decisions involved in managing for sustainability. Moreover, the developmental stages encourage measurement of progress without forcing quantitative measures. Thus, AISHE provides for potential cross-institutional and intra-institutional comparison (which is beginning to happen).

A significant weakness of AISHE is that the criteria are somewhat abstract and difficult to comprehend. However, the creators of AISHE are developing assistance tools, examples, reference lists, and a training program to make the criteria more tangible and comprehensible. In fact, the Dutch government is funding an AISHE team of consultants to assistant colleges and universities attempting to implement sustainability initiatives (Roorda, 2002). Another potential weakness is that AISHE does not explicitly include indicators about motivations for pursuing sustainability. In other words, it seems possible to use the tool without explicitly addressing the reasons for moving a campus in a particular direction. Overall, however, AISHE is an excellent example of a process-oriented approach to sustainability assessment. The consensus-building approach to designing AISHE created a flexible platform upon which to stimulate and operationalize sustainability in higher education. Therefore, AISHE has the potential for global reach and appeal.

Environmental Report and Workbook (England)

To assist "those within universities who are responsible for implementing environmental policy", the Higher Education Funding Council for England developed an environmental report (1998a) and workbook (1998b). The workbook – which includes over 130 self-assessment questions – guides colleges and universities through a legislative and environmental review. The greatest strengths of this effort include its strategic foci on: baseline data, best practices, policy, management systems (including creating responsibility and information systems), conditions for success, and meeting (English) legal requirements. The self-assessment worksheet included in the workbook can help college or university personnel rate, plan and prioritize environmental management. However, the effort is focused on operations, and sustainability is rarely mentioned and is never used as a goal-setting target. Regulatory compliance and eco-efficiency are stressed, to the detriment of more systemic changes. Moreover, the self-assessment format leaves little room for comparisons between institutions or aggregate measures of progress, and motivations are largely ignored.

Greening Campuses (Global)

The primary goal of "Greening Campuses" (Chernushenko, 1996, p. vi) is to be "a comprehensive source of information and strategies designed as much for institutions already grappling with environmental issues as it is for those that have barely begun to do so." Greening Campuses is a practical manual (which comes on a diskette) created through a partnership between the United Nations Environment

Programme, the Association of Community Colleges of Canada and the International Institute for Sustainable Development. The manual begins with a call to action as well as definitions of sustainability. The sustainability orientation continues throughout the manual. A major strength of Greening Campuses is its comprehensive, process orientation. Each of the many topics is addressed by clearly identifying: the problem and potential solutions; common obstacles and how to avoid them; costs, benefits and opportunities; priorities for action; and best practices. Thus, Greening Campuses creates a systematic, holistic framework for action toward sustainability that incorporates specific, prioritized recommendations as well as examples of institutions further along the path to sustainability. Moreover, Greening Campuses raises profound issues about social and ecological sustainability. For example, the "Facilities Design and Construction" section recommends beginning the design process by asking the question: "Is this facility needed?" However, Greening Campuses fails to provide an adequate way to calculate and compare progress toward sustainability. In addition, the manual focuses on Canadian community colleges, not to the exclusion of other institutions, but enough to hamper the usefulness for other types of campuses. Moreover, many of the resources in the manual are outdated. Overall, Greening Campuses (Chernushenko, 1996) is an excellent resource for campus environmental decision-makers developing action strategies, but falls short as a measurable and comparable assessment tool.

Campus Ecology (Global)

Students and others across the U.S. and world have used the book "Campus Ecology" (Smith and the Student Environmental Action Coalition, 1993) extensively to conduct environmental audits. The cross-functional and comprehensive focus was unique at the time the book was published. Although these topics are addressed largely through an eco-efficiency lens, the emergence and integration of social and economic topics into the debate can be seen through the inclusion of "environmental justice" and "investment policies". The major strength of "Campus Ecology" is its practicality as a clear, coherent framework for assessment: frame the problem, design assessment questions, gather data, identify best practices, develop recommendations and strategies, and find resources for implementation. Moreover, "Campus Ecology" encouraged the thought about processes, life-cycle analysis, and sense of place that is reflected in the more progressive current tools. Although this tool is no longer "state-of-the art", it far exceeded its goal of being a starting point for student environmental assessments and has become a basis for cross-institutional sustainability assessments.

Environmental Performance Survey (Canada and the U.S.)

To assist the University of Calgary and other institutions in implementing environmental management systems, Herremans and Allwright (2000) designed a survey to answer the question: What drives good environmental performance at

North American colleges and universities? This survey was sent (1998-1999) to at least the largest two colleges or universities in each province and state as well as to Talloires Declaration[4] signatories. Fifty (12 Canadian/38 U.S.) out of the 251 institutions in the sample (20%) returned the survey, which takes a cost-centered approach to environmental management, focusing not on quantitative data, but on four managerial "elements": focus, commitment, capability and learning. The strengths of Herremans and Allwright's effort come from their process-orientation, simplicity and compatibility with established environmental management systems. Moreover, this effort addresses and categorizes environmental posture and behavior in a holistic manner. However, the results are limited almost solely to operations, largely ignoring systemic cross-functional, cultural changes required for movement toward sustainability.

Campus Sustainability Selected Indicators Snapshot and Guide (U.S.)

The New Jersey Higher Education Partnership for Sustainability developed its "Campus Sustainability Selected Indicators Snapshot and Guide" with the goal of becoming a "simplified and workable" approach to sustainability assessment. For each of the 10 categories of indicators, each campus provides a "snapshot" (rating sustainability on a 1 to 7 scale) as well as a ranking of priorities. Campuses fill out a more detailed indicators guide for the highest priority items. The strength of the Partnership's effort is in providing a quick, prioritized overview of environmental facets of campus operations. However, this effort is narrowly focused on eco-efficiency in operations (e.g. lighting retrofits) – devoting little attention to sustainability and cross-functional initiatives – although institutions are asked to rank the "sustainability" of these efforts. There is little reference to processes, motivations or other important decision-making parameters. Moreover, there is no way to benchmark sustainability initiatives across campuses.

"Grey Pinstripes with Green Ties" Business School Survey (U.S.)

In 1998, the World Resources Institute (WRI) surveyed the top 67 MBA programs in the U.S. (50 respondents) on environmental courses, institutional support and faculty research (Finlay, Bunch, & Neubert, 1998). While the results of the survey are not directly relevant to this chapter, this survey represents a model for collecting digestible curriculum and research-based campus data. The results are portrayed in "quartiles", which allow stakeholders to assess and benchmark institutions without forcing quantitative comparisons. Moreover, WRI's assessment captures programs and reputations, and includes environmental courses as well as environmental modules in core courses. However, WRI's survey is not sustainability-specific (i.e. does not distinguish between sustainability and environmental issues), lacks information on decision-making processes, and does not include operations (nor

---

[4] The Talloires Declaration – created in 1990 – asks colleges and universities to work individually and collaboratively toward sustainability. An institution is a "signatory" if the president signs this "pledge". For more information about the Talloires Declaration, see www.ulsf.org.

service to a significant degree). WRI conducts follow-up surveys called "Beyond Grey Pinstripes" (Finlay & Samuelson, 1999) every two years.

Environmental Management System Self-Assessment Checklist (U.S.)

The Campus Consortium for Environmental Excellence – which consists of U.S. environmental safety officers – developed its Environmental Management Self-Assessment Checklist (2000) to "help campuses identify the strengths and weaknesses of its current EMS (environmental management system)." The 33-part questionnaire is technical, process-oriented, based on ISO 14001, and directed at campus environmental health and safety professionals. The strength of this tool is as a "rapid self-assessment" which helps campuses visually focus on environmental management processes. The four-part scale for each question follows a "plan, do, check, act" framework in five major areas: environmental policy, planning, implementation and operations, checking and corrective action, and management review. However, the checklist does not reflect sustainability, focusing on eco-efficiency in operational areas such as compliance, documentation, policies and procedures.

CONCLUSIONS

The ten campus sustainability assessment tools reviewed in this chapter (Table 1) vary in purpose, scope, function and state of development. They also differ greatly in how well they meet the assessment tool criteria outlined at the beginning of this chapter. Unsurprisingly, most assessments excel in capturing baseline data on environmental and sustainability performance through an eco-efficiency framework. Surprisingly, many tools also excel at gathering process-oriented information on how campuses are beginning to manage for sustainability. This orientation may reflect an emerging recognition of the complexities and important of process in developing sustainability management systems. Moreover, most tools create a foundation for strategic planning through identifying important issues as well as methods to set and achieve prioritized sustainability goals.

Most assessment tools do not provide mechanisms for comparing campus efforts against other institutions or national/international averages. This may reflect the fear of sustainability rankings discussed at the end of this section. Moreover, most assessments do not address the rationale for "why" initiatives began and are maintained (i.e. motivations), thus failing to provide input into effective advocacy strategies. Many tools focus on operational efficiency, although theory and practice point to the need for sustainability integration across all functional areas. Finally, many analysts and assessment tools do not effectively communicate methods and results, although this situation is likely to change as the tools are used more extensively and demand grows for outcomes.

In addition to lessons about sustainability measurement, cross-institutional assessment tools provide valuable insight into essential attributes of sustainability in higher education through their structure and content. An analysis of included and excluded factors reveals the following parameters:

1. *Decreased Consumption/Throughput*: All assessment tools reflect the need for campuses to decrease usage of energy, water, and other materials and inputs. Tools that orient toward sustainability incorporate goals of adjusting throughput to ecosystem carrying capacities.

2. *Centrality of Sustainability Education*: While elective courses on sustainability are necessary and commendable, state-of-the-art assessment tools recognize that sustainability education needs to be incorporated into core curricula and courses in many disciplines. Curricula on sustainability must include active and service learning about the home institution as well as larger ecological and social issues. Moving towards institutional sustainability means that all students should be exposed to environmental and interrelated social issues.

3. *Cross-Functional Integration*: Strong assessment tools emphasize progress on issues that incorporate teaching, research, operations and service, such as land stewardship and ecological building design. Incorporating multiple functions ensures attention to the interrelated environmental, economic and social aspects of sustainability.

4. *Cross-Institutional Integration*: Leading institutions in sustainability and leading assessment tools reach across institutional boundaries through initiatives and cross-campus comparisons. For example, assessments of campus investments as well as outreach and employment of graduates address the crucial function that colleges and universities play in social development through promoting or hindering sustainability. Moreover, campuses help each other by sharing successes, constraints and opportunities.

5. *Incremental and Systemic Progress*: Recognizing that sustainability is a long-term and difficult goal and process, the tools reflect a two-prong approach. First, campuses should pursue incremental steps to move toward eco-efficiency (e.g. water conservation). The weaker assessment tools stop with incremental steps while the stronger tools incorporate the simultaneous second prong, systemic changes, which include incentive and reward structures, mission and goals statements, procedures, annual reports and other organizational decision-making processes.

The most useful cross-institutional assessment tools reflect the larger transition in thought from environmental management (eco-efficiency) to management for sustainability. Of course, assessment approaches also inevitably reflect the biases of their creators and users. These tools are rapidly emerging, yet there are no well-defined "quality controls" or criteria for assessment. Therefore, a strong effort must be made to ensure that these tools meet the needs of colleges and universities, and provide valid results to all stakeholders. This chapter is designed to begin this process.

Several assessment tools strive to become an international standard, which raises the question of whether a "universal tool" is desirable and feasible. This assessment approach has several clear benefits. First, an international model based on a single assessment tool would allow unprecedented opportunities for comparisons and standardization. Currently, analysts are developing assessment tools on somewhat parallel tracks, and working toward an international standard could minimize these development efforts by combining strategies, resources and approaches. Ideally, a universal tool would make sustainability assessment less cumbersome and idiosyncratic. However, there is no agreement over whether such an international consensus-based approach is necessary to gather and share knowledge. The current approach – in which countries, regions and individual campuses develop or tailor tools for their own needs – is succeeding in gathering piecemeal data, if not comparable and verifiable results. Moreover, developing a "universal tool" would be a painstaking process, which would take longer than many stakeholders are willing to wait for results. In addition, the desirability of a "universal tool" is debatable as contextually important information is likely to be overlooked. What constitutes a sustainable system in one community, country or culture may not correlate with the needs and desires of a different stakeholder group. Therefore, scholars and practitioners need to carefully consider the necessity, feasibility and desirability of a "universal assessment tool".

A second major question/challenge for the future is: Should analysts numerically rank and publicly report on college and university progress toward sustainability? This question has divided analysts perhaps more than any other assessment issue. In the U.S., magazines such as *U.S. News & World Report* provide rankings of colleges and universities on a wide variety of academic, social and economic criteria. Institutions are deeply affected by these rankings and work hard to improve their relative positions. However, there are no ranking systems for environmental issues or sustainability. Clearly, "sustainability rankings" would provide digestible information to students, parents, administrators and other stakeholders on the relative position of campuses on sustainability. While these rankings would perhaps be most relevant on a national level, they would also provide an international perspective on which colleges and universities are taking the lead on sustainability. However, most assessment tools and analysts have shied away from rankings due to widespread resistance from administrators and others. There is no clear way to arrange campuses on a sustainability scale, yet lack of coherent criteria has not stopped campus rankings on other important issues, such as diversity, academic prowess, social life, etc. Therefore, scholars and practitioners need to either help shape a ranking system or provide a clear rationale for why ranking is not appropriate. This controversial potential next step in campus sustainability assessment will have far reaching practical and theoretical implications since it appears be important to the major "client" of higher education: students.

REFERENCES

Bandy II, G. (1998). *Sustainability Booklet*. Houston, TX: The University of Texas-Houston Health Sciences Center.

Campus Consortium for Environmental Excellence (2000). *Environmental management system self-assessment checklist (Version 1.0)*. Boston, MA: Campus Consortium for Environmental Excellence.

Chernushenko, D. (1996). *Greening campuses: Environmental citizenship for colleges and universities*. Winnipeg, Manitoba, Canada: International Institute for Sustainable Development.

Cortese, A.D. (1999a). Education for sustainability: The need for a new human perspective. *Second Nature*. Retrieved March 12, 2000 from the World Wide Web: http://www.secondnature.org.

Cortese, A.D. (1999b). Education for sustainability: The university as a model of sustainability. *Second Nature.* Retrieved March 12, 2000 from the World Wide Web: http://www.secondnature.org.

Cortese, A.D. (2001). Education for sustainability: Accelerating the transition to sustainability through higher education. *Second Nature.* Retrieved April 21, 2001 from World Wide Web: http://www.secondnature.org.

Creighton, S.H. (1998). *Greening the ivory tower: Improving the environmental track record of universities, colleges, and other institutions.* Cambridge, MA: The MIT Press.

Eagan, D.J., & Keniry, J. (1998). *Green investment, green return: How practical conservation projects save millions on America's campuses.* Vienna, VA: National Wildlife Federation's Campus Ecology Program.

Eagan, D.J., & Orr, D.W. (Eds.). (1992). *The campus and environmental responsibility.* San Francisco: Josey-Bass.

Filho, W.L. (Ed.). (1999). *Sustainability and university life.* New York: Peter Lang.

Filho, W.L. (2000). Dealing with misconceptions on the concept of sustainability. *International Journal of Sustainability in Higher Education,* 1(1), 9-19.

Filho, W.L. (Ed.). (2002). *Teaching sustainability at universities: Toward curriculum greening.* New York: Peter Lang.

Finlay, J., Bunch, R., & Neubert, B. (1998*). Grey pinstripes with green ties: MBA programs where the environment matters.* Washington, DC: World Resources Institute.

Finlay, J., & Samuelson, J. (1999). *Beyond grey pinstripes: Preparing MBAs for social and environmental stewardship.* Washington, DC: World Resources Institute/Aspen Institute's Initiative for Social Innovation through Business.

Fussler, C. (1996). *Driving eco innovation: A breakthrough discipline for innovation and sustainability.* Washington, DC: Pitman Publishing.

Herremans, I., & Allwright, D.E. (2000). Environmental management systems at North American Universities: What drives good performance? *International Journal of Sustainability in Higher Education, 1*(2), 168-181.

Higher Education Funding Council for England. (1998a). *Environmental report (98/61).* London: Higher Education Funding Council for England.

Higher Education Funding Council for England. (1998b). *Environmental workbook (98/62).* London: Higher Education Funding Council for England.

Keniry, J. (1995). *Ecodemia: Campus environmental stewardship at the turn of the 21st century.* Washington, DC: National Wildlife Federation.

McIntosh, M., Cacciola, K., Clermont, S., & Keniry, J. (2001). *State of the campus environment: A national report card on environmental performance and sustainability in higher education.* Reston, VA: National Wildlife Federation.

Monteith, J., & Sabbatini, R. (1997). The evolving role of sustainability on the new campus of California State University. In: R. Koester (ed.), *Greening of the Campus II: The Next Step* (pp. 56-60). Muncie, IN: Ball State University.

National Wildlife Federation. (2001). Campus environmental report card. *National Wildlife Federation.* Retrieved February 17, 2001 from World Wide Web: http://www.nwf.org/campusecology/stateofthecampusenvironment/index.html.

O'Connor, J.C. (1995). Toward environmentally sustainable development: Measuring progress. In T. C. Trzyna (Ed.), *A sustainable world: Defining and measuring sustainable development* (pp. 87-114). Sacramento, CA: International Center for the Environment and Public Policy.

Onisto, L. (1999). The business of sustainability. *Ecological Economics,* 29, 37-43.

Penn State Green Destiny Council. (2000). *Penn State indicators report 2000: Steps toward a sustainable university.* State College, PA.

Roorda, N. (2000). Auditing sustainability in engineering education with AISHE. In *ENTREE 2000, EEE Network.* Brussels, Belgium.

Roorda, N. (2002). Assessment and policy development of sustainability in higher education with AISHE. In W. L. Filho (Ed.), *Teaching sustainability at universities: Towards curriculum greening* (pp. 459-486). New York: Peter Lang.

Smith, A.A., & The Student Environmental Action Coalition (1993). *Campus ecology: A guide to assessing environmental quality and creating strategies for change.* Los Angeles: Living Planet Press.

U.S. Interagency Working Group on Sustainable Development Indicators (1998). *Sustainable development in the United States: An experimental set of indicators.* Washington, DC: U.S Interagency Working Group on Sustainable Development Indicators.
University Leaders for a Sustainable Future (1999). *Sustainability assessment questionnaire (SAQ) for colleges and universities.* Washington, DC: University Leaders for a Sustainable Future.
Wackernagel, M., & Rees, W. (1996). *Our ecological footprint: Reducing human impact on the Earth.* British Columbia: New Society Publishers.

BIOGRAPHY

Dr. Michael Shriberg is Assistant Professor and Program Director for Environmental Studies at Chatham College (Pittsburgh, PA, U.S.A.). Michael received his Ph.D. and M.S. from the University of Michigan School of Natural Resources & Environment's Resource Policy & Behavior program. He has worked in the corporate and environmental consulting sectors. His current research explores organizational factors which determine why and how some campuses are emerging as sustainability leaders while most campus lag. He is also using this environmental management and organizational change model to assess prospects for sustainable business development. Michael is currently the North American Editor for the International Journal of Sustainability in Higher Education. He has lectured at conferences and campuses across the U.S. and world, and received research fellowships from organizations including the Frederick E. Erb Environmental Management Institute, the U.S. Environmental Protection Agency and the University of Michigan Rackham Graduate School.

CHAPTER 7

THE PROBLEMATICS OF SUSTAINABILITY IN HIGHER EDUCATION: A SYNTHESIS

Peter Blaze Corcoran & Arjen E.J. Wals

As the previous chapters illustrate, there is no one way of developing sustainability in higher education. There is not even one way of viewing sustainability. As has been pointed out in Part One, sustainability has a variety of meanings depending on the users, their backgrounds, interests and values, and the context in which it is used. Attempts to define and measure sustainability in order to contain it and provide it with a universal meaning have been manifold but have, by and large, failed. On the contrary, Dobson's research showed that in the mid-nineties three hundred definitions for sustainability and sustainable development were available, up from just a few in the late eighties (Dobson, 1996). After a decade of work to bring meaning to these terms there is perhaps less coherence and more divergence. The likelihood of arriving at some common understanding seems more remote than ever (Wals & Jickling, 2002).

One can argue that this "ill-definedness" (van Weelie & Wals, 2002) is as much a distinguishing property of sustainability as it is of the post-modernity that has tightened its grip on the social sciences and society. Postmodern social science has elevated the value of local knowledge, contextual meaning, and diversity of perspectives, thereby increasing the number of actors involved in decision and meaning-making. No longer is meaning pre-determined by a small group of so-called experts, instrumentally handed down and externally evaluated. Instead, meaning is often co-created in a specific context through a collaborative, and relatively open-ended, process that involves a broad range of stakeholders.

In a postmodern world, pathways towards sustainable universities are unlikely to develop without friction, controversy, and conflict. After all, we live in a pluralistic society, characterized by multiple actors and diverging interests, values, perspectives, and constructions of reality (Wals & Heymann, *in press*). The ill-defined and uncertain nature of working towards sustainable living and the complex and contextual nature of higher education itself, does not allow for universally applicable recipes for implementing sustainability in higher education. University boards cannot rely on the exclusive use of economic incentives, rules, standards, and regulations to enforce sustainability in higher education. At the same time, reliance on the instrumental use of education, training, and communication to promote or even force one particular view of sustainability, is problematic as well, particularly

Peter Blaze Corcoran & Arjen E.J. Wals (Eds.), Higher Education and the Challenge of Sustainability: Problematics, Promise and Practice, 87-88.

in higher education where critical and autonomous thinking should perhaps be emphasized the most.

When recognizing that sustainability is an ill-defined concept that derives meaning in a specific context with the involvement of multiple stakeholders, an important question is raised as to how one deals with the inevitable tension between the divergence of interests, values, and worldviews on the one hand - and the need for the shared resolution of issues that arise in working on sustainability in higher education on the other. This leads us to the role of multiple perspectives in exploring sustainability in higher education. We believe that pluralism of thought, when applied constructively, can be a driving force for reaching solutions to sustainability issues in higher education. It is this pluralism of thought that can lead to creative solutions to complex challenges. In Part Two, we begin to cultivate such pluralism while exploring the promise of sustainability from a variety of vantage points.

REFERENCES

Dobson, A. (1996). Environmental Sustainabilities: An Analysis and a Typology. *Environmental Politics*, 5(3), 401-428.

Wals, A.E.J. and Heymann, F.V. (2004). *Learning on the edge: exploring the change potential of conflict in social learning for sustainable living*. In: A. Wenden (Ed.) Working toward a Culture of Peace and Social Sustainability. New York: SUNY Press.

Wals, A.E.J. and Jickling, B. (2002). "Sustainability" in Higher Education from doublethink and newspeak to critical thinking and meaningful learning. *Higher Education Policy*, 15, 121-131.

Van Weelie, D. and Wals, A.E.J. (2002). Making biodiversity meaningful through environmental education. *International Journal of Science Education*, 24(11), 1143-1156.

PART TWO

PROMISE

CHAPTER 8

THE PROMISE OF SUSTAINABILITY IN HIGHER EDUCATION: AN INTRODUCTION

Arjen E.J. Wals & Peter Blaze Corcoran

As we see in Part One, sustainability can mean many things. Some are uncomfortable with this notion and will argue that when something is so inclusive it can mean anything and therefore becomes meaningless. They may be looking for clearly defined concepts that can be operationalized and translated into teachable products that, when fully mastered, will result in a better behaviors and lifestyles. A universal understanding of a concept is seen as a pre-requisite for making it a focal point for learning. There are also those who will argue that sustainability does have a clear meaning. There are many definitions of sustainability and as long as one is carefully selected and applied it can be a useful concept for teaching and learning for a better world.

We have taken the position that the multiple meaning of sustainability are not a weakness but a strength. The fact that it is ill-defined allows people to give it their own meaning as is appropriate for their own context. The process of giving meaning within a context is meaningful learning. Clearly there are different imaginable educational responses to sustainability.

In Part Two we introduce the reader to a variety of educational responses from the vantage point that a pluralism of perspectives can be a driving force for reaching solutions to sustainability issues in higher education. As we write in the synthesis of Part One, it is this pluralism of thought that can lead to creative solutions to complex challenges.

When responding to the challenge of sustainability in higher education, the emergence of conflicting perspectives or frames (Kaufman & Smith, 1999) is both inevitable and, when properly managed, desirable. Conflicting sustainability-related frames often result from a divergence of the frames people use to describe and interpret their reality. The selective elevation of particular frames and the (ab)use of power - the approach of 'singularism '- rarely lead to satisfying long-lasting results. Singularism tends to lead to two types of outcomes: a distributive outcome or a false consensus outcome (van Woerkum and Aarts, 1998). A distributive outcome results when, for instance, ecological frames are privileged over economic frames, or, as is more common, when economic frames are privileged over ecological ones. When this happens, a hardening of positions occurs and results are translated in terms of loss - gain or winners - losers. Heymann and Wals (*in press*) speak of a false consensus outcome when a third party neutral or mediator tries to mould the

Peter Blaze Corcoran & Arjen E.J. Wals (Eds.), Higher Education and the Challenge of Sustainability: Problematics, Promise and Practice, 91-95.

different frames into a seemingly shared reality to arrive at a solution that on the surface appears to be satisfactory to all involved but fails to address the deeper underlying issues.

Instead, the approach of pluralism appears to have more promise in the exploration of sustainability in higher education. Conflict resolution strategies based on pluralism are based on the premise that conflict is inherent to the interaction between dissimilar and diverse societal groups. The idea of pluralism is founded on the creation of respect and openness for being different, in other words, for being sensitive to frames that differ from one's own. In Part Two, we present contributions from a variety of perspectives on the role of sustainability in higher education and the role of higher education in society. In doing so, we hope to move beyond the pioneering and, therefore, important contributions made earlier by those closely connected to the environmental education field (i.e. Kormondy and Corcoran, 1998; Collett and Karakashian, 1996; Filho, 1999; 2002). The idea being that these multiple perspectives will provide a variety of insights and raise a number of questions that do not have clear cut answers but require a struggle to resolve their underlying tensions and conflicts. This struggle and the way it is facilitated or managed, we believe, is crucial for the emergence of a deep sustainability supported by multiple stakeholders involved in academia and which is continuously under scrutiny and re-construction. The vantage points for exploring sustainability in higher education we selected are from: environmental education, environmental justice, deep ecology, ecofeminism, transformative learning, natural resource management, whole systems thinking, and enriched disciplinary education.

In Chapter 9 Daniella Tilbury claims that Environmental Education for Sustainability has a critical and transformative contribution to make towards conceptualizing, motivating, and managing change in higher education. She interprets innovation as a political, economic, and social process, which challenges existing stakeholder relationships and redefines professional and institutional priorities. Tilbury identifies the distinctive contribution of environmental education to the challenge of sustainability in higher education. With specific reference to recent higher education initiatives, it critically reviews the major trends and developments in environmental education and gives three examples of how this can work. Tilbury explores the role environmental education can play, not only within, but also through higher education as a catalyst of change towards sustainability.

In Chapter 10, Julian Agyeman and Craig Crouch make the case that justice in its many forms should become a core part of sustainability education in university curricula. Taking as its starting point that sustainability is about improving the quality of human life in a just and equitable manner, while living within the limits of ecosystems, the chapter maps some of the thematic and content-based material that occurs at the nexus of justice and sustainability. These include unequal distribution of environmental benefits and detriments; community participation, advocacy and activism; biodiversity, biopiracy and cultural diversity, and environmental human rights. Agyeman and Crouch discuss the pedagogical and learning process implications involved in using this perspective. The implications include an emphasis on experiential learning, participatory research, teamwork, reflection and discussion. Such an approach, they argue, will serve to develop within students an improved ability to analyze societal problems and to realistically envision more positive, just, and sustainable futures.

In Chapter 11, Harold Glasser invites us to consider what kind of world we might have if we were constantly engaged by both reason and joy in the pursuit of exploring the connections among knowledge, values, and concrete actions. Glasser, drawing on Arne Naess' philosophy of deep ecology, explores the relationship of our values to our praxis. Might mindfulness of a wider web of environmental costs and consequences lead us to make changes in our behavior? He identifies four philosophical notions to reconsider our value priorities and to relate them to our actions: "deep questioning", consideration of "vital needs", the principle of "universalizability", and the concept of "beautiful action". He suggests these tools can inspire a sustainability mission for higher education that is based on resolving the paradoxical disjunction between people's stated environmental concern and un-environmental everyday behavior.

In Chapter 12, Annette Gough argues that women have a distinctive contribution to make to sustainability policy, pedagogy, and research. Gough discusses research into the gaps and silences present in policies, pedagogy, and research in sustainability from a feminist perspective and sketches possibilities for new directions when ecofeminist pedagogies and research methodologies are considered in the development of sustainability in higher education. She is convinced we can do many types of research and teaching which put the social construction of gender at the center of the inquiry, and that we need more stories from women's lives relating to environments that we can use in the development of sustainability in higher education. Her chapter discusses strategies for achieving this goal. Gough's contribution is a strong reminder of the powerful contribution women can make to shaping sustainability in higher education and points out that their omission will severely diminish such possibilities.

In Chapter 13, Edmund O' Sullivan develops the idea of a tranformative vision for higher education and its implications for sustainability. His fundamental assumption is that higher education must embark upon a discussion of this transformative vision if we are to have an education that holds the core values of sustainability. He places this question in a cosmological context and argues that universities founder for lack of a comprehensive context. He further advocates, using the suggestion of Thomas Berry, that "the university should be a place where the universe story is encountered." He the uses the Earth Charter as an example of a foundational document for establishing such a context and commitment for transforming higher education.

In Chapter 14, Röling tells us that the very "success" of industrial society has made humans a major force of nature creating the anthropogenic transformation of the earth that is fraying the web of life on which all organisms depend. Human behavior now is seen as the greatest threat to human survival. This means that we must seriously begin to ask the question how we can manage ourselves to create a sustainable society. The market largely fails in this respect and there are no technological fixes. There is, he argues, a need for a third way to getting things done: the interactive way, which focuses on sustainability as the emergent property of human interaction. Sustainability is seen to be the outcome of such processes as conflict resolution, negotiated agreement, social learning, distributed cognition, alleviating social dilemmas, building social capital, collective decision making, and concerted action. The chapter summarizes the essence of the approach and discusses its applications in higher education.

Whole systems design theory and practice which maximize stakeholders' involvement in management, engagement in core values and assessment in organizational change, lend themselves well to moving higher education towards sustainability, according to James Pittman. In Chapter 15, he suggests several critical elements integral to success in organizational change for "living sustainably through higher education." He further contends that such change has the potential to reach success far beyond previous scientific management models of organizational development.

Part Two concludes by looking the intellectual challenges of sustainable development from a disciplinary, rather than a multidisciplinary or transdisciplinary, perspective. In Chapter 16, Geertje Appel, Irene Dankelman, and Kirsten Kuipers outline a project in Dutch universities in which experts from various disciplines produced overviews on the fundamental linkages and applications of their fields to sustainable development. The reviews were based on interviews, literature study, and primary source material. The authors analyze the process of preparation, dissemination, and follow-up. They critique this methodology of enriched disciplinary education for sustainable development.

We trust the reader will relate the diverging frames presented in the various chapters to one another and also to her or his own perspective on sustainability as it relates to higher education. After all, this is what meaningful learning is: a change process resulting from a critical analysis of one's own norms, values, interests, and constructions of reality (deconstruction), exposure to alternative ones and the construction of new ones (reconstruction) (Wals & Heymann, 2004). Such a change process is greatly enhanced when the participants in the innovation process are mindful and respectful of other perspectives. In addition, there needs to be room for new views that broaden the realm of possibilities; in other words, when there is space for dialogue rather than just for the mere transmission or exchange of points of view. This is why we include multiple voices in Part Two of this book on sustainability in higher education, although we acknowledge that other voices are not represented. The voices that follow are well-articulated, thought-provoking, and promising in leading the way for sustainability in higher education.

REFERENCES

Collett, J. and Karakashian, S. (Eds.) (1996). *Greening the College Curriculum: A Guide to Environmental Teaching in the Liberal Arts.* Washington DC: Island Press.

Filho, W. (Ed.) (1999). *Sustainability and University Life.* Frankfurt Am Main: Peter Lang Scientific Publishers.

Filho, Walter Leal (Ed.) (2002). *Teaching Sustainability at Universities.* Peter Lang Publishing: Frankfurt am Main, Germany.

Kaufman, S. and Smith, J. (1999). *Framing and Reframing in Land Use Conflicts. Journal of Architecture, Planning and Research,* Special Issue on Managing Conflict in Planning and Design 16(2), 164-180.

Kormondy, E.J. and Corcoran, P.B. (1998). *Environmental Education: Academia's Response*. Troy, Ohio: NAAEE.

Wals, A.E.J. and Heymann, F.V. (2004). *Learning on the edge: exploring the change potential of conflict in social learning for sustainable living*. In: A. Wenden (Ed.) Working toward a Culture of Peace and Social Sustainability. New York: SUNY Press.

Woerkum, C.J. and Aarts, N.M.C. (1998). *Communication between farmers and government over nature: a new approach to policy development*. In: Roling, N.G., and Wagemakers, M.A.E. (Eds.) Facilitating sustainable agriculture: participatory learning and adaptive management in times of environmental uncertainty. Cambridge University Press, Cambridge, UK, p. 272-280.

CHAPTER 9

ENVIRONMENTAL EDUCATION FOR SUSTAINABILITY: A FORCE FOR CHANGE IN HIGHER EDUCATION

Daniella Tilbury

INTRODUCTION

Sustainability is becoming increasingly pertinent to higher education. Forward-looking higher education institutions are establishing relevant links between widely accepted policies on environmental protection, social justice, economic development and the way they run their institutions and provide learning experiences for students (see Forum for the Future et al., 1999; Benn, 1999; Calvo, Benayas & Guitierrez, 2002). These institutions are linking learning, innovation and competitiveness to sustainable development. Some recognise that a country's future prosperity rests on its people's ability to address sustainability issues and have embraced sustainability at its core – rethinking their missions, developing visions and strategic plans across the University (see Forum for the Future et al., 1999). Indeed, over one thousand university presidents and vice-chancellors have signed the Halifax Declaration (1991), Swansea Declaration (1999) Copernicus Charter (1994), Talloires Declaration (1999), Kyoto Declaration (1993) and Lunenburg Declaration (2001), committing their institutions to change towards sustainability.

Others, resisting systemic change but increasingly influenced by internal and external stakeholders, are responding through discrete initiatives or pilot projects which are impacting in some administrative and/or curriculum areas within their institutions (e.g. Fien, Heck & Ferriera, 1997; Bowdler et al., 2001; Tilbury, Reid & Podger, 2002). These projects are financed by intergovernmental agencies such as EU and UNESCO (e.g. Fien & Tilbury, 1996; Geli & Junyent, 2002); national and state government agencies (e.g. Watkin et al., 1995; Tilbury, Reid & Podger, 2002) and NGOs (e.g. Forum for the Future et al., 1999). Student bodies and other on-campus groups are also contributing by promoting sustainable living projects and influencing University practice through student union politics (e.g. Bowdler et al., 2001).

However, many of these efforts have focused on actions to minimise the ecological footprints of universities. This is being achieved through reducing levels of energy consumption, opting for more sustainable waste management practices and putting in place environmental managements systems to monitor impacts. A

Peter Blaze Corcoran & Arjen E.J. Wals (Eds.), Higher Education and the Challenge of Sustainability: Problematics, Promise and Practice, 97-112.

number of these initiatives have also involved students in learning about and/or managing this innovative practice (Campus Earth Summit, 1995; Calvo, Benayas & Guitirrez, 2002).

It is now being recognised that a next and more critical step needs to be taken to address sustainability through higher education. This requires educating about and for sustainability through the taught curriculum (see Richardson & Ali-Khan, 1995; Alabaster & Blair, 1996; Benn, 1999; Bowdler et al., 2001). Calls to restructure higher education courses towards Environmental Education for Sustainability are being supported by the corporate sector, which seeks graduates with the personal and professional knowledge, skills and experience necessary for contributing to sustainability. Corporate stakeholders, attending a recent University-Industry Summit, argued that every student, regardless of specialism, should have opportunities to learn about and for sustainability in higher education (Tilbury & Cooke, 2002).

However, this form of education commonly referred to as Environmental Education for Sustainability cannot simply be integrated into existing curricula. Many attempting this task are encountering a number of challenges at the conceptual, planning and management level (UNESCO NIER, 1996; Yeung, 1996; Tilbury & Turner, 1997). This chapter defines the conceptual aspects of Environmental Education for Sustainability and describes how higher education institutions are beginning to grapple in practice with a process that has 'learning for change' at its core aim.

The chapter argues that Environmental Education for Sustainability is an innovative and interdisciplinary process requiring participative and holistic approaches to the curriculum. It cannot be inserted into existing teaching and learning structures. Environmental Education for Sustainability has a transformative agenda that requires, and often leads to, professional, curriculum as well as structural change (Robottom, 1987; Tilbury, 1998a & b). Innovation not integration lies at the heart of Environmental Education for Sustainability (Tilbury, 1998a). The chapter identifies the challenges that this transformative agenda presents to institutions in higher education. It sees this learning process as the next stage in the challenge towards sustainability and argues that only by engaging with Environmental Education for Sustainability, will higher education institutions contribute to building social capacity for change towards sustainability. It predicts that the institutions themselves will be the subject of change (and not just a vehicle for change) as teachers and students engage in making changes for a better world.

ENVIRONMENTAL EDUCATION FOR SUSTAINABILITY

At Johannesburg, UNESCO explicitly recognised the critical role that formal and higher education play in providing opportunities for social learning and change towards sustainable development (UNESCO, 2002 p. 7). Their WSSD document *'Education for Sustainability: From Rio to Johannesburg'* (UNESCO, 2002) called for socially critical forms of learning which could help us transform the world we live in. To achieve an improved environment and quality of life, it argued, we need active and knowledgeable citizens as well as informed decision makers capable of making the right choices about complex and interrelated economic, social and environmental issues facing the world today (UNESCO, 2002, p. 7). It identified a need for an education that questioned our current mental models and assumptions that underpin them and called for 'deeper, more ambitious ways of thinking about education, one that retains a commitment to critical analysis while fostering creativity and innovation' (p. 8). This interpretation of education has been promoted by environmental educators such Saul (2000) who calls for culturally critical perspectives and by Huckle (1983; 1996; 1997) who argues that only through asking socially critical questions can we progress towards a more sustainable future.

In fact, the transformative goals outlined in the above document mirror those contained in earlier environmental education texts that have informed practice over the last thirty years. It was the 1972 United Nations Conference on Human Environment that formally acknowledged the emerging field of environmental education and recommended it be promoted in all countries. The United Nations International Environmental Education Programme (IEEP, 1975-1995) which resulted from this recommendation did much to inspire and promote environmental education practices that questioned thinking and assumptions as well as action for change. This program was the first to introduce concepts of sustainability into higher education. The 1977 Tbilisi Declaration gave momentum to these early commitments to environmental education, although translating these into practice, as the UNESCO (2002) document recognises, has proved to be a considerable challenge. Current education practice, UNESCO argues, 'falls far short of what is required' (UNESCO, 2002, p. 9).

Thinking Critically, Thinking Culturally

> Sustainable Development is more about new ways of thinking than about science and ecology. While sustainable development involves the natural sciences, policy and economics; it is primarily a matter of culture' (UNESCO, 2002, p. 8).

Environmental educators have argued for over a decade that formal and higher education must engage learners in critical reflection - a process needed to interpret the root causes of environment and development problems, to challenge bias, support rational decision-making and to examine personal and political contributions to change (Fien, 1993; Tilbury, 1993; 2001; Huckle, 1996; 1997; Sterling, 1996). Critical reflection arose out of the ideas of Jurgen Habermas. The social-cultural theory he advocated drew on both Weber and Marx but shifted the focus from social relations of production to social interaction and the nature of language and morals. Huckle (1997) interprets Habermas' principal claim to be that interaction has become distorted by the rise of positivism and instrumental reasoning which promotes science as meta-narrative and value-free knowledge. This, according to Habermas, fosters a distorted and incomplete understanding of our relations with one another and the rest of the world. Critical theory seeks to reveal this distorted or incomplete rationality and empower people to think and act in genuinely rational and autonomous ways (Fien,1995; Huckle, 1997).

Saul (2000) is critical of the how this process is practiced, and argues that models of critical rationality currently used blinds learners to cultural complexities. Teaching critical rationality, he argues is not enough. Echoing, Habermas' initial thinking, Saul argues that we need to teach learners that often conflicts are not only about rational arguments, but about the clash of cultural values and perspectives (Saul, 2000). Saul contends that:

> 'Environmental problems result from environmental practices and environmental
> practices are cultural activities.....we need to teach how culture works, because cultural
> differences frame what are seen as rational arguments.' (Saul, 2000, p. 7)

To achieve sustainable development we need critical reflective models which will help learners 'not only think critically but also culturally' (Saul, 2000, p. 8). Values clarification is a process that can help learners uncover the layers of assumptions and deconstruct socialised views. It can help them engage in a critical review of their own environmental and political values as well as help them comprehend that other complex cultural perceptions exist (Tilbury, 2002). It has been used extensively in environmental education but originates from the global studies and development education movement of the 1970s that developed alternative and interactive approaches for teaching for a better world (Tilbury, 1993). Values clarification resists the reduction of complex situations into simplified binary oppositions that often develops when controversy arises. It can develop learners who are aware and critical of cultural perceptions and processes that lead us to unsustainable development.

Critical thinking and values clarification approaches help learners take the first steps towards learning for change. Experience and competence in action taking is also critical if learners are to transform their thinking into practice. Action research offers opportunities for learners to develop critical thinking, values clarification, action and management skills to address sustainability at the practical level.

Participatory Action Research

Participatory action research is a process, rooted in the critical theory paradigm, which engages learners in practical issues of power, politics and participation. It is a process used by environmental educators to assist communities to engage in environmental problem-solving and social change (see Gough & Robottom, 1993; Stapp & Wals, 1994; Allen, 2000). It can empower individuals and communities by providing them with support and experience in planning and managing change (Tilbury, 2002).

Four basic themes underpin action research approaches: i) collaboration through participation; ii) acquisition of knowledge; iii) social change; and iv) empowerment of participants (Hillcoat, 1996). A key assumption underlies action research - that effective social change depends on the commitment and understanding of those involved in the change process.

Through the action research process, the researcher engages in a spiral of cycles consisting of phases of planning, acting, observing and reflecting (Masters, 1995). The goal of the action researcher is to increase the closeness between the actual problems encountered by practitioners in a specific setting and the theory used to explain and resolve the problem. The second goal, which goes beyond the other two approaches, is to assist practitioners in identifying and making explicit fundamental problems by raising their collective consciousness (Holter et al., 1993).

Participatory action research resulted from a Group Dynamics Movement which used action research to address racial prejudice and for social reconstruction after World War II. Lewin (1947) was the most distinguished researcher of this group. He used action research as a form of experimental enquiry based upon groups experiencing problems. Since then, action research has evolved into a socially critical process which promotes a critical consciousness - which exhibits itself in political as well as practical action to promote change.

Core Components

The terms 'critical reflection', 'values clarification' and 'participative action research' have become core components of Environmental Education for Sustainability (see Sterling et al., 1992; Fien & Trainer, 1993; Gough & Robottom, 1993; Huckle & Sterling, 1996; Huckle, 1997; Robottom, 1987; Fien & Tilbury, 1996; Hesselink et al., 2000; Tilbury, 1993; 2001a; 2001b). These approaches provide opportunities for students; to engage in critically reflecting upon the basis of their socio-cultural values and assumptions; to identify how they are conditioned and confined by the socio-cultural structures they are operating in and, more significantly, to build their capacity as agents of change. They are perceived as critical to addressing capacity building and education needs for sustainable development.

Potentially any subject is a resource for learning for change towards sustainability through environmental education (Alabaster & Blair, 1996). It solely requires that these key learning components form part of the curriculum in higher education. Curriculum developers and lecturers from Languages, Social Sciences, Sciences and the Arts can address these generic skills and provide learning experiences which students need to contribute to sustainability at the institutional, professional and personal level.

CHANGE IN HIGHER EDUCATION

'If it is to fulfil its potential as an agent of change towards a more sustainable society, sufficient attention must be given to education as the subject of change itself.' (Sterling, 1996, p. 18)

Sterling (1996) argues that education itself must be transformed if it is to be transforming and sees environmental education for sustainability as the catalyst for this change. The paradigm of learning promoted by environmental education for sustainability offers possibilities for changing the very core of teaching, administration and management in Universities (Ali-Khan, 1990). It will leave its mark in higher education as it begins to involve teachers and students in education for social change.

Environmental Education for Sustainability can activate and/or support change at a number of levels and has a particular contribution to make towards conceptualising, motivating and managing change. However, to understand Environmental Education for Sustainability contribution to higher education, one needs to understand how it perceives the change process as well as how it conceptualises sustainability.

Conceptualising Change

Environmental Education for Sustainability promotes holistic and systemic change (Sterling, 1996) recognising that efforts need to be made simultaneously at a number of levels for meaningful change to occur (Tilbury, Reid & Podger, 2002). This process facilitates change by confronting the complexity, tensions and relationships within a system as well as by offering an integrated response. Sterling (1996, p. 23) describes the approach to change promoted by environmental education for sustainability as 'holistic but human' in scale:

> '....recognising that all educational dimensions, such as curriculum, pedagogy, structure, organisation and ethos are mutually affecting and need to be seen as consistent whole.'

Those engaged in change through Environmental Education for Sustainability would ask questions such as; 'how do people, institutions and communities interact- what is the hidden and operational curriculum?; how do we engender a sustainability ethos that is both lived and critically reflected upon?; how do we change towards democratised classrooms and decision-making; how do we change our management structures to sustain the changes?; how can higher education become a learning centre for the whole community?; how do we establish meaningful links with community stakeholders? (adapted from Sterling, 1996, p. 36).

Motivating and Managing Change

> Since 1992, an international consensus has emerged that achieving SD is essentially a process of learning' (UNESCO, 2002, p. 7).

Environmental Education for Sustainability, interprets sustainability as a process rather than a concept to be implemented. This process is also seen as multi-dimensional (political, economic and social) and ultimately seeking cultural change. It places learning as the driving force for change and at the heart of a sustainable society (Huckle & Sterling, 1996; Dovers, 2002).

Learning is vital to motivate and manage meaningful change. Connor and Dovers recently undertook an empirical study in an attempt to construct models of institutions for sustainability (2002, p. 10). Recognising the role of learning and supporting adaptive management approaches to change, their research report argues that:

> '...sustainability is an ideal and not something likely to be fully achieved any time soon. It is a matter for ongoing social consideration at the most serious level, and requires

mechanisms to accumulate experience and knowledge of decision-making so that
learning may proceed into the far future'

Environmental Education for Sustainability sees both sustainability and change
towards sustainability as engaged and participative rather than passive and promotes
democratic ownership of change (Sterling, 1996). It provides an effective way of
rooting change by engaging stakeholders in changes towards sustainable
development. The Johannesburg Declaration (UN, 2002) and Agenda 21 (UN, 1992)
saw stakeholder engagement as a critical first step. Connor and Dovers (2002)
remind us that engaging stakeholders meaningfully requires more than just
consultation, particularly when institutional change is a goal. They call for
opportunities for stakeholders to: vote in decision-making; contribute to knowledge
generation; be given ownership so that they themselves contribute to change (p. 9).
Ali Khan (1996) interprets meaningful change as involving internal and external
stakeholders in constructing alternative futures and repositioning higher education
with its community. Environmental education for sustainability can offer a process
for facilitating change within an institution as it directly addresses these dimensions
of power, politics and participation for change.

INNOVATION AND CHANGE: THE ROLE OF ENVIRONMENTAL EDUCATION FOR SUSTAINABILITY

As argued above, Environmental Education for Sustainability can be a force for
change through and within higher education, when interpreted as more than the
greening of the university or the integration of the environmental perspective into
the curriculum. It has argued the need to educate for cultural change not just campus
or curriculum change.

There are different approaches to motivating and managing cultural change
within higher education. However, critical to the success of innovation in this area is
the need to recognise the multi-faceted nature of change

Summarised below are three Environmental Education for Sustainability projects
taking place within one university. Although the projects take different approaches
to motivating and managing cultural change towards sustainability, all three
recognise the multi-faceted nature of change and the need to engage stakeholders in
defining and managing change. The projects tackle professional, curriculum and
institutional innovation simultaneously.

Project 1 – MU Summit in Environmental Education for Sustainability- Change on Higher Education

When interpreted as a process of learning which enhances generic skills, rather than a content to be taught, Environmental Education for Sustainability is relevant to all fields of learning. The key issue is often that few higher education staff is not familiar with Environmental Education for Sustainability as a process of learning. Many see it as environmental content to be taught rather than a process of inquiry and therefore struggle to see the relevance of this concept to their field of knowledge. Only a handful of Universities confronting the challenge of sustainability have invested in professional development to support staff that is questioning how sustainability is relevant to their teaching and research activities.

The issues identified above were raised at a recent Summit on 'Building Capacity for a Sustainable Future: Environmental education for Sustainability' held at Macquarie University in Sydney, Australia (Tilbury & Cooke, 2002). The event was an initiative of the *National Environmental Education Council* (NEEC), an advisory body established, by the Commonwealth Government in July 2000, to advise the Federal Minister for the Environment and Heritage on priority environmental education needs. The Summit which brought together University and Industry leaders is one in a series being held throughout Australia to help address emerging questions in the education for sustainability arena.

At the University-Industry Summit, the business and industry speakers spoke about the sustainability challenges they face and the difficulties of recruiting graduates who have the necessary skills to address these needs. The speakers referred to developments in environmental management and triple bottom line reporting and spoke of how sustainability requires these initiatives to establish greater links with processes of risk assessment, social equity and corporate responsibility. The suggestion was that there had been little engagement in higher education teaching with these more complex issues.

Many speakers identified the need for universities to develop graduates with creative and futures thinking skills necessary for initiating and managing changes towards sustainability. The importance of teaching students how to think and reflect critically, regardless of their area of specialism, was also highlighted. There was agreement that sustainability should not just be the concern of environmental disciplines. There were calls for reconceptualizing how and where sustainability is taught within higher education. All students, it was argued, need to be exposed to sustainability. There was also the suggestion that sustainability should be compulsory to students undertaking professional courses and that accountancy, law and business management faculties should assess student competencies in this area.

Many generic skills were seen as necessary to address the sustainability challenge within the corporate sector. Well-developed communication skills were perceived as vital to effectively promote changes towards sustainability from within the 'tent'. The ability to problem-solve and cope with uncertainty, innovation and risk was also cited as necessary for graduates wishing to contribute positively to business and industry within the area of sustainability. Some members of the audience highlighted the need to explicitly identify these generic skills within the position descriptions used to recruit graduates. This, it was argued, would raise their profile amongst students and strengthen University responses to education for sustainability.

There was general agreement, amongst those present at the Summit, that there is lack of graduates with the necessary skills needed to address the sustainability challenges within the corporate sector. Innovative programs, which helped develop professional skills in this area, were valued but were considered to be rare. As a result the following recommendations were made:

1. Universities need to provide students with the *critical, creative and futures thinking skills* to develop innovative and alternative solutions to sustainability issues.
2. Universities need to provide students with *needs assessment and action-oriented skills* needed to motivate, manage and measure change towards sustainability.
3. Universities need to provide students with the *interpersonal and intercultural skills* needed to redefine relationships amongst stakeholders (directors and board executives, workforce, legislators and government agencies, clients, community). This is key to addressing the challenge of sustainability.
4. Universities need to provide students with the confidence and skills to deal with *complexity and uncertainty.*
5. Universities need to review their curriculum to ensure the effective development and assessment of *generic skills* in education for sustainability (outlined in 1.-4.) across all faculties.
6. Universities need to involve a range of stakeholders (including business and industry) in overseeing this curriculum review process.
7. Universities need to increase opportunities for students to learn through engaging with real and specific problems or tasks.
8. Universities need to help create/facilitate active networks between business professionals and future graduates. Knowledge on sustainability comes through dialogue and through reflecting upon experiences. The networks need to incorporate diverse experiences at the international and well as national level.
9. Universities need to develop and strengthen partnerships with business and industry which will allow for increased opportunities for:
   a) students to engage directly with industry e.g. through work placements
   b) industry speakers to contribute to undergraduate and postgraduate courses
   c) across the disciplines.
   d) the university to support in-house training for graduates
   e) faster industry uptake of innovative new approaches to sustainability
   f) developed from student projects
10. Universities need to offer opportunities for professional development of staff in education for sustainability.
11. Universities need to ensure that all graduates, regardless of specialism have opportunities to learn about and for sustainability.

These recommendations were sent to Australian Vice-Chancellors Committee, senior management in Universities within New South Wales, as well as to educators grappling with the issue of how to integrate sustainability knowledge and skills into their teaching programs. The recommendations provide a case for redefining priorities within an institution, for engaging closely with external stakeholders to construct a new curriculum and for supporting those planning innovation to meet sustainability needs within higher education.

A major outcome of the Summit was the setting up of an Industry Advisory Group on Sustainability which currently informs the development of University-Industry activities at Macquarie University. The Advisory Group is composed of a representative from the National Environmental Education Council; Commonwealth Bank of Australia, Clayton Utz Law; Environment Institute of Australia; Integral Energy; Ecosteps Consultancy; Environs Local Government Australia; Australian Institute for Corporate Citizenship; BP Australia.

This initiative is an example of how external stakeholders are providing the motivation and often, financial support for institutional change towards sustainability. Many stakeholders expect to be involved in defining, planning and/or managing this change process. Environmental Education for Sustainability can offer opportunities to bring together stakeholders – internal and external – and offers approaches for constructing future scenarios and repositioning higher education within the community.

Project 2 - Masters in Sustainable Development – Change through Higher Education

Critical praxis is a core component of the Masters in Sustainable Development offered by the Graduate School of the Environment. Through a number of short courses, the program engages students in processes of critical reflection, values clarification and action research. Its objective is to develop the professional and personal skills needed to make a contribution to sustainable development. The course encourages the learner to explore notions of culture, identity and sustainability through multidimensional frameworks.

The Masters course adopts a broad definition of culture which highlights human activity and social meaning. In '*Introduction to Sustainable Development*' students explore, through a range of case studies, how culture can be a key determinant of sustainable development. In the workshops they are given opportunities to reflect critically upon their cultural assumptions, values and social relations and how these inform their decisions, influence lifestyle choices and actions. Their assignments require them to identify and reflect upon the socio-cultural context within which the Major Groups (defined by Agenda 21) operate and the implications of these frames for sustainable development strategies.

In '*Education for Sustainable Development*' students explore participatory approaches to learning and capacity building. They experience and reflect upon a democratic pedagogy that recognises the individual within the group, promotes social engagement and cooperation. In this course students engage with critical education approaches such as values clarification and critical reflection – and are required to keep a journal to explore the value and limitations of these potentially transformative processes. Their assignments require them to design and facilitate a workshop, which encourages the learner to ask critical questions, reflect upon new knowledge and make and enact choices about their futures.

'Ecotourism for Sustainable Development' encourages students to question the existing and potential contribution of this economic activity to sustainable development. Through a number of case studies, students consider whether ecotourism protects or threatens communities with environmental and cultural erosion. They experience a range of community and stakeholder participation processes for the planning, management and monitoring of ecotourism activities. The course is based on active, collaborative and reflective pedagogical approaches.

'Action Research for Sustainable Development' also engages students with transformative approaches, which empower citizens for changes towards sustainable development. This process sensitises students to the role of power, politics and participation in socio-cultural change for sustainable development. The course introduces students to research paradigms and the ontological and epistemological assumptions that underpin them. Learners explore the contribution of critical theory and action research to sustainable development. Through the assignments student develop practical knowledge and skills in action research planning, data collection and analysis techniques and evaluation. The assignments in this course are designed to further develop student's skills in critical reflection, values clarification and collaborative action.

This Masters in Sustainable Development attracts national and international students who have an interest in environment and/or development. The students participate in the core courses outlined above and then choose another four options to further develop their knowledge and skills in planning and facilitation, social and cultural impact assessment, values-clarification, critical thinking, interpretation, communication and decision-making in sustainable development.

The course provides learners with experience of processes that enable socially critical pedagogy and democratic problem-solving associated with critical theory. Through these, students explore different kinds of practical and theoretical knowledge to decide what people can, might, do to attain sustainable development. The Masters promotes critical praxis - a pedagogy that integrates reflection and action. Critical praxis was developed by Freire (1972) to raise the consciousness of learners to dominant ideological interests present in their socio-cultural environment and to engage them in reflective action (praxis).

The course is educating students about and for change. Not surprisingly a significant number of students who have successfully completed the Masters course have returned to their communities and workplaces and are developing programs to address sustainability. These programs have and are being developed in NGOs, local government sector, environmental consultancy firms, energy and water utility companies, building and planning authorities; landscape design organisations; aid agencies, amongst others. The environmental education for sustainability approaches developed by the course have provided the knowledge and skills for students to motivate and facilitate social change.

Project 3: Action Research for Change in Curriculum and Graduate Skills Towards Sustainability – Change in Higher Education

This two year pilot project funded jointly by Environment Australia and Macquarie University is investigating ways post-graduate lecturers interpret sustainability and creativity as graduate attributes and involve their students in exploring these ideas. It is concerned with identifying ways action research, grounded in the critical theory paradigm, enables lecturers through reflective praxis to transform thinking and action towards sustainability. The project adopts an innovative approach combining two research methodologies to explore sustainability across disciplines .

The phenomenographic component of the project aims to investigate and identify the range of qualitatively different ways that teachers of post-graduate students understand sustainability, creativity and innovation. Identification is significant, as it will inform the development of curriculum, and provide a theoretical focus for learning through action research activities. Phenomenography can be seen to richly describe the object of study through an emphasis on describing the variation in the meaning that is found in the participants' experience of the phenomenon and are reported in the form or hierarchically related categories. (Tilbury, Reid & Podger, 2002). Phenomenography examines the experience of each participant and recognises that each person's experience is an *internal* relation between the subject and the object, in other words, between the participant and the phenomenon. However, it is the structure of the variation across the *group* that emerges through iterative readings of descriptions of the experience that is the object. This is an innovative component of this project as academics' thinking on sustainability and creativity as higher order graduate attributes has never been explored.

As mentioned previously, action research combines critical intellectual discourse with practical action. It can be used as a collaborative research tool, which is often represented as a four-phase cyclical process of critical enquiry of praxis - plan formulation, action, outcome observation and reflection. The project uses this process to engages lecturers in critical research into contextual epistemology, pedagogy and ontology - transforming thinking and action towards sustainability. Participants carry out individual research involving implementation of plans, collection of data through observation, evaluation and reflection, and revision of plans with focus groups. The four tasks involve action researchers in i) exploratory sessions in action research, sustainability and education for sustainability; (ii) mentoring on the development of individual research plans, identifying the research focus, critical questions and plans for action; (iii) enacting plans (iv) group workshops to review and discuss research plans throughout the research process; (iv) post research evaluation (Tilbury, Reid & Podger, 2002). The project, thus, involves both the individual and individuals collaborating as a group in order to explore creativity and sustainability within the teaching context and to subsequently contribute to change towards sustainability.

Lecturers from across the University are participating in this project. Although it is still early days, initial feedback suggests that these lecturers are beginning to sow the seeds of change within the curriculum and are challenging some of the management and administration structures of the institutions. This project, informed and supported by the MU Industry Advisory Group (see project 1) and co-facilitated by past students of the Masters in Sustainable Development (see project 2) has great potential for changing higher education itself.

CONCLUSION

A number of key anchor points underpin the ideas articulated in this chapter:

1. Environmental Education for Sustainability has a transformative agenda. The chapter defines Environmental Education for Sustainability as a process that a) questions current mental models and the assumptions that underpin them, and; b) promotes new ways of thinking through fostering creativity and innovation. Critical reflection, values clarification and participatiory action research are identified as core components of this learning process. They are seen as critical to capacity building for sustainable development. The chapter argues that potentially any subject is a resource for learning for change towards sustainability through environmental education (Alabaster & Blair, 1996). It solely requires that these core learning components form part of the curriculum in higher education.

2. Environmental Education for Sustainability can activate and/or support change at a number of levels and has a particular contribution to make towards conceptualising, motivating and managing change. The chapter has argued that this process facilitates change by confronting the complexity, tensions and relationships within a system as well as by offering an integrated response. It promotes holistic and systemic change (Sterling, 1996) recognising that efforts need to be made simultaneously at a number of levels for meaningful change to occur (Tilbury, Reid & Podger, 2002).

3. Innovation not integration lies at the heart of Environmental Education for Sustainability. The chapter argues that once engaged with this learning process, higher education itself will also be the subject of change, as teachers and students engage in making changes for a better world. Those engaged in change through Environmental Education for Sustainability would ask questions such as; 'how do people, institutions and communities interact- what is the hidden and operational curriculum?; how do we engender a sustainability ethos that is both lived and critically reflected upon?; how do we change towards democratised classrooms and decision-making; how do we change our management structures to sustain the changes?; how can this higher education become a learning centre for the whole community?; how do we establish meaningful links with community stakeholders? (adapted from Sterling, 1996 p. 36).

4. Environmental education for sustainability can offer a process for facilitating change within an institution as it directly addresses these dimensions of power,

politics and participation for change. Internal and external stakeholders are already playing a critical role in shifting the organisational culture and redefining professional and institutional priorities. The chapter argues that Environmental Education for Sustainability can offer a methodology that brings together stakeholders – internal and external – and offers approaches for constructing future scenarios and repositioning higher education within the community.

5. Learning is a critical component of the process of change towards sustainability. Moving towards sustainability requires mechanisms to accumulate experience and knowledge of decision-making so that learning may proceed into the future (Connor & Dovers, 2002).The paradigm of learning promoted by Environmental Education for Sustainability offers possibilities for changing the very core of teaching, administration and management in Universities (Ali-Khan, 1990).

REFERENCES

Alabaster T., and Blair, D. (1996). Greening the University. In: Huckle, J. and Sterling, S. (Eds.) *Education for Sustainability.* London: Earthscan Publications.

Ali-Khan, S. (1990). *Greening the Curriculum.* Committee of Directors of Polytechnics, London.

Ali-Khan, S. (1996). 'A Vision of a 21st-Century Community Learning Centre' In: Huckle, J. and Sterling, S. (Eds.) *Education for Sustainability.* Earthscan, London pp. 222-228.

Benn, S. (1999). *Education for Sustainability: Integrating Environmental Responsibility into Higher Education Curricula.* Sydney: Institute of Environmental Studies, University of New South Wales.

Bowdler, L., Bowly, N., Cooke, K., McPherson, S., Morris, M., Podger, D. (2001). *Audit of Education for Sustainable Development Provision in the Division of Environment and Life Sciences at Macquarie University.* Sydney: Macquarie University (Unpublished).

Calvo, S. Benayas, J. and Guiterrez, J. (2002). 'Learning for Sustainable Environments: Greening Higher Education in Spain'. In: Tilbury, D., Stevenson, R. Fien, J. and Schreuder, D. (Eds.) *Education and Sustainability: Responding to the Global Challenge.* Gland: IUCN.

Campus Earth Summit (1995). *Blueprint for a Green Campus: The Campus Earth Summit Initiatives for Higher Education.* Yale University: Heinz Family Foundation.

Connors, R.D. and Dovers, S. (2002). *'Institutional Change and Learning for Sustainable Development'* Working Paper 2002/1. Centre for Resource and Environmental Studies, Australian National University.

COPERNICUS secretariat (1994). *COPERNICUS: The University Charter for Sustainable Development.* Geneva: COPERNICUS. http://www.copernicus-campus.org/index.html

COPERNICUS secretariat (2001). *The Lunenburg Declaration* Geneva: COPERNICUS. http://www.copernicus-campus.org/index.html

Dovers, S. (2001). *Institutions for Sustainability.* Melbourne: Australian Conservation Foundation.

Fien, J. (1993). *Environmental Education: A Pathway to Sustainability.* Geelong: Deakin University.

Fien, J., Heck, D. and Ferreira, J. (1997*). Learning for a Sustainable Environment: A professional development guide for teachers.* Brisbane: Griffith University.

Fien, J. and Trainer, T. (1993). Education for Sustainability. In: Fien. J. (Ed.). *Environmental Education: A Pathway to Sustainability.* Geelong: Deakin University Press, pp.11-23.

Fien, J. and Tilbury, D. (1996). *Learning for a Sustainable Environment: An Agenda for Asia and the Pacific.* Bangkok: UNESCO Asia Pacific Centre for Educational Innovation for Development BKA/96/M/252-500.

Forum for the Future, Department of Environment, Transport and Regions and Higher Education Funding Council for England (1999). *Learning for Life: Higher Education and Sustainable Development.* London: Forum for the Future.

Geli, A., and Junyent, M. (2002). *Reorienting Higher Education Studies Towards Sustainability: Designing Interventions and Analysing the Process.* Spain: University of Girona (unpublished).

Gough, A. and Robottom, I. (1993). Towards a socially critical environmental education: water quality studies in coastal school, *Journal of Curriculum Studies*, 25 (4), 301-316.

Hillcoat, J. (1996). Action Research. In: Williams, M. (Ed.) *Understanding Geographical and Environmental Education: The Role of Research.* London: Cassells.

Huckle, J. (1983). *Geographical Education, Reflection and Action* Oxford: Oxford University Press

Huckle, J. (1996). Realizing Sustainability in Changing Times. In: Huckle, J. and Sterling, S. (Eds.) *Education for Sustainability.* London: Earthscan.

Huckle, J. (1997). Towards a critical school geography. In: Tilbury, D. and Williams, M. (Eds.) *Teaching and Learning Geography.* London: Routledge, pp.241-244.

Huckle, J. and Sterling, S. (1996). *Education for Sustainability.* London: Earthscan.

International Association for Universities (1991). *The Halifax Declaration* http://www.unesco.org/iau/ftsd_halifax.html

International Associations for Universities (1993). *The Swansea Declaration* http://www.unesco.org/iau/tfsd_swansea.html

International Associations for Universities (1993). *The Kyoto Declaration* http://www.unesco.org/iau/tfsd.html#THEKYOTO

Lewin, (1947). Group Decisions and Social Change. In: Newcomb, T., and Hartley, E. (Eds.) *Readings in Social Psychology.* New York: Henry Holt, pp.330-344.

Richardson, S. and Ali-Khan, S. (Eds.) (1995). *The Environmental Agenda: Taking responsibility promoting sustainable practice through higher education curricula.* London: Pluto Press.

Robottom, I. (1987). The dual challenge of professional development in environmental education. In: A. Greenall (Ed.) *Environmental Education: Past, Present and Future.* Canberra: AGPS.

Saul, D. (2000). Expanding environmental education: thinking critically, thinking culturally. *Journal of Environmental Education*, 31(2), 5-7.

Stapp, W. and Wals, A.E.J. (1994). An action research approach to environmental problem-solving: theory, practice and possibilities in environmental education. In: Bardwell, L.V., Monroe, M.C. and Tudor, M.T. (Eds.). *Environmental Problem Solving: Theory, Practices and Possibilities in Environmental Education.* Troy, OH: NAAEE, pp. 49-66.

Sterling, S. and the Environment, Development and Training Group (1992). *Good Earthkeeping: Education and Training for a Sustainable Future.* London: EDTG.

Sterling, S. (1996). Education in Change. In: Huckle, J and Sterling, S. (Eds.). *Education for Sustainability.* London: Earthscan, pp.18-39.

Tilbury, D. (1993). *Environmental Education: a model for teacher education.* Unpublished, PhD Thesis, University of Cambridge CHECKB

Tilbury, D. and Turner, K. (1997). Environmental Education in Europe: Philosophy into practice. *International Journal of Environmental Education and Communication*, 16(2), 2-14.

Tilbury, D. (1998a). The Role of Research in Initiating and Sustaining Developments in Teacher Education. *International Research in Geographical and Environmental Education*, 7(3), 239-264

Tilbury, D. (1998b). Sustaining educational innovation through research: A European environmental education project. *Journal of the European Environment*, 4(3), 23-32.

Tilbury, D. (2001). Sustaining Environmental Education at the University Level: Experiences from the LSE Research Project. *Journal of Environmental Education*, 17, 14-18.

Tilbury, D. (2002). Active Citizenship: Empowering People as Cultural Agents of Change Through Geography. In: Gerber, R. and Williams, M. (Eds.). *Geography, Culture and Education.* Kluwer Geolearn Series.

Tilbury, D. and Cooke, K. (2002). *Environmental Education for a Sustainable Future: A University and Industry Summit.* Report prepared for Environment Australia accessible at www.gse.mq.edu.au/summit.

Tilbury, D., Reid, A. and Podger D. (2002). *Action Research: Graduate Skills Towards Sustainability* Sydney: Macquarie University, (Unpublished) accessible at www.gse.mq.edu.au/action

United Nations Conference on Environment and Development (UNCED) (1992). *Agenda 21: Programme of Action for Sustainable Development: Rio Declaration on Environment and Development.* New York: UN.

UNESCO (2001). *Education and Public Awareness for Sustainable Development: Report of the Secretary General.* UN E/CN 17/2001-advance unedited copy.

UNESCO (2002). *Education for Sustainability, From Rio to Johannesburg: Lessons Learnt from a Decade of Commitment.* Report presented at the Johannesburg World Summit for Sustainable Development, Paris: UNESCO.

UNESCO-NIER (1996). *Learning for a Sustainable Environment: Teacher Education and Environmental Education in Asia and the Pacific: Final Report.* Tokyo: National Institute for Educational Research (NIER).

UNESCO-UNEP. (1976). The Belgrade Charter. *Connect,* 1(1), 1-9.

UNESCO-UNEP. (1978). The Tbilisi Declaration. *Connect,* 3(1), 1-8.

UNESCO-UNEP. (1988). *International Strategy for action in the filed of Environmental Education and Training for the 1990's.* Moscow, Paris and Nairobi: UNESCO-UNEP.

United Nations (2002). *World Summit for Sustainable Development: Implementation Plan.* Johannesburg: United Nations.

University Presidents for a Sustainable Future (1999). *The Talloires Declaration* http://www.ulsf.org/about/tallo.html

Watkin, G. Wanklyn, M. and Wylie, V. (1995). *The Environmental Agenda-Taking Responsibility: Promoting Sustainable Practice through Higher Education Curricula.* London: Pluto Press in association with WWF-UK.

BIOGRAPHY

Dr. Daniella Tilbury is an Associate Professor in Environmental Education and Sustainable Development at the Graduate School of the Environment, Macquarie University, Sydney. She is currently the Chair in Education for Sustainable Development for the IUCN Commission in Education and Communication (CEC) and a member of The Earth Charter International Advisory Committee on Education. She is also an Australian representative on the OECD Environment and Schools Initiative (ENSI) programme. Daniella's PhD study undertaken at the University of Cambridge developed a grounded theory of change in environmental education within higher education. Since then she has lectured in environmental education at the postgraduate and undergraduate level in many universities across the globe. She organised and facilitated the IUCN Seminar 'Engaging People in Sustainability' held in Johannesburg at the World Summit for Sustainable Development.

CHAPTER 10

THE CONTRIBUTION OF ENVIRONMENTAL JUSTICE TO SUSTAINABILITY IN HIGHER EDUCATION

Julian Agyeman & Craig Crouch

INTRODUCTION

In this chapter, we explore the inextricable, yet often unrecognized links between environmental justice[1] and sustainability in the context of sustainability education. The current, predominant orientation of sustainability discourses in higher education is one of *environmental sustainability*. At the university level, themes associated with sustainability are therefore usually taught in departments of environmental science and environmental studies, emphasizing ecology, resource management, and environmental economics.

This pedagogical approach means that aspects of sustainability having to do with justice and equity are, if at all, dealt with in departments and disciplines traditionally outside environmental studies such as sociology, anthropology and law. The domination of *'sustainability as science'*, together with its polarization against *'sustainability as justice and equity'* is not only an inaccurate representation of the reality of sustainability issues, but also imparts a distorted picture to students. As a result, it is possible to graduate from university programs with credentials implying expertise in sustainability issues with a full understanding of the science of sustainability, but not the fundamental justice and equity issues that are inseparable from holistic considerations of sustainability.

Using case studies, we make the case that the development of pedagogy and curricula around sustainability in higher education should take account of the centrality of justice and equity issues as well as the natural sciences, and we put forth environmental justice as an analytic lens, conceptual tool and logical companion to sustainability. We explore the current reality of sustainability discourses, environmental justice discourses, examples of dissociation between the two, and make suggestions as to how the latter might inform the former in the context of sustainability studies in higher education. Pedagogical implications of this perspective, which we touch on in our case studies include, among others, emphases on experiential learning, participatory research, teamwork, reflection and discussion,

---

[1] In this chapter, we will use 'environmental justice' as an overarching phrase which incorporates issues of social justice as we see the delineations between environmental justice and social justice as somewhat arbitrary.

Peter Blaze Corcoran & Arjen E.J. Wals (Eds.), Higher Education and the Challenge of Sustainability: Problematics, Promise and Practice, 113-130.

contextualized learning, and straddling boundaries not only between disciplines, but between theory and practice.

SUSTAINABILITY DISCOURSES

Environmental perspectives

While many current sustainability discourses do not take more than a passing account of the linkage between environmental degradation and social justice and equity, it is worthwhile exploring what *is* often meant by those who see sustainability primarily though its environmentalist lens. Redclift (1987) argues that the 'limits to growth' debates of the 1970s in general, and the 1972 UN Stockholm Conference in particular, shifted the collective environmental thinking of the day towards its current *environmental* sustainability-focused paradigm. The current, prevailing definition of sustainable development comes from a 1987 World Commission on Environment and Development (WCED) report known as the Brundtland Report.

Perhaps the most significant contribution of the Brundtland definition of sustainability is the so-called 'futurity principle,' that is, the incorporation of notions of the welfare of future generations into our current moral sphere. It argued that "sustainable development is development that meets the needs of the present without compromising the ability of future generations to meet their own needs." (WCED, 1987, p. 43) By 1991, the International Union for the Conservation of Nature (IUCN) had also updated its 1980, conservation-based definition of sustainability. It was rewritten to suggest that sustainable enterprises are those based on " improving the quality of human life while living within the carrying capacity of supporting ecosystems." (IUCN, 1991, p. 10).

However, Kula (1998) suggests the roots of current sustainability debates are much older than the 1970s and 80s. Concepts of sustainability in forestry are found as early as the mid-nineteenth century when forest managers such as Von Thunen, in 1826, and Faustmann, in 1849, wrote of opportune harvesting times to ensure that forest yields were sustained over time and not depleted for the sake of short term gain.

It has been the relatively recent recognition of the scope of our impact on the natural world, and thus our ability to alter the opportunities for future generations, that has created an intellectual climate in which the concept of sustainability, long known in agriculture, forestry and fisheries, has been applied to the human enterprise more broadly. According to Kula (1998, p. 152), "the sustainable development debate is essentially about the claims of future generations which have been brought to prominence by environmental problems that have no precedent in human history." He argues that our natural capital must be the central element in our definitions of sustainability. He designates a proportion of this as "critical capital" which is necessary for human survival: the ozone layer, the carbon cycle, biodiversity, etc. This, the so-called "constant natural capital stocks criterion" (Rees, 1995), must be kept constant from generation to generation for two reasons.

The first is irreversibility. This is based on the recognition that when they are exhausted, such natural capital stocks are irreplaceable. The second is that we do not fully understand the workings of our ecosystem(s) and thus do not know the myriad effects of depleting one of these stocks from the biosphere. But Kula also warns that sustainability may be an anthropocentrist trap. While the notion of preserving the integrity of natural systems for future generations is an ethical position, it is not necessarily an environmental one. If the motivation behind sustainability is solely the well-being of future humans, then those aspects of the environment that serve no utilitarian function to humans may fall outside of a sustainability ethic.

Harris (2000, p. 1) suggests that a "thoroughgoing application of the principles of sustainability seems to imply a basic change in patterns of agriculture and industrial growth, an emphasis on non-renewable energy and energy efficiency, an integration of population policy into macroeconomic policy, a sustainable management policy for natural resources, and new instruments of macroeconomic measurement, as well as new social and institutional structures." Yet despite at least two decades of sustainability rhetoric, what we find is development programs that are much the same as they were before the concept of sustainability came to global prominence. Harris (2000, p. 3) argues that regardless of attempts at implementation of sustainable development, "when the existing social institutions are such as to deny many people access to the power and knowledge they need to affect the development process, sustainable development is not possible." Only when issues of power, knowledge and institutional structure are addressed can we move toward what is being called SAEJAS, 'socially and environmentally just and sustainable development'.

Social and political perspectives

Harris's (2000) thoughts on the inclusion of social issues leads into a discussion of the relevance of justice and equity to sustainability. Campbell (1996, p. 301) has argued that "in the battle of big public ideas, sustainability has won: the task of the coming years is simply to work out the details, and to narrow the gap between its theory and practice". Yet before narrowing the gap between theory and practice, the theory itself must reflect the centrality of social justice and equity. With regard to that centrality, there is a discernable North-South difference in interpretations of sustainability. As Jacobs (1999, p. 33) argues "in Southern debates about sustainable development the notion of equity remains central…In the North, by stark contrast, equity is much the least emphasized of the core ideas[2], and is often ignored altogether". Our position on sustainability (and sustainability education) is twofold. First, sustainability is primarily a political, rather than a solely technical enterprise (Agyeman & Evans, 1995). Second, an understanding of justice and equity is an integral component of any realistic attempt at achieving sustainability globally, nationally and locally.

Several writers have taken this position as the foundation for their work in sustainability: "like other political terms (democracy, liberty, social justice, and so

---

[2] The others being environmental protection and participation.

on), sustainable development is a contestable concept'" (Jacobs, 1999, p. 25); "the emerging sustainability ethic may be more interesting for what it implies about politics than for what it promises about ecology" (Hempel, 1999, p. 43); and "sustainability will be achieved, if at all, not by engineers, agronomists, economists and biotechnicians but by citizens" (Prugh, Costanza & Daly, 2000, p. 5). These statements make the case that sustainability is ultimately about politics and citizens, and, as Agyeman and Evans (1995, p. 36) argue "any attempt to 'technicise' sustainability is doomed to failure".

Regarding the second point, Agyeman, Bullard and Evans (2003, p. 2) argue that "sustainability is clearly a contested concept, but our interpretation of it places great emphasis upon precaution: on the need to ensure a better quality of life for all now, and into the future, in a just and equitable manner, whilst living within the limits of supporting ecosystems". They also endorse the point made by Middleton and O'Keefe (2001, p. 16) that "unless analyses of development begin not with the symptoms, environmental or economic instability, but with the cause, social injustice, then no development can be sustainable".

These definitional and perspectival shifts have lead to a surge in material in recent years dealing with the concepts of sustainability and sustainable development. This has given rise to competing and conflicting views over what the terms actually mean and what are the most desirable means of achieving the goal of sustainable communities. As multifaceted and contested as the concept of sustainability has become, for the purposes of education, Huckle and Sterling (1996) argue that the goal of 'education for sustainability' is precisely to encourage students to reflect upon these many perspectives. An understanding of these conflicts and contests will enable citizens to engage with society's efforts to move toward a sustainable path in a more informed capacity.

ENVIRONMENTAL JUSTICE DISCOURSES

Definition

The political movement, academic field, and philosophy of environmental justice, we suggest, offers a set of conceptual tools for thinking about the role of justice and equity within sustainability discourses. Environmental justice is the name given to the intersection of social justice and environmental issues, both in theory and in practice. Though defining environmental justice is not an easy task. Like sustainability, there are many, sometimes conflicting, interpretations at this intersection.

Generally speaking, concerns over racial and socioeconomic bias in the siting of LULUs (locally unwanted land uses), the enforcement of environmental laws and access to the benefits of clean environments have mobilized concerned citizens into what came to be known as the environmental justice movement. As quantitative and qualitative empirical research substantiated many of these citizen claims, an academic field of environmental justice emerged to further address these claims and to develop a theoretical basis for what began as a grassroots movement. In response to this political pressure and in keeping with these findings, municipal, state and the federal government have taken up the environmental justice cause in their respective ways. For example, the Commonwealth of Massachusetts (Commonwealth of Massachusetts, 2002, p. 2) uses the following definition in its Environmental Justice Policy: "environmental justice is based on the principle that all people have a right to be protected from environmental pollution and to live in and enjoy a clean and healthful environment. Environmental justice is the equal protection and meaningful involvement of all people with respect to the development, implementation, and enforcement of environmental laws, regulations, and policies and the equitable distribution of environmental benefits."

In particular, environmental justice has traditionally dealt with the degree to which minority populations bear a disproportionate burden of the products of environmental degradation. In 1983, the Government Accounting Office released a report (GAO, 1983) indicating that African-Americans comprised the majority populations in three of the four communities of the south-eastern US where hazardous waste landfills were located. The landmark 1987 United Church of Christ study 'Toxic Wastes and Race in the United States' showed that certain, predominantly communities of color, are at disproportionate risk from commercial toxic waste. This finding was confirmed by later research (Adeola, 1994; Bryant & Mohai, 1992; Bullard 1990a, 1990b; Mohai & Bryant, 1992). It also led to the coining of a term by Benjamin Chavis, which became the rallying cry of many: environmental racism. The notion of environmental *racism* – that certain minority populations are forced, through their lack of access to the decision-making and policy making process, to live with a disproportionate share of environmental burdens, and thus suffer disproportionately a public health and quality of life burden – led to calls for environmental *equity*, which typically focused on sharing environmental burdens equally across all communities.

The movement soon evolved a broader set of environmental *justice* concerns that dealt with the political economy of the environment. Environmental justice activists claim that the path-of-least-political-resistance approach to, for example, siting toxic waste facilities, functions to the detriment of minorities, and, moreover, that the resulting disproportionate burden is therefore an *intentional* result (Portney, 1994). Cole and Foster (2001, p. 15-16) argue that environmental justice "both expresses our aspiration and encompasses the political economy of environmental decision-making." They continue, "most important in our concept of environmental justice is the element of democratic decision making, or community self-determination. Current environmental decision making processes have not been effective in providing meaningful participation opportunities for those most burdened by environmental decisions." Environmental justice, therefore, has at its heart notions of self-determination, righting a wrong, and correcting an unjustly imposed burden. It is not simply a 'not-in-my-backyard' (NIMBY) approach to environmental decision-making.

Its issue focus has broadened from LULUs such as waste facility siting, toxic storage and disposal facilities (TSDFs) and other issues such as lead contamination, pesticides, water and air pollution, workplace safety, and transportation to issues such as sprawl, smart growth (Bullard et al., 2000), and 'climate justice'

(International Climate Justice Network, 2002). The environmental justice movement has also evolved beyond a sole focus on the experiences of people of color (Bullard, 1994). As Cutter (1995, p. 113) notes, "environmental justice ... moves beyond racism to include others (regardless of race or ethnicity) who are deprived of their environmental rights such as women, children and the poor." As the Massachusetts policy implies, environmental justice is about increasing minority and low-income access to environmental "goods" as well as tackling the disproportionate impact of environmental "bads." Environmental justice is thus being linked with discourses on rights, and being able to experience quality environments and environmental quality. (Adeola, 2000; Agyeman, 2001; UNECE, 1999).

In effect, the environmental justice movement has reframed environmentalism. At the first national People of Color Environmental Leadership Summit in October 1991, Dana Alston attempted to define this new environmental agenda from a social justice and equity standpoint. She addressed the degree to which the traditional environmental movement failed to take account of equity and social justice concerns in the way they frame their issues. She explained that "the issues of the environment do not stand alone by themselves. They are not narrowly defined. Our vision of the environment is woven into an overall framework of social, racial and economic justice. The environment, for us, is where we live, where we work, and where we play. ...we refuse narrow definitions." (quoted in Gottlieb, 1995, p. 5) With those words, Alston was questioning the central premises of environmental politics.

Alston spoke from a perspective on environmental issues that dates back at least to Rachel Carson's 1962 *Silent Spring*. With the onset of pollution-based environmental concerns came concerns about human quality-of-life issues, versus earlier natural resource protection perspectives on the environment. Gottlieb (1995, p. 7) claims that history has documented resource management and regulatory approaches to environmentalism and largely ignored social-movement-based quality-of-life concerns. He suggests, "defining contemporary environmentalism primarily in reference to its mainstream, institutional forms, such a history cannot account for the spontaneity and diversity of an environmentalism rooted in communities and constituencies seeking to address issues of where and how people live, work and play." In short, he argues that the reality of environmental concerns are much more social-justice oriented than historical recordings of the movement suggest, with the environment being a core concept within a broader set of social justice issues that date back to concerns over urbanization during the era of industrialization.

Contestation and critique

Among the most critical of the environmental justice movement is the work of Christopher Foreman (1998). He sees the major weaknesses of environmental justice as: a) lack of empirical evidence supporting claims of disproportionate environmental impacts and regulatory enforcement; b) attention being directed away from environmental problems that are providing the greatest risk; c) exacerbating such problems as inefficient policy making; and d) unrealistic goals such as "zero tolerance" for environmental risks. He goes on to add that the environmental justice movement is unable to adequately respond to these shortcomings due to its lack of a unified political front, and an "unwillingness to face politically inconvenient facts about environmental health risks." (Foreman 1998, p. 4)

Regarding the distribution of risk from toxic chemicals, waste disposal facilities, air and water pollution, Nichols (1994) also argues that the "zero-tolerance" position, which the environmental justice movement often promotes, hides the very real tradeoffs between environmental protection and other priorities that necessarily take place in environmental policy making. If the question of *which* problems are to be addressed is not then a function of rational decision-making, which Nichols (1994) suggests a "risk-based" program would accomplish, then such decisions would instead be highly politicized.

However, if as Nichols (1994) and Foreman (1998) would have us believe, further politicization is sufficient rationale for rejecting an environmental justice framework and adopting a risk-based regime, we may be left wanting. Winner (1986, p. 138-39) shows us that "the arena in which discussions of risk take place is [itself] highly politicized and contentions. Powerful social and economic interests are invested in attempts to answer the question, How safe is safe enough? Indeed, the very introduction of 'risk' as a common way of defining policy issues is itself far from a neutral issue." At the very least, Winner suggests, using "risk" as a framework for environmental policy favors the status-quo because it narrowly defines the terms of a given debate to the single risk in question and does not allow a broader debate about the societal issues that may have enabled the risk to manifest itself in the first place. In addition, defining these issues in terms of "risk" opens up a degree of uncertainty not seen when similar issues are defined in terms of "hazard" or "threat." Under a "risk" regime, one must prove the chance of the potential harm as well as the magnitude of the potential harm, both of which rely heavily on quantitative scientific knowledge and high degrees of uncertainty. With the burden of proof ordinarily lying with the plaintiff, we can see where a "risk" regime favors inaction.

Internationalizing environmental justice

At the international level, environmental justice issues are being played out in many arenas. Some, like those in the USA, are based around toxics, or other LULUs. But there are other arenas which are not (yet?) to the fore in the USA. Rixecker and

Tipene-Matua (2002, p. 259), in their analysis of bioprospecting and the Maori people's resistance to (bio)cultural assimilation, argue that

> "extraction of such information [about natural resources] and life forms for the purpose of economic profit has been called biological prospecting. It is regarded as part of the ongoing colonization of local and indigenous peoples around the world, and it has [also] been labeled biopiracy (Shiva, 1997). Where colonizers once exploited the land and degraded the culture, incurring an 'ecological debt', current science and technology now continue this by exploiting and stealing the remaining treasures of indigenous peoples - their genes of their flora and fauna, traditional ecological knowledge, spiritual integrity and their relationship with the human and nonhuman world."

Looking at these and other international dimensions shows that environmental justice cannot be considered a US, or solely contemporary phenomenon. Ecological debt can also be seen as *historical* environmental injustice, whereas biopiracy and bioprospecting can be seen as *contemporary* (and *future*) environmental injustices.

Another international level issue being played out in the environmental justice arena is that of human rights. Agyeman et al. (2003, p. 11) note that "environmentalists are also increasingly questioning the justice and equity implications of other international agreements, especially those related to trade or economic development. There is great (and under-researched) potential for the notions of environmental justice, human rights and sustainability to permeate environmental regimes and international policy and agreements." It is being increasingly recognized that one of the best ways to protect environmental rights, and thereby the rights of the most marginalized people, is to uphold the basic civil, human and political rights of the individual (Sachs, 1995; Anderson, 1996).

The environmental justice paradigm

Increasingly, environmental justice can be seen as a conceptual framework which represents the socio-cultural realities of contemporary environmental politics. The social and political context of environmental policy and of environmentalism more broadly has changed significantly in recent years. Ringquist (2003, p. 49) argues that the "environmental justice movement has the potential to broaden the base of support for the traditional environmental movement, and it may reinvigorate and refocus the forces of progressive politics behind environmental concerns. In short, it has the potential to change the face of environmental politics and policy." This is clearly seen in Taylor's (2000) work on the ideological pillar of the environmental justice movement: the Principles of Environmental Justice. They were the main outcome of the October 1991 People of Color Environmental Leadership Summit held in Washington DC. They are a set of seventeen criteria along which to develop and evaluate policies for environmental and social justice (see Appendix 1). The principles of environmental justice are "a well developed environmental ideological framework that explicitly links ecological concerns with labor and social justice concerns" (Taylor, 2000, p. 538).

The US environmental justice movement "appropriated...the preexisting salient frames of racism and civil rights" (Taylor, 2000, p. 62). This, she argues led to the development of the 'Environmental Justice Paradigm' (EJP) which "is most clearly articulated through the Principles" (537), and "is the first paradigm to link environment and race, class, gender, and social justice concerns in an explicit framework" (542). When compared with the New Environmental Paradigm (NEP) of Dunlap and Van Liere (1978) which Milbrath (1989, p. 118) describes as "a new set of [environmental] beliefs and values", Taylor (2002, p. 542) notes that "the EJP has its roots in the NEP, but it extends the NEP in radical ways...The EJP builds on the core principles of the NEP; however, there are significant differences...vis a vis the relationship between environment and social inequality. The NEP does not recognize such a relationship; consequently it has a social justice component that is very weak or non-existent." This weakness or non-existence, we argue, is precisely why environmental justice should be more closely linked to sustainability and sustainable development in higher education. Goldman (1993, p. 27) may well have been thinking the same way when he said that, "sustainable development may well be seen as the next phase of the environmental justice movement".

Resistance to incorporating an environmental justice perspective in sustainability programs in higher education may be for many reasons. The very term 'environmental justice', for those who have heard of it, may be perceived as 'overtly political', or dealing with sensitive issues such as race. It may also be due, in part, to the body of literature, such as that of Foreman (1998) and Nichols (1994), which refutes the claims of environmental justice advocates. Debates over whether empirical evidence supports or refutes the claims made by environmental justice advocates, regarding for example siting decisions, or risk analyses are an important contribution to relevant policy-making and scholarship, as well as to the efforts of grassroots advocates themselves. However, these debates, like those advocated over the contestation of sustainability by Huckle and Sterling (1996), are more appropriately used to enliven the intersection of social justice and environmental issues, rather than as tools to undermine awareness raising and action at this intersection.

For the sake of university programs in sustainability, these varying perspectives on the efficacy of the environmental justice project constitute the beginning of an intellectual conversation, not the end. That the finer points of environmental justice, like those of sustainability, are still open to debate, and in some cases are unanswerable, in no way diminishes the problematic fact that many students are being trained to work in the '*sustainability as science*' realm, but are ill-equipped to negotiate the realities of the '*sustainability as justice and equity*' realm of the sustainability enterprise.

For example, the highly charged political environment associated with the unequal distribution of the products of environmental degradation is a reality graduates of sustainability programs will face regardless of whether the unequal distribution is perceived or actual, intentional or unintentional. Employing environmental justice as an organizing framework for such programs (see the University of Vermont case below) will enable students to attend to these and other issues, including the efficacy of the various positions, in an intellectually robust way that appreciates the inherent social justice dimensions of sustainability issues.

EMPIRICAL EVIDENCE OF THE DISSOCIATION OF JUSTICE AND SUSTAINABILITY

Despite the well-articulated theoretical linkages between justice and sustainability, there remains dissociation at the practical level. In municipalities and local governments, but less so in community organizations (Agyeman, Bullard & Evans, 2003), there is a growing body of evidence of the rifts between sustainability issues and their social justice components.

To find empirical evidence of the dissociation between social and environmental justice and sustainability at both the theoretical and practical levels, we can look to the growing post-Earth Summit (1992) interest in local municipal sustainability projects which often come under the banner of Local Agenda 21 (LA21), or 'Communities 21' in the USA. Warner (2002, p. 37), in an internet-based survey found that "more than 40 percent of the largest cities (33 of 77) in the United States had sustainability projects on the web, but only five of these dealt with environmental justice on their web pages." By 'dealt with', he means 'makes mention of,' and he categorizes the scale of 'making mention' from the lowest: '*education*,' i.e. background information for users of the site, through '*policy*,' i.e. a stated policy commitment to environmental justice, to '*implementation*,' i.e. integration of environmental justice into sustainability.

He continued, "few communities were building environmental justice into local definitions of sustainability. Only five local sustainability projects made these connections: Albuquerque, New Mexico; Austin, Texas; Cleveland, Ohio; San Francisco, California and Seattle, Washington." Only one city reached 'implementation': San Francisco, the others were all at the 'education' level, except Cleveland, which reached the 'policy' level. Warner's (2002, p. 38) conclusion was that "while environmental justice seemed to be having an impact on mainstream environmental organizations and on government agencies, this did not apparently extend to groups working on sustainability projects."

Similarly, Portney (2003, p. 57) argues that "most cities that have sustainability indicators do not explicitly use social or environmental equity." Yet a search of the database of the well respected training and consulting organization 'Sustainable Measures' (http://www.sustainablemeasures.com) under '*race*' yields eight sustainability indicators in different US and Canadian cities, '*ethnicity*' yields two and '*low income*' yields fifteen[3]. What does this say about the integration of environmental justice into sustainability concerns in the US? It says that it is ad hoc at best, rather than fundamental or foundational.

In addition, a 1999 research report by the Washington DC based Environmental Law Institute, entitled *Sustainability in Practice,* analyzed 579 applications to the US Environmental Protection Agency's (EPA) 1996 'Sustainable Development Challenge Grant Program'. Less than 5% of applications had 'equity' as a goal, and, interestingly, less than 1% addressed 'international responsibility' (Friends of the

---

[3] Examples include: Racial and ethnic representation in Legislature; Unemployment rate by ethnicity; Occupational distribution of women and minorities; Number of homeless people; Number and value of business loans in low income area; Percent of jobs that pay a livable wage for a family of two; Number or percent of residents receiving welfare assistance.

Earth Scotland, 2000). Finally, Tuxworth (1996, p. 285), regarding the Local Agenda 21 surveys 1994-1996 among local governments in the UK, notes that "areas of policy work where the influence of sustainable development lies between 'no part' and 'minor influence' are predominantly at the social end of the spectrum" including "social services, welfare/equal opportunities work, and anti-poverty strategies."

CASE STUDIES OF GOOD PRACTICE

Despite this broader evidence of dissociation between sustainability and environmental justice, there are many specific examples of good practice, both within and outside higher education. Here we explore four university programs, one in Scotland, the others in the US, that are making the connections between sustainability and environmental justice.

As the programs are all relatively recent, there is little literature on the implications of an environmental justice perspective for teaching and learning in sustainability. This may be due to the fact that *'sustainability as justice and equity'* is harder for most to teach, in terms of both pedagogy and resources, than *'sustainability as science'*. As we mentioned earlier, the 'overtly political' perception of environmental justice, dealing as it does with issues of race and poverty, may work against it. However, Kaza (2000) who looks at teaching ethics through environmental justice, perhaps has more to say than many.

Queen Margaret University College certificate of higher education in environmental justice

As part of Friends of the Earth Scotland's (FoES) 'Campaign for Environmental Justice,' the organization recently completed a three-year project developing links with community organizations across Scotland. Some 2,600 people were involved in providing information about their needs in relation to the future work of FoES. One of the explicit demands recorded in project evaluation feedback was a Level 1 accredited course in environmental justice, which would offer community-based education and support for practical projects. Accordingly, FoES approached Queen Margaret University College, Edinburgh (QMUC), to validate a program (Certificate of Higher Education in Environmental Justice) which will be carried out by part-time study and delivered largely by FoES staff.

The overall aim is to provide a flexible program of part-time study which enables learners to apply the concepts and approaches of environmental justice to community development in their own localities. More specifically, the objectives of the program are: to produce certified individuals qualified to undertake the role of an agent of environmental justice for the community in which they live; to develop in students an understanding of the ways in which relevant theories and concepts may be applied in the real world; and to provide access to Higher Education for individuals who otherwise would not find it available.

Among the modules constituting the certificate program are, 'Principles of sustainable development,' 'Principles of environmental justice,' 'Principles and practice of community development,' 'Planning and environmental law,' 'Citizenship, social movements and political change,' and 'Communications and the media.' The combination of subjects in the program reflects the reality of community activists' work in defending or developing their community and its environment. Thus the program is interdisciplinary and combines theoretical and practical elements in a way that will appear cohesive from the perspective of the student. The nature of the student is the crucial consideration in identifying pedagogical approaches. Each learner will bring a unique combination of skills and experience to the program, and will seek to apply concepts and theories in the context of their own work within the community.

University of Michigan's degree concentration in Environmental Justice

Within the University of Michigan's School of Natural Resources and Environment, Professor Bunyan Bryant has managed to bring a robust social justice and equity perspective to bear on environmental issues. He and his colleagues have created an environmental justice degree concentration in the school based on the constellation of courses already in existence. Among others, they include: Domestic and international issues in environmental justice, Applications of environmental justice, Theoretical approaches to environmental justice, and Research methods in environmental justice.

Pulling on disciplinary traditions in public health, urban planning, and ecology, the program is attempting to bridge the social-environmental divide by undertaking environmentally related community research, with an eye toward informing policy decisions and discouraging environmental racism. A strong theme running throughout the program is one of participatory research, wherein students are encouraged to go into surrounding communities and engage community-members in research designs that are appropriate to their particular needs. Integrating these hands-on experiences into the curriculum in effect bridges rifts between research and practice, and between academia and community, by producing knowledge that is relevant to local communities.

Wellesley University's Environmental Studies concentration in Environmental Justice

In Massachusetts, Wellesley University is experimenting with an interdepartmental undergraduate major in environmental studies. In its second year, the major offers concentrations in one of four possible strands: Environmental justice, Environmental philosophy and ethics, Environmental policy and economics, and Environmental science. In keeping with the developmental trajectory new programs often take, the environmental studies program at Wellesley was built on resources that were pre-existing in the institution. Wellesley had strong social science and humanities programs which supported the development of their environmental studies program, endowing it with a robust social science perspective. It is, in effect, a major that reflects what is on campus.

In addition to an institutionalized social science perspective, Wellesley is an institution that values multidisciplinary studies. Many areas of study on campus follow a multidisciplinary model. Having that as the pedagogical and conceptual underpinning of the environmental studies program during its development brought together multiple disciplinary constituencies under the environment banner, rather than a center of gravity in the natural sciences. Finally, Wellesley has already instituted programs that have a strong ethics and justice concerns, such as their Peace and Justice program, Africana studies, etc. The goal from inception was to define the environment broadly and incorporate as many of the different perspectives on environmental issues as the school could offer.

University of Vermont's Teaching Ethics Through Environmental Justice

Stephanie Kaza, an environmental educator at the University of Vermont uses environmental justice issues to produce an empowering curriculum. Kaza utilizes the liberation theology of Gerard Fourez to raise critical issues. Fourez developed a four step model which: assesses dominant social norms and names the promulgating agents; notes how these norms serve those in power; develops the process of 'conscientization' and finally assists in the articulation of a structural ethics to address (white) privilege. Kaza and her colleagues use Fourez's model to immerse students in environmental justice at the predominantly white and wealthy University of Vermont. She notes that through this experience, students recognize their own *denial;* they get firsthand experience of *inequity*; they become aware of their own *complicity*, and finally, they witness *resistance*. An interesting exercise would be to utilize this model to teach an identical syllabus at a very diverse institution, and to share experiences and observations with Kaza and her colleagues.

CONCLUSION

In this chapter, we've explored what we consider to be the inextricable links between two contested public policy areas: environmental justice and sustainability. In addition, we've tried to relate these to what we perceive to be a dominant '*sustainability as science*' formulation in much sustainability education, arguing for a balance between this, and the less traveled path of '*sustainability as justice and equity*'. We have, in support of this, cited four programs which we feel achieve something near this equilibrium. Our argument of course, is that students emerging from these "model" programs will be better equipped than their '*sustainability as science*' counterparts, to deal with the political realities of today (and tomorrow), wherever in the world they choose to live and work.

Moreover, we have argued that environmental justice and sustainability are "logical companions". The former reminds the latter of the pivotal nature of issues of justice and equity; the latter reminds the former of the importance of futurity and vision. Environmental justice is one of the few intellectual domains where both the causes and effects of environmental degradation are seen as social and cultural phenomena. As Conway et al. (1999) remind us, there is nothing wrong with the environment; environmental degradation is first and foremost a human problem and its solutions are more likely to be found in the intellectual spheres that deal with human issues than those that deal exclusively with the natural sciences.

If sustainability educators in higher education want to more accurately reflect the political *and* scientific realities of sustainability issues as they are occurring in the world outside academia, they would do well to incorporate a '*sustainability as justice and equity*' perspective. The only vehicle for this is the growing body of theory and practice we call environmental justice. This incorporation we see as *the* major challenge for sustainability educators.

REFERENCES

Adeola, F.O. (2000). Cross-National Environmental justice and Human Rights Issues - A Review of Evidence in the Developing World. *American Behavioral Scientist*, 43(4), 686-706.

Adeola, F.O. (1994). Environmental Hazards, Health, and Racial Inequity in Hazardous Waste Distribution. *Environment and Behavior*, 26 (1), 99-126.

Agyeman, J. (2000). *Environmental justice: from the margins to the mainstream?* London: TCPA.

Agyeman, J. (2001). Ethnic minorities in Britain: short change, systematic indifference and sustainable development. *Journal of Environmental Policy and Planning*, 3(1), 15-30.

Agyeman, J., Bullard R. and Evans, B. (2003). *Just Sustainabilities: Development in an Unequal World.* London/Cambridge, MA: Earthscan Publications Ltd/MIT Press.

Agyeman, J. and Evans, B. (1995). Sustainability and Democracy: Community Participation in Local Agenda 21. *Local Government Policy Making*, 22(2), 35-40.

Anderson, M (1996). Human Rights Approaches to Environmental Protection: An Overview in Boyle, A and Anderson, M *Human Rights Approaches to Environmental Protection*. Oxford: Clarendon Press.

Bryant, B. and Mohai, P. (1992). *Race and the Incidence of Environmental Hazards: A Time for Discourse*. Boulder, CO, Westview Press.

Bullard, R (1990a). *Dumping in Dixie: Race, Class, and Environmental Quality* Boulder, CO: Westview.

Bullard, R (1990b). Solid waste sites and the Black Houston Community. *Sociological Inquiry* 53 (spring), 273-288.

Bullard, R. (1993). Anatomy of Environmental Racism, in Hofrichter, R. (Ed) *Toxic Struggles: The Theory and Practice of Environmental Justice*. Gabriola Island, British Columbia: New Society Publishers.

Bullard, R. (Ed) (1994). *Unequal Protection: Environmental Justice and Communities of Colour*. San Francisco: Sierra Club Books.

Bullard, R, Johnson G, and Torres A, (2000). *Sprawl City: Race, Politics, and Planning in Atlanta*. Washington DC. Island Press.

Campbell, S (1996). Green Cities, Growing Cities, Just Cities. Urban Planning and the Contradictions of Sustainable Development. *Journal of the American Planning Association*, 62(3), 296-312.

Campbell, C. and Heck, W. (1997). An Ecological Perspective on Sustainable Development. In Muschett, D. (Ed.) *Principles of Sustainable Development*. Delray Beach, FL: St. Lucie Press.

Cole, L. and Foster, S. (2001). *From the Ground Up: Environmental Racism and the Rise of the Environmental Justice Movement*. New York: NYU Press.

Commonwealth of Massachusetts (2000). *Draft Environmental Justice Policy of the Executive Office of Environmental Affairs*. Boston: EOEA. (12/7/00 version).

Conway, J., Keniston, K. and Marx, L. (1999). *Earth, Air, Fire, Water: Humanistic Studies of the Environment*. Boston: University of Massachusetts Press.

Cutter, S. (1995). Race, Class and Environmental Justice. *Progress in Geography*, 19(1), 111-122.

Dunlap, R and Van Liere, K (1978). The New Environmental Paradigm *The Journal of Environmental Education*, 9(4), 10-19.

Environmental Law Institute (1999). *Sustainability in Practice*. Washington DC: ELI.

Foreman, C. 1998). *The Promise and Peril of Environmental justice.* Washington, DC: The Brookings Institute.

Friends of the Earth Scotland (2000). *"Harris Superquarry: A Briefing from Friends of the Earth".* Edinburgh: Friends of the Earth.

General Accounting Office (1983). *Siting of Hazardous Waste Landfills and their Correlation with Racial and Economic Status of Surrounding Communities.* Washington, DC: GPO.

Goldman, B (1993). *Not just prosperity: Achieving sustainability with environmental justice.* Washington DC: National Wildlife Federation.

Gottlieb, R. (1995). *Forcing the Spring: The Transformation of the American Environmental Movement.* Washington, DC: Island Press.

Harris, J. (2000). *Rethinking Sustainability: Power, Knowledge and Institutions.* Ann Arbor: University of Michigan Press.

Hempel, L.C. (1999). Conceptual and analytical challenges in building sustainable communities, in Mazmanian, D.A. & Kraft, M.E. (Eds.). *Towards Sustainable Communities: Transition and Transformations in Environmental Policy.* Cambridge: MIT Press.

Huckle, J. and Sterling, S. (1996). *Education for Sustainability.* London: Earthscan Publications Ltd.

International Climate Justice Network (2002) Bali principles of climate justice. http://www.corpwatch.org/campaigns/PCD.jsp?articleid=3748 Accessed 3/31/03.

IUCN. (1991). *Caring for the Earth: A Strategy for Sustainable Living.* Gland: IUCN.

Jacobs, M (1999). Sustainable development as a contested concept in Dobson, A. (ed) *Fairness and futurity.* Oxford. OUP pp. 21-45.

Kaza, S (2002). Teaching ethics through environmental education. *Canadian Journal of Environmental Education* 7(1), 99-109.

Kula, E. (1998). *The History of Environmental Economic Thought.* New York: Routledge.

Middleton, N and O'Keefe, P (2001). *Redefining Sustainable Development.* London: Pluto Press.

Milbrath, L (1989). *Envisioning a Sustainable Society: Learning Our Way Out.* Albany. SUNY Press.

Mohai, P, and Bryant, B (1992). Environmental Injustice: Weighing Race and Class as Factors in the Distribution of Environmental Hazards. *University of Colorado Law Review,* No. 63, pp. 921-932.

Nichols, A. (1994). Risk-Based Priorities and Environmental Justice. In: Finkel, A. et al. (Eds.) *Worst Things First? The Debate over Risk-Based Environmental Priorities.* Washington DC: Resources for the Future.

Portney, K. E. (1994). Environmental justice and Sustainability: Is There a Critical Nexus in the Case of Waste Disposal or Treatment Facility Siting? *Fordham Urban Journal,* Spring, 827-839.

Portney, K (2003). *Taking sustainable cities seriously.* Cambridge MIT Press.

Prugh, T, Costanza, R and Daly, H (2000). *The Local Politics of Global Sustainability.* Washington DC, Island Press.

Rees, W (1995). Achieving sustainability: reform or transformation? *Journal of Planning Literature,* 9(4), 343-361.

Redclift, M. (1987). *Sustainable Development.* London: Routledge.

Rixecker, S and Tipene-Matua, B (2002). Maori Kaupapa and the Inseparability of Social and Environmental justice: An Analysis of Bioprospecting and a People's Resistance to (Bio)cultural Assimilation. In Agyeman, J., Bullard, R. and Evans, B. (Eds) *Just Sustainabilities: Development in an Unequal World.* London/Cambridge, MA: Earthscan/MIT Press.

Ringquist, E (1999). Environmental justice: normative concerns and empirical evidence. In: Vig, N. and Kraft, M. (eds.) *Environmental Policy: New Directions for the Twenty-First Century.* Washington, DC: Congressional Quarterly Press.

Sachs, A. (1995). *Eco-Justice: Linking Human Rights and the Environment.* Worldwatch Paper 127. Washington DC: The Worldwatch Institute.

Shiva, V. (1997). *Staying Alive.* London: Zed Books.

Shutkin, W. (2001). *The Land That Could Be: Environmentalism and Democracy in the Twenty-First Century.* Cambridge, MA: MIT Press.

Taylor, D (2000). The Rise of the Environmental Justice Paradigm. *American Behavioural Scientist,* 43 (4), 508-580.

Tuxworth, B (1996). From Environment to Sustainability: Surveys and Analysis of Local Agenda 21 Process Development in UK Local Authorities. *Local Environment,* 1(3), 277-297.

United Church of Christ Commission for Racial Justice (1987). *Toxic Wastes and Race in the United States.* New York: United Church of Christ Commission for Racial Justice.

United Nations (2002). *World Summit on Sustainable Development*, [Website]. Available: http://www.johannesburgsummit.org/.
United Nations Conference on Environment and Development (1992). *Agenda 21*. Geneva: United Nations.
United Nations Economic Commission for Europe (1999). *Convention on Access to Information, Public Participation in Decision Making and Access to Justice in Environmental Matters*. Geneva: UNECE.
Vig, N. and Kraft, M. (2003). *Environmental Policy: New Directions for the Twenty-First Century*. Washington DC: Congressional Quarterly Press.
Warner, K. (2002). Linking Local Sustainability Initiatives with Environmental Justice. *Local Environment*, 7(1), 35-47.
Winner, L (1996). The Whale and the Reactor: A Search for Limits in an Age of High Technology. Chicago: University of Chicago Press.
World Commission on Environment and Development (1987). Our Common Future. Oxford: Oxford University Press.

BIOGRAPHIES

Julian Agyeman is Assistant Professor of Environmental Policy and Planning at Tufts University, Boston-Medford. His interests are in the links between sustainability and environmental justice, community involvement in local environmental and sustainability policy and the development of sustainable communities. He is founder, and co-editor of the international journal 'Local Environment'. His most recent book 'Just sustainabilities: development in an unequal world ' (Earthscan/MIT Press) places issues of justice and equity to the fore in sustainability debates.

Richard Craig Crouch is a Doctoral Candidate at the Harvard Graduate School of Education. After receiving his Bachelor's Degree from UC Berkeley, he was an intern in the White House office of Vice President Al Gore. He then worked at the Center for Ecoliteracy in Berkeley, California. He also worked on the 1996 Clinton/Gore campaign before receiving his Master's Degree at Harvard.

APPENDIX

PRINCIPLES OF ENVIRONMENTAL JUSTICE (1991)*

WE, THE PEOPLE OF COLOR, gathered together at this multinational People of Color Environmental Leadership Summit to begin to build a national and our lands and communities, do hereby re-establish our spiritual interdependence to the sacredness of our Mother Earth; to respect and celebrate each of our cultures, languages and beliefs about the natural world and our roles in healing ourselves; to insure environmental justice; to promote economic alternatives which would contribute to the development of environmentally sage livelihoods; and, to secure our political, economic and cultural liberation that has been denied for over 500 years of colonization and oppression, resulting in the poisoning of our communities and land and the genocide of our peoples, do affirm and adopt these Principles of Environmental Justice:

1. Environmental justice affirms the sacredness of Mother Earth, ecological unity and the interdependence of all species, and the right to be free from ecological destruction.
2. Environmental justice demands that public policy be based on mutual respect and justice for all peoples, free from any form of discrimination or bias.
3. Environmental justice mandates the right to ethical, balanced and responsible uses of land and renewable resources in the interest of a sustainable planet for humans and other living things.
4. Environmental justice calls for universal protection from nuclear testing, extraction, production and disposal of toxic/hazardous wastes and poisons and nuclear testing that threaten the fundamental right to clean air, land, water, and food.
5. Environmental justice affirms the fundamental right to political, economic, cultural, and environmental self-determination of all peoples.
6. Environmental justice demands the cessation of the production of all toxins, hazardous wastes, and radioactive materials, and that all past and current producers be held strictly accountable to the people for detoxification and containment at the point of production.
7. Environmental justice demands the right to participate as equal partners at every level of decision making including needs assessment, planning, implementation, enforcement and evaluation.
8. Environmental justice affirms the right of all workers to a safe and healthy work environment, without being forced to choose between an unsafe livelihood and unemployment. It also affirms the right of those who work at home to be free from environmental hazards.
9. Environmental justice protects the right of all victims of environmental injustice to receive full compensation and reparations for damages as well as quality health care.

10. Environmental justice considers governmental acts of environmental injustice a violation of international law, the Universal Declaration On Human Rights, and the United Nations Convention on Genocide.

11. Environmental justice must recognize a special legal and natural relationship of Native Peoples to the U.S. government through treaties, agreements, compacts, and covenants affirming sovereignty and self-determination.

12. Environmental justice affirms the need for urban and rural ecological policies to clean up and rebuild our cities and rural areas in balance with nature, honoring the cultural integrity of all of our communities, and providing fair access for all to the full range of resources.

13. Environmental justice calls for the strict enforcement of principles of informed consent, and a halt to the testing of experimental reproductive and medical procedures and vaccinations on people of color.

14. Environmental justice opposes the destructive operations of multinational corporations.

15. Environmental justice opposes military occupation, repression and exploitation of lands, peoples and cultures, and other life forms.

16. Environmental justice calls for the education of present and future generations which emphasizes social and environmental issues, based on our experience and an appreciation of our diverse cultural perspectives.

17. Environmental justice requires that we, as individuals, make personal and consumer choices to consume as little of Mother Earth's resources and to produce as little waste as possible; and make the conscious decision to challenge and re-prioritize our lifestyles to insure the health of the natural world for present and future generations.

*Source: Principles of Environmental Justice (October, 1991). Ratified at the First People of Color Environmental Leadership Summit. Washington DC. October, 1991.

CHAPTER 11

LEARNING OUR WAY TO A SUSTAINABLE AND DESIRABLE WORLD: IDEAS INSPIRED BY ARNE NAESS AND DEEP ECOLOGY

Harold Glasser

The challenge of today is to save the planet from further devastation which violates both the enlightened self-interest of humans and nonhumans, and decreases the potential of joyful existence for all (Naess, 1995, p. 226).

INTRODUCTION

The world is shaped by the ways in which humans choose to live. Nature, as the material world and all its collective objects and phenomena, is embedded in our genes and our cultures. Our cultures, through their artifacts and actions, reflect our shifting attitudes toward nature. Our changing attitudes toward nature guide our shaping of it. And this refashioned nature, in turn, reshapes our cultures.

Many of today's environmental problems—global warming, stratospheric ozone reduction, biodiversity loss, pollution, water shortages, desertification and salinization, invasive species—are the unintended, unforeseen (but not necessarily unforeseeable) consequences of a failure to adequately appreciate the two-way character of this relationship. Humans can be positive and creative forces on this planet, but we must learn to live—and find joy living—in ways that celebrate and cherish the full richness and diversity of the Earth, both cultural and biological. This is, perhaps, the most vital challenge before humanity.

Humans have co-evolved with the planet's other life forms and all of their terror, splendor, utility, and wonder. As ecologist E.O. Wilson contends, we have an inborn affinity for them—"an affiliation evoked, according to circumstance, by pleasure, or a sense of security, or awe, or even fascination blended with revulsion" (Wilson, 1994, p. 360). We are *a part* of nature, but it is a nature that we continue to dominate and diminish through our collective efforts to refashion it in our own image. These efforts to distance ourselves from nature, to de-wild and de-sacralize it through continual marginalization and homogenization, may, however, only end up diminishing and de-humanizing us. We might, metaphorically speaking, be sewing the seeds of our own destruction with Round-up Ready Soybeans™, Bt Potatoes, and biopharming. The fatuousness of pursuing this de-wilding strategy to its logical end—setting ourselves *apart* from nature—should have been made vividly clear by

Peter Blaze Corcoran & Arjen E.J. Wals (Editors), Higher Education and the Challenge of Sustainability: Problematics, Promise and Practice, 131-148.
© 2004 *Kluwer Academic Publishers. Printed in the Netherlands.*

Biosphere 2's manifest dependence on "external" ecosystems services—Biosphere 1[1]. And yet, the essence of our humanity, which is somehow tied to the flourishing of the Earth's richness and diversity, still seems under-appreciated. The future is not yet drawn, but our everyday actions, in ways likely unclear to us now, are adumbrating its contours.

Concern over the health, status, and character of the world, our inheritance and legacy to all the world's future inhabitants, is on many minds and in many hearts. The primatologist Jane Goodall describes her sense of loss for a once fecund and diverse forest and despairs over the cycle of poverty that accompanies its destruction (Goodall, 1994, p. 21):

> A little while ago I drove along a road in Tanzania that once ran through miles of forest. Twenty years ago there were lions and elephants, leopards and wild dogs, and a myriad of birds. But now the trees are gone and the road guided us relentlessly, mile after mile, through hot, dusty country, where crops were withered under the glare of the sun and there was no shade. I felt a great melancholy, and also anger. This anger was not directed against the poor farmers who were trying to eke out a livelihood from the inhospitable land, but against mankind in general. We multiply and we destroy, chopping and killing. Now, in this desecrated area, the women searching for firewood must dig up the roots of the trees they have long since cut down to make space for crops.

A quite different view of the "state of the world," from a macro perspective, is offered by the Danish political scientist Bjørn Lomborg (2001, pp. 351-2), the latest author in the tradition of "doomslayer" Julian Simon:

> We are actually leaving the world a better place than when we got it and this is the really fantastic point about the real state of the world: that mankind's lot has vastly improved in every significant measurable field and is likely to continue to do so.... [C]hildren born today—in both the industrialized world and developing countries—will live longer and be healthier, they will get more food, a better education, a higher standard of living, more leisure time and far more possibilities—without the global environment being destroyed.

How do we reconcile the conflicting views of Lomborg and Goodall on the state of the world? Are their descriptions of "two" worlds simply the result of viewing a non-homogeneous world from two radically different scales? Or are they the result of fundamentally different perceptions of the same world, based on radically

---

[1] Biosphere 2 was created to explore the potential of developing self-sustaining environments separate from planet Earth—Biosphere 1. A 1.26 hectare greenhouse was constructed in the Arizona desert near Tucson. The greenhouse, which was designed to be hermetically sealed, was populated with a "representative" sampling of the Earth's major ecosystems, including grasslands, marshlands, ocean complete with coral reef, and a tropical rainforest. In mid-1991 four men and four women were sealed into Biosphere 2 for a two year adventure. A series of unforeseen events rapidly occurred. Oxygen levels dropped precipitously, levels of nitrous oxide spiked, carbon dioxide levels fluctuated dramatically, ecosystems failed, and over-harvesting ensued. In a short time, nineteen of twenty-four vertebrate species were extinct along with all of the nonhuman pollinators. Additional oxygen had to be added from Biosphere 1 (before the two years were out). The Biospherians also began turning the tropical rainforest into a farm and started smuggling in rations from Biosphere 1. While the experiment was ostensibly a failure, it presented us with a sobering clarification of both the significance of *in situ* ecosystem services and how little we understand about what is required to sustain them and us.

different values and distinctly different criteria and standards for judging what is important?

Contemporary sustainability discussions, with their many diverse, competing, and sometimes conflicting perspectives, represent nothing less than impassioned conversations on the human prospect and the fate of the planet. The term "sustainability," in its widest sense, can be conceived as a heuristic device for introducing, exploring, and peeling back the many layers of the human problématique. From a meta-perspective, three fundamental questions are suggested. First and foremost, "What makes life worth living—what truly nourishes and fulfills us?" Second, "How *should* we, as individuals and societies, approach the paradox of using but not abusing nature?" And finally, "What is our place on Earth—what are our responsibilities, duties, and obligations toward humans and nonhumans alike?" Exploring these three questions and relating them to the many interpretations of sustainability can help us make sense of the past, claim the present, and plan for the future.

Sustainability is an elusive and inescapably normative term—it involves our values and subjective perceptions about the state of the world, technology, economics, and the value of all life. Characterizations of sustainability are inextricably tied to our views on the existence, or lack thereof, of real physical carrying capacity limits and social or human behavioral constraints that might affect our abilities to acquire and process knowledge, make wise judgments, govern, manage, and plan. For instance, when considering issues such as climate change or biodiversity loss, do we emphasize adaptation and change or conservation of structural characteristics?

Any thoughtful consideration of sustainability demands a careful examination of four key questions. First, *What are we trying to sustain*—the human race, a viable economy, unrestricted technological development, our lifestyle, biodiversity, cultural diversity, ecosystem services, particular species? Second, *For whom*—all living humans, some living humans, future generations, all life? Third, *For how long*—till the sun dies, thousands of years, decades, the weekend? And finally, *Who decides*—who is making the decisions for the *whom*? How did they come to be in this role, and what values and standards will they use to make their decisions?

Two related questions also warrant consideration. First, does "to sustain" merely mean to keep in existence, to persist—regardless of the state of existence? A culture or species could be "kept alive" in a museum or zoo—just as with Ishi or a snow leopard[2]. Is this acceptable? Or by "sustain" do we mean something more akin to the flourishing of cultures and species? Second, does the particular "sustainable" world under consideration constitute a "desirable" world? Would we be spiritually or

---

[2] Ishi was the lone survivor of the Yahi tribe of Northern California. He lived in the "wild" until 1911, when he was found, emaciated and enervated, in the corral of a slaughterhouse. Ishi was delivered to the University of California's Museum of Anthropology, where he spent the remainder of his life (some five years) under the care of the anthropologist's Thomas Waterman and Alfred Kroeber. Snow leopards (*Panthera uncia*) are solitary and very rare members of the *Felidae* family that are native to the mountains of Central Asia. Unfortunately, their beautiful, creamy gray and black spotted fur is a valuable commodity. They also suffer from habitat loss and fragmentation. Strangely, while they undergo increasing threats in the wild, they continue to breed successfully in zoos around the world.

aesthetically satisfied with a sustainable but much less diverse world of plastic trees and nature DVDs? Must the "sustainable" state also be socially just? What is the cost of going from here to there (or not going)—how are past relationships and traditions altered in the transition? Can it embrace serendipity and is it prepared for the unexpected? As with any complex multicriteria problem there will be competing interests and conflicts regarding the weighing of priorities. Are there provisions for addressing these in an ethically responsible and equitable manner?

This is a representative selection of the many questions that can and should be raised when taking an open-minded, meta-perspective on the sustainability debate. I have coined the term "ecocultural sustainability" to refer to a state and process that is both desirable and ecologically-sound. In my view, realizing a state of ecocultural sustainability requires that we, at a minimum, can support over successive generations: (1) the flourishing of rich cultural and biological diversity; (2) forms of governance that are democratic, open, transparent, and socially just; (3) sufficient, bioregionally-sound, and respectful economies; and (4) accountable and creative economies that keep their ecocultural wake in-check by both learning from and working with nature *and* limiting the total life-cycle costs (social, environmental, and financial) of production and consumption.[3]

But is such concern about the future and the environment only limited to academics in writing chairs? Where does the public stand on these issues? Do they believe that significant environmental problems exist? How do they see the future? What is their stated willingness to trade-off standard of living for quality of life? Are they knowledgeable about the issues and do they possess a sophisticated understanding of them? What are their environmental values?

PUBLIC PERCEPTION VERSUS ACTION: A PARADOX OF EPIC PROPORTIONS

In the United States, the public has been surveyed on their perception and knowledge of environmental issues since the late 1960s (Dunlap, 1992; NEETF, 2000, 1998). These surveys indicate a widespread sentiment that environmental quality is deteriorating at all levels—from the local to the global. These surveys also suggest that the public perceives environmental deterioration as posing a growing threat to human health and well-being. They also suggest that the public is willing to make trade-offs to garner improved environmental quality. When posed with a hypothetical choice requiring a trade-off between "economic development" and "environmental protection," seventy-one percent of those surveyed chose "environmental protection" (NEETF, 2000).

Conventional wisdom suggests that the environment is a luxury good—that such views will be limited to the citizens of wealthy, industrialized nations who can afford to be more concerned about environmental problems than citizens of less

---

[3] This definition of ecocultural sustainability is meant to exist as an ideal, as a state to strive for. It is, however, operationalizable and it is capable of being used in practice to establish a series of objectives and indicators for guiding design, policy, and decision-making *and* monitoring our progress toward or away from these objectives.

economically advantaged, so-called "developing" nations. Gallup's "Health of the Planet" survey, the largest environmental opinion survey ever conducted, suggests that conventional wisdom desperately needs revision (Dunlap et al., 1993a,b; Bloom, 1995). The survey, conducted in 1992, covered twenty-four nations, eleven classified as high income by the World Bank and the remaining thirteen representing high-medium, low-medium, and low income countries. The goal of the survey was to compare citizens' views on the seriousness of environmental problems and gauge their support for environmental protection.

The survey results indicate that concern about environmental problems, while widespread throughout the surveyed countries, is actually more significant among citizens of the "developing" countries. These citizens, who are often more directly dependent on the environment for food, water, fuel, and raw materials for building and clothing, believe environmental problems affect their health now and pose a greater threat for the future. They generally view their nation's environmental quality as worse than those in the wealthy, industrialized countries. And while citizens of the industrialized nations all view their nation's environmental quality as much better than the world average, only one-half of the "developing" countries view their nation's environmental quality as better than the world average. Given the tremendous economic disparities, it is astonishing that in nine of the "developing" countries surveyed, a majority of respondents stated a willingness to give environmental protection priority, even at the risk of slowing economic growth. In addition, in half of the "developing" countries surveyed a majority of respondents stated a willingness to pay higher prices to protect the environment. Finally, when asked, who is "more responsible for today's environmental problems in the world," citizens of the rich and poor nations alike were both willing to assume significant responsibility for the Earth's environmental troubles.

Perhaps most surprising of all, are the non-anthropocentric, non-instrumental expressions of environmental concern. Despite the absence of instrumental gains, a variety of public opinion surveys demonstrate that a growing majority of lay people view the more-than-human world as intrinsically valuable—as having value in its own right—and deserving of moral consideration (Kempton et al., 1995; Dunlap et al., 1993a). The statement, "Plants and animals do not exist primarily to be used by humans" yielded a 69% approval rating in the "Health of the Planet" survey performed by Dunlap et al. (1993a). This viewpoint has also been supported by significant numbers of high-level policy-makers in Norway (Naess, 1986b; 1987) and high level European policy makers in the field of global warming (Glasser et al., 1994). In a small, but unique survey on the perceptions of nature by young adolescents from urban and suburban Detroit metropolitan area schools, Wals (1994, p. 136) concludes that "[a]lthough all students are anthropocentrically concerned about pollution and other environmental issues, there are some students who express concern about the vanishing of nature areas as a result of human activity. These students seem to say that nature and the species that are part of nature have a right to exist on their own." Sadly, while global concern for the environment appears strong, basic knowledge of environmental issues, at least in the U.S., appears woefully inadequate (NEETF, 2000).

We are left with a paradox of epic proportions. Concern over the health, status, and character of the world and stated support for the environment have generally not translated into effective action. Reading over a decade's worth of *State of the World* and *World Resources Institute* reports, it is difficult for me to be as sanguine as Lomborg about the current direction of the world. As sustainability discussions have become more prevalent, many of the planet's vital statistics have shown increasingly downward trends.

We are overmining ancient water supplies; desertifying and salinizing or paving over once productive agricultural lands; overharvesting forests and fisheries; proliferating the planet with toxic wastes and endocrine disrupters; and creating nuclear wastes that must be isolated from living systems for more than ten millenia. We are also now the primary driving force behind a warming climate with more intense storms. In the wake of our production, consumption, and waste generation spree are extirpated cultures, languages, and species as well as increasingly vulnerable communities and degraded, fragmented ecosystems.

In short, we are building on a long-standing, hubris-filled pattern of planning and living that has had little regard for the environmental (and resulting social) consequences of our actions—a pattern that has been implicated in the collapse of societies from ancient Sumer and Rome to Easter Island. What can account for this disjunction between our stated concerns and our environmentally destructive actions?

DEEP ECOLOGY: A POTENTIAL STRATEGY FOR WORKING OUR WAY THROUGH THE PARADOX

The term *deep ecology* was introduced by Norwegian philosopher and mountaineer Arne Naess (1912–) in 1972 at the third World Future Research Conference in Bucharest.[4] Naess coined the terms *deep ecology* and *shallow ecology* to juxtapose what he regarded as two radically different approaches for problematizing (*Problematisieren*) and responding to the ecological crisis.[5] Deep ecology calls for expanding our sphere of concern to all living beings—charismatic or dull, gargantuan or tiny, sentient or not. This *wide-identification* is characterized by the perception that all life is interdependent; common goals bind all living beings to the life process. In its most expansive form, wide-identification is the intuition that nature's interests and our own coincide. The purpose of deep ecology as an ecophilosophical approach is to encourage and help individuals to weave together their ultimate beliefs (including wide-identification), their life philosophy, and other descriptive and prescriptive premises about the world and ecological science into

---

[4] For a discussion of this inaugural presentation of deep ecology, see Naess (1973).
[5] For a more detailed, mature version of deep ecology, see Naess (1986a). For a concise overview of deep ecology that chronicles its key evolutionary changes and identifies the distinctions among the "deep ecology approach to ecophilosophy," the "deep ecology movement," and Naess's "Ecosophy T" (a particular deep ecological total view), see Glasser (2001).

systematic conceptual structures for relating to the world—ecologically inspired *total views* or *ecosophies*.[6]

The "shallow," currently more influential approach to environmentalism is identified with treating the symptoms of the ecological crisis, such as pollution and resource degradation. Its central concern is the health and prosperity of people in the economically privileged countries. This reform-oriented approach is grounded in technological optimism, economic growth, and scientific management, not in ultimate premises that plumb the relationship between humans and nature. A core premise is "*all* environmental problems are manageable"—nature is a puzzle to be deciphered by human ingenuity and manipulated, albeit more efficiently, for human benefit. From this perspective, remedy for environmental problems is limited to economic, technological, and managerial reforms. This effort to palliate human impacts, rather than probe and address their underlying causes, favors a search for "technical" solutions to what are more likely social, political, and ethical problems. By truncating the realm of problematizing, the shallow approach, perhaps inadvertently, prunes the set of conceivable social changes to a feeble incrementalism.

The "deep" approach, on the other hand, while in no way discounting the exigency of addressing pollution and resource degradation, adopts a broader, long-term, more skeptical stance. Doubtful about technological optimism, critical of limitless economic growth, and decidedly against valuing nature in purely instrumental terms, it asks if the shallow approach's proposed solutions take into consideration the complexity and insidiousness of the problems they hope to rectify. Drawing on a wide diversity of philosophical or religious ultimate premises, which acknowledge that every living being has value in itself, the deep approach sees the flourishing of nature and culture as fundamentally intertwined. Nature is viewed as mentor, standard, and partner rather than vassal.

A key premise is that environmental management is much more about managing the habits and desires of humans than attempting to control nature. Remedy for environmental problems is sought by identifying and responding to the complex "root" causes of the ecological crisis, dedicating special attention to protect the wild and free from thoughtless human interference. Taking less for granted, the deep approach calls for the public questioning of every practice, assumption, and value that propels the ecological crisis.

By juxtaposing these two, almost caricatured, perspectives, Naess employs a technique of Gandhian nonviolent communication designed to confront core disagreements. The central premise is that society's potential to overcome the ecological crisis rests on guiding discussion and debate to its root causes. One of the primary root causes, Naess asserts, is the widespread disjunction between people's core beliefs and actions. People, in general, neither comprehend how their practices and everyday lifestyle choices harm the environment, nor recognize how these consequences may be in direct conflict with their core beliefs—this is the primary weakness of the shallow approach. A crucial, underlying hypothesis of the deep

---

[6] For a detailed discussion of deep ecology as an ecophilosophical approach, with special attention to its policy implications, see Glasser (1996).

approach is that teasing out the presumed inconsistencies between an individual's actions and their fundamental beliefs, while possibly engendering serious ancillary conflict along the way, will ultimately generate progress toward ecocultural sustainability.

Naess argues that humans act as if we have total views whether or not we make such structures explicit. Because our decisions regarding society and nature are guided by our total views, Naess maintains that we should attempt to articulate them. By making these structures explicit, we expand both our opportunities for fruitful debate and interchange and our possibilities for creating policies that are consistent with our collective ultimate beliefs. Total views are dynamic and tentative, as well as adaptive and revisable. They are always fragmentary and incomplete, but this is no justification for abdicating our responsibility for attempting to articulate them. In fact, the goal of ecocultural sustainability hinges on our ability to integrate description with prescription in a manner that relates ethics, norms, rules, and practice. This integration is necessary, in part, because norms like "Respect for all living beings!" cannot prescribe behavior in particular situations. "Respect!" as a norm, neither implies that behavior towards all creatures should be equal nor that some creatures should not be eaten. "Respect!" removed from social context and the vicissitudes of life, cannot dictate conduct.

This focus on praxis (responsibility and action) separates the deep ecology approach from more descriptive inquiries into environmental philosophy that focus on axiological questions, such as extending "rights" to certain nonhumans or grading intrinsic value. The ontologically inspired deep ecology approach attempts to counter the perception of fundamental people/environment and spiritual/physical cleavages. Its primary strategy for overcoming the ecological crisis is to help individuals avoid pseudo-rational thinking.

Naess argues that many regrettable environmental decisions are made in a state of "philosophical stupor," where narrow concerns are confused with, and then substituted for, more fundamental ones. In proposing the deep/shallow contrast, Naess applies his research on empirical semantics, philosophy of science, the inquiring skepticism of Sextus Empiricus, Spinoza, and Gandhian nonviolent communication. His technical semantic distinction is directed at our *level of problematizing*—the extent to which we can, and do, coherently and consistently trace our views, practices, and actions back to our ultimate beliefs or bedrock assumptions.

In relating this notion of persistently asking deeper questions to the ecological crisis, Naess broadens his concept of "depth." In the context of deep ecology as an ecophilosophical approach, depth refers to both the general level of problematizing we employ in seeking out the underlying, coevolving causes of the ecological crisis *and* the extent of our willingness to consider an expansive array of social and policy responses, even if they necessitate changes that constitute a radical departure from the status quo.

Rather than calling for a new environmental ethic or a radical change in fundamental values, Naess's approach to ecophilosophy centers on transforming practice and policy by challenging us to develop more thoroughly reasoned, consistent, and ecologically inspired *total views*. Some will take issue with the core premise underlying this goal. They counter by asking: Can thoroughly reasoned and consistent positions based on existing wide-identifying ultimate norms actually help to generate policies and actions that conserve the earth's full richness and diversity? Widening identification may serve to moderate technological hubris by rekindling humility, but it cannot eliminate conflicts between humans and nature. Individuals will still choose to fell trees, dam rivers, drive cars, eat animals, use toxic chemicals, procreate, and pollute. Naess's hope, however, is that they will be more mindful of the costs and web of consequences that emerge from these various courses of action.

Expanding our concern to others does not, in any way, imply a consequent disregard or decrease of concern for each other, quite the contrary. As a farmer learns to listen to the land and work it well, the land has a way of bestowing well-being on the farmer. By inspiring love for life, encouraging accountability, and promoting methodical reasoning that integrates our feelings and emotions, the practice of forming a total view may work similarly.

THE IMPORTANCE OF EFFECTIVE CHANGE STRATEGIES: FIRST- AND SECOND-ORDER CHANGE

If we place any credence in the public opinion surveys on peoples' perceptions of the environment and we generally embrace the deep/shallow ecology contrast, then the broad-scale persistence of environmentally destructive habits cannot be attributed solely to a basic lack of awareness of environmental degradation. I repeat, the essential problem of continued environmental degradation is not a failure to recognize that there is a problem. Rather, it is what to do about it and how to go about doing it. Concern clearly has not yet translated into effective action.

Why do people say they care about something and then seemingly act hypocritically—pursuing ends that appear counter to previously stated concerns? Explanations abound. Sometimes people are disingenuous or their concerns are superficial.[7] Sometimes people do not recognize how their actions can generate undesirable effects or serious norm conflicts. Sometimes people fail to appreciate the interconnectedness of problems such as consumption, economic growth, population growth, poverty, sprawl, poor health, ennui, and environmental degradation. These issues are exacerbated by the fact that the effects of the environmental problems we create are often separated from us in space and time, uncertain, and indirect—environmental problems often appear unrelated to the actions that engender them, in part because they represent the composite result of thousands or millions of similar actions by others. Sometimes society and government exacerbate these problems by creating perverse incentives (Weizsäcker et al., 1998) and social traps (Costanza, 1987). Other explanations also exist. Sometimes goals are in conflict—people may act "rationally," but simply view other objectives or considerations as more meaningful or more pressing at the time. Finally, sometimes people recognize the importance of change and actually attempt

---

[7] Our previous discussion of the public's perception of environmental problems and their stated willingness to trade-off standard of living for improved environmental quality demonstrates that the publics concern for the environment is serious and consequential.

it, but get stymied along the way. Change is difficult. This last explanation, which to some extent encompasses many of the others, is the central focus of this section.

In the 1960s and 1970s Paul Watzlawick and his colleagues at the Mental Research Institute in Palo Alto, California, performed ground-breaking research on how problems arise and persist in some cases, yet are resolved in others (Watzlawick et al., 1974). A key insight from this research is that there are two fundamentally different types of change. "First-order" change occurs *within* a given system, which itself remains unchanged. Examples of first-order change include recycling, pollution reduction, and standards to promote increased fuel efficiency. However beneficial any of these changes might be, none of them individually, nor even all of them collectively, can bring an end to the ecological crisis. From the perspective of deep ecology, the shallow ecology approach is constrained to create first-order changes. "Second-order" change, on the other hand, results in a transformation of the system itself. It represents a radical break or logical jump from the status quo, and thus its practical manifestations may appear outlandish, irrational, or paradoxical. An interesting aspect of second-order change is that it is often not necessary or important to deeply understand the fundamental cause of the problematic behavior.[8] The crucial issue is to find a strategy for breaking the problematic behavior pattern and engender a new, more appropriate behavior pattern. A transition from our present state of unsustainablity to a state of global sustainability would constitute a second-order change.

Watzlawick and colleagues identified three primary strategies for mishandling change (1974, p. 39). The first is what I refer to as "solution by denial." It involves cases where some form of action is necessary, but none is taken. An example is Bjørn Lomborg's (2001) highly publicized, rosy prognostications about the "real state of the world," which are intended to convince us that environmental quality and quality of life are generally improving and that the environmental community has exaggerated claims to the contrary. If Lomborg's conclusions, which have been widely lauded in the media, were embraced uncritically, the public might be lulled into believing that there are no significant environmental challenges or that the poverty, disease, injustice, and biodiversity loss that spring from them are a mere fabrication of a cynical and self-serving environmental community. Lomborg's cornucopian conclusions, however, are not drawn from a sound blending of science and values, nor are they based on a sophisticated and thorough statistical evaluation of existing data. They are based on what I call the "Don't worry, be happy" theory of human progress. According to this view, things have been good in the past, so they will only get better in the future—the Earth is ever fecund and resilient to assault and, in any case, humans are creative beyond all imagination. What Lomborg is peddling is a sort of "environmental somnambulance"—a strategy for dis-solving the pressing environmental problems before us by sleepwalking our way through

---

[8] This important and subtle point represents an area where Naess and I may disagree. In my view, the deep ecology approach may help us to resolve the ecological crisis indirectly by focusing our attention on critical issues and stimulating second-order change, but without our ever truly identifying the "root" causes of the ecological crises.

them. He is preying on our seemingly insatiable desire for gratification and our general tendency not to postpone the same for the future or for future generations.

The second strategy for mishandling change is when change is attempted for a problem that is essentially irresoluble or a problem that does not exist. Lomborg and many others argue that pursuing climate mitigation strategies would be an example of this form of mishandling change. Their argument is that the likely effects are uncertain and probably not very severe in any case (at least for wealthy humans). Mitigation measures would be expensive and preclude other much more beneficial investments. Taking a "wait and see" approach and adapting, if necessary, is seen as the only rational strategy from this perspective.

The third and final strategy for mishandling change occurs when change is initiated at the wrong level. Two forms exist. A second-order change may be initiated for a first-order problem or a first-order change strategy may be applied to a second-order problem. In either case an error of logical type is committed and an endless, irresoluble cycle is established. In complex, difficult, or subtle situations, solutions may be initiated that not only do not produce the desired change, but actually exacerbate the problem. This is the insidious danger of the shallow ecology approach, which, on first principles, appears sensible, if not particularly effective, but harmless at worst. The shallow ecology approach represents an example of applying a first-order change strategy to a second-order problem. The general disjunction between people's stated environmental concerns and their actions, which continue to perpetuate a state of global unsustainability, clearly demonstrates the necessity for a second-order approach to this problem. To change the world, we must change ourselves. But how should we proceed?

WHY DOES THE PARADOX PERSIST? A DEEP ECOLOGY INSPIRED CRITIQUE OF ORTHODOX HIGHER EDUCATION

To move forward, we must begin working our way through the paradox. We must consider the insights regarding change theory and reexamine the chasm between our stated environmental concerns and our generally unsustainable lifestyles. Now is the time to ask: What is higher education's role in perpetuating the paradox? And how might it help resolve the paradox?

A facile, although not particularly insightful, answer to the first question is that higher education, however well-intentioned, neither prepares us to recognize the distinction between first- and second-order change, nor helps us to contemplate the consequences of our everyday actions. From the standpoint of deep ecology, both of these explanations have merit, but they reveal only part of the story. As with most admonishments, they also belie many commendable efforts to do otherwise. More significantly, however, these explanations only probe the surface.

I offer the following deep ecology inspired critique of orthodox higher education in the United States to highlight three core issues that lie below the surface. The first is higher education's tendency to promote alienation from the non-human world. By focusing on dissecting, subduing, and transcending nature, we have, perhaps inadvertently, come to define ourselves in opposition to it. A consequence of our anthropocentrism is ennui and estrangement from the world from which we sprang—the world we depend on for sustenance and meaning. Like a root-bound plant, this restriction of a key source of sustenance binds us in a perpetual state of immaturity.

The second issue is higher education's emphasis on producing and regurgitating objective information. This emphasis undervalues the importance of giving information meaning. It also undervalues the importance and excitement of learning how to seek out meaningful information. Amassing facts about the world or creating abstract models, which describe how subsystems in nature work, does not imply that we have an intimate knowledge of the world or the ability to restore what we have impoverished. Academe's effort to neatly separate facts from values leaves us ill-prepared to process and make sense of information. We are left unable to use the information we have wisely—in service of the planet and people.

The third issue is higher education's promotion of passivity in relation to subjective, real-world problems. With regard to environmental problems, higher education prepares us to, at best, document nature's decline or improve our understanding of the causes of the decline. Practical problem solving is generally viewed as mundane and unsuitable for scholars whose primary purpose is contributing to the growth of knowledge. Advocating for or against is seen as compromising one's objectivity as a scholar and is usually looked upon with suspicion, mistrust, and disapproval.

In summary, orthodox higher education and the intellectual tools and skills that it offers provide us with little protection from ourselves. The orthodox approach to higher education is shallow because it does not outfit us with the skills, tools, and vision to probe the depths of our predicament or guide our way to a sustainable future. Two examples underscore the depth of the crisis. Despite high levels of environmental concern, surveys that explore knowledge of environmental issues in the U.S. demonstrate that a profound environmental illiteracy persists (NEETF, 2000; 1998; Kempton et al., 1995). These surveys also indicate that there is very little difference in environmental literacy levels between college graduates and those with a high school education or less (NEETF, 1998). While higher education institutions may be uniquely suited to help usher in a transition to a more sustainable world, a recent study by the National Wildlife Federation demonstrates that few have taken the initiative to actively incorporate sustainability considerations into all aspects of their research, operations, outreach, and teaching (Glasser, 2002).

Finally, in relation to the question of exploring the relevance of second-order change, we must appreciate that academia has a significant investment in perpetuating the status quo. It protects this investment by promoting an ideology of disciplinary idolatry, anthropocentrism, infallibility, invulnerability, and—if all else fails—adaptability. After more than three decades of creating international declarations for environmental sustainability in higher education and nine declarations later, we are still creating new declarations for environmental sustainability in higher education. However insightful the Stockholm, Tbilisi, Tailloires, Swansea, Thessoloniki, and other declarations may be, their ability to facilitate second-order change has been limited (Wright, 2002). Change, especially second-order change, can be slow and difficult. The need for new, second-order learning strategies cannot be more apparent.

SECOND-ORDER LEARNING FOR SECOND-ORDER CHANGE: PEDAGOGY FOR A SUSTAINABLE AND DESIRABLE WORLD

The deep ecology vision for a sustainable and desirable world calls for considerable social, economic, technological, and ideological change.[9] It acknowledges the inadequacy of applying first-order learning to facilitate second-order change. This vision sees academia as a place to: integrate reason and emotion (Naess, 2002), "test drive" new ideas, be captivated by the fervor of learning, develop our appreciation for nature, and learn skills and values that will prepare us to be a positive and creative force on Earth. Distinguishing between preparing people for a job and preparing them for life, it highlights the significance of promoting emotional maturity (Naess, 2002). While not attempting to dictate a particular set of values (except, some form of wide-identification), deep ecology offers three "tools" for helping us to reconsider our value priorities and more consistently relate them to our lifestyles and everyday actions.

The first tool, "deep questioning," has already been discussed at length. The second is the idea of "vital needs." It is meant to help us contemplate the relationship between quality of life and standard of living. Beyond a certain point (that of satisfying our vital needs), the acquisition of more things does not generally lead to more satisfaction. Furthermore, the downstream consequences of satisfying these "non-vital needs" often stand in the way of others satisfying their own vital needs. While everyone will have a different concept of what constitute vital needs for their own circumstances, the goal is to have each of us consider the acquisition of every new thing in light of this concept. The third is the "principle of universalizability." It is meant to help us reflect on the equity implications of our actions and acquisitions. When applied to any act or acquisition, we are to ask ourselves two questions. First, is it both possible and feasible for anyone in the world to perform the same act or acquire the same good or service? And second, what would happen if everyone in the world actually did act in the same fashion or acquire the same good or service?

A final concept that may play a key role in facilitating second-order change, perhaps by serendipity, is Naess's concept of "beautiful action" (1993). Naess's concept is an elaboration of Kant's distinction between "dutiful acts," which are dictated by respect for the moral law, and "beautiful acts," which result from inclination. By tapping into our innate affinity for nonhumans (biophilia), we may begin to reduce our dependence on prescriptive laws and regulations and cultivate a global society committed to ecocultural sustainability.[10]

---

[9] See the Eight Points of the Deep Ecology Platform, particularly Point Six (Naess, 2002, p. 108-109).

[10] For a significant start along this path, see for instance the policy strategies used in the green planning efforts of the Netherlands and New Zealand (Johnson, 1997).

But how can this seemingly disjointed collection of tools and concepts from deep ecology help us with the central question of creating a pedagogy for a sustainable and desirable world? How can deep ecology help us create a learning community that engages the whole person—and the entire academic community—in the challenge of second-order change through second-order learning. Deep ecology can do this by helping us to see sustainability as an outcome and process as well as a catalyst for institutional innovation.

Our earlier discussion on the disjunction between the public's stated environmental concern and their unsustainable actions demonstrated that the paradox does not arise from a simple ignorance of environmental problems— although knowledge of environmental issues does not run particularly deep. The central problem also cannot be attributed to an unwillingness to change—although it is often not clear what to do and we often pursue ineffective change strategies. The solution to the paradox lies in activating and deepening our existing concerns, helping to make our actions more consistent with them, and choosing more effective change strategies, not in creating entirely new values for environmental concern. In order to do so we need to make the consequences of our actions more vivid, critical trade-offs more transparent, and value conflicts more real. We can do this by going back to our roots and creating a new core mission for higher education.

ROOTS—Research, Operations, Outreach, and Teaching for Sustainability:
A Deep Ecology Inspired Mission for Higher Education

1. Help nurture a sense of wonder and a passion for life-long learning that integrates reason and emotion and stimulates our imaginations.
2. Inspire positive attitudes toward nature.
3. Create opportunities for regular and direct contact with nature.
4. Provide more thorough, sophisticated, and realistic models of nature and models of how the environment functions (this includes understanding the many ways in which ecosystem services provide for our sustenance).
5. Prepare everyone to consider and explore the impacts of everyday actions—on themselves, their families, their communities, and those distant from them in space and time, including nonhumans (this includes understanding how population, consumption, technology, and values interplay to generate impacts).
6. Encourage open-mindedness and non-dogmatism in relation to discussion and problem solving.
7. Develop the skills for wise or mindful decision-making (this includes developing skills in questioning the "taken-for-granted" assumptions about the world and society that currently perpetuate unsustainability). Help prepare people to distinguish between: needs and wants, quality of life and standard of living, benefits and drawbacks of new technologies, etc.
8. Break down barriers of disciplinary idolatry and encourage true interdisciplinary and transdisciplinary thinking. Infuse the entire curriculum, across all colleges, with a discussion of sustainability questions—from the impacts of globalization to the potential effects of a proposed application or policy.

9. Inspire a sense of responsibility and activeness in relation to pressing social and environmental problems. Promote planetary CPR—creativity, prudence, and responsibility in service of people and the planet.
10. Support real-world problem-solving in service of people and the planet.
11. Use the campus and the community as living laboratories. Create opportunities for research on: education for sustainability, sustainable living, ecological engineering, ecological design, ecological economics, sustainability indicators, green building, green business, green planning, sustainable agriculture, renewable energy, industrial ecology, life-cycle analysis, and sustainable water, fisheries, and forest management.
12. Make academic institutions models of sustainability in all aspects of their functioning. Create a Campus Environmental Impact/Sustainability Committee and a Campus Environmental Impact/Sustainability Committee Mission Statement. Perform regular campus sustainability assessments and use these to refine and update campus policies.[11]

The deep ecological vision of higher education entreats us to wake up from our philosophical stupor. It asks us to develop our skills of self-criticism to new, unimagined heights; to distinguish between knowledge and wisdom; and to be active in relation to problems. It helps us to draw conclusions that are consistent with both our core values and a deeper, more informed understanding of the state of the world. And it does this not by demanding obeisance to a particular set of values, but by cultivating our own proclivities for love of life and helping us to integrate our core values with our lifestyles. In doing so, it helps us to resolve the paradox of the disjunction between people's everyday actions and stated environmental concern.

CONCLUSION

> Whatever our job we need to integrate life theory and life practice, clarify our value priorities, distinguish life quality from mere standard of life, and contribute in our own way to diminish unsustainability (Naess, 1992, p. 303)

Naess is fond of saying that he is pessimistic about the twenty-first century, yet optimistic about the twenty-second. If society is to initiate a paradigm shift toward sustainability and meet Naess's goal, people today will need to do much more than simply muddle along becoming less unsustainable. A world that is less unsustainable is neither sustainable nor a positive vision for the future. Ultimately, we must shift our focus from preventing the destructive, which is a vacuous goal, to promoting the good. By drawing on our strengths as a species—ingenuity, sympathy, optimism, love of wisdom, potential to reason, capacity for transformation—we can inspire joy and hope and stimulate much more powerful, positive motivations.

    The practice of creating total views is a strategy for helping us to stay in-touch with how our lifestyles and everyday actions shape the world. Active engagement in the process can help us appreciate the significance and magnitude of the problems

---

[11] For more information on campus sustainability assessment (CSA), including rationale, trends, best-practice and a searchable database with data on more than 1,100 CSA projects, see the work of the Campus Sustainability Assessment Project (which I direct) at: http://csap.envs.wmich.edu/

we create, as well as inspire sensitivity to the dilemma of trying to engineer our way out of them. By highlighting the importance of wide-identification and pointing out its existence in most of the world's religions and many of its philosophical traditions (if only as minority views), deep ecology offers a "middle way" between seeking out entirely new ultimate norms and ethics and presuming that no fundamental changes from the status quo are required.

In the end, deep ecology opens more questions than it answers. Like a koan, this is its allure, frustration, and promise. As a pedagogical strategy, it is subversive. It inculcates a profound open-mindedness and non-dogmatism that builds our potential for creating positive, second-order change.

Imagine that we (and enough others) do believe that the planet was not made for us alone—that we appreciate natures' use value, cherish it as a source of inspiration, and respect its right to live and flourish. Consider what kind of world we might encourage if our actions and lifestyles integrate reason and emotion and are tempered by the intuition that humans are not the center of existence. The idea of humans as "plain planetary citizens," but ones with unique capabilities and profound responsibilities, is not new. It is a thread that appears throughout history in the writings of Buddha, Asoka, St. Francis of Assisi, Aldo Leopold, Arne Naess, and many others. It just has not taken hold—yet. Human influence on the planet may ultimately be judged, not by our potential to transform it in our own image—which is considerable—but by our presence of mind and spirit, both as individuals and as a species, to exercise creativity, compassion, and restraint in service of the planet and ourselves. By choosing to reshape our relationship to the world, the world can be reshaped in positive ways. Our ability to bring about such change is only limited by our imaginations and our desire to learn how to be the change we wish to see.

REFERENCES

Bloom, D. E. (1995). International Public Opinion and the Environment. Science, 269(21 July), 354-358.

Costanza, R. (1987). Social Traps and Environmental Policy. BioScience, 37, 407-412.

Dunlap, R. E., Gallup, Jr., G. H., & Gallup, A. M. (1993a). Health of the Planet. Princeton, N.J.: George H. Gallup International Institute.

Dunlop, R. E., Gallup, Jr., G. H., & Gallup, A. M. (1993b). Of Global Concern: Results of the Health of the Planet Survey. Environment, 35(9), 7-15, 33-39.

Dunlap, R. E. (1992). Trends in Public Opinion Toward Environmental Issues: 1965-1990. In: Riley E. Dunlap & Angela G. Mertig (Eds.), American Environmentalism: The U.S. Environmental Movement 1970-1990 (pp. 89-116). Washington, D.C. and London: Taylor and Francis.

Glasser, H. (2002). Murky Grades on Campus Sustainability: A Survey Reveals a Widespread Unwillingness to Make the Environment a High Priority. AGB Trusteeship, 10(2), 34-35.

Glasser, H. (2001). Deep Ecology. In: N. J. Smelser & P. B. Bates (Eds.), International Encyclopedia of Social and Behavioral Sciences Vol. 6. (pp. 4041-4045). Oxford: Pergamon.

Glasser, H. (1996). Naess's Deep Ecology Approach and Environmental Policy. Inquiry, 39(2), 157-187. Reprinted in N. Witoszek & A. Brennan (Eds.), Philosophical Dialogues: Arne Naess and the Progress of Ecophilosophy (pp. 360-390). Lanham, Maryland: Rowman and Littlefield.

Glasser, H., Craig, P., & Kempton, W. (1994). Ethics and Values in Environmental Policy: The Said and the UNCED. In: J. van der Straaten & J. van den Bergh (Eds.), Toward Sustainable Development: Concepts, Methods, and Policy (pp. 80-103). Washington, D.C.: Island Press.

Goodall, J. (1994). Digging up the Roots. Orion (Winter), 20-21.

Johnson, H. D., & Brower, F. b. D. R. (1997). Green Plans: Greenprint for Sustainability. Lincoln and London: University of Nebraska.

Kempton, W., Boster, J. S., & Hartley, J. A. (1995). *Environmental Values in American Culture.* Cambridge, Massachusetts: MIT Press.

Lomborg, B. (2001). *The Skeptical Environmentalist: Measuring the Real State of the World.* Cambridge, U.K.: Cambridge University Press.

Naess, A., Haukland, W., McKibben, F., & Glasser, H. (2002). *Life's Philosophy: Reason and Feeling in a Deeper World* (R. Huntford, Trans.). Athens and London: University of Georgia Press.

Naess, A. (1995). Self Realization: An Ecological Approach to Being in the World. Reprinted in Sessions, 1995. In G. Sessions (Ed.), *Deep Ecology For the 21st Century: Readings on the Philosophy and Practice of the New Environmentalism* (pp. 226-239). Boston: Shambala. This essay was originally given as a lecture, March 12, 1986, at Murdoch University, Western Australia, sponsored by the Keith Roby Memorial Trust.

Naess, A. (1993). Beautiful Action. Its Function in the Ecological Crisis. *Environmental Values,* 2(1), 67-71.

Naess, A. (1992). Sustainability! The Integral Approach. In O. T. Sandlund & K. Hindar & A. H. D. Brown (Eds.), *Conservation of Biodiversity for Sustainable Development* (pp. 303-310). Oslo: Scandinavian University Press.

Naess, A. (1987). *Ekspertenes Syn På Naturens Egenverdi (Expert Views on the Intrinsic Value of Nature).* Trondheim: Tapir Forlag.

Naess, A. (1986a). The Deep Ecology Movement: Some Philosophical Aspects. *Philosophical Inquiry, 8,* 10-31. Reprinted in G. Sessions (Ed.). (1995). *Deep Ecology for the Twenty-First Century* (pp. 64-84). Boston: Shambhala.

Naess, A. (1986b). Intrinsic Nature: Will the Defenders of Nature Please Rise? In M. E. Soulé (Ed.), *Conservation Biology: The Science and Scarcity of Diversity* (pp. 504-515). Sunderland, Massachusetts: Sinauer Associates.

Naess, A. (1973). The Shallow and the Deep, Long-Range Ecology Movement. A Summary. *Inquiry, 16,* 95-100. Reprinted in G. Sessions (Ed.) (1995). *Deep Ecology for the Twenty-First Century* (pp. 151-155). Boston: Shambhala.

National Environmental Education and Training Foundation & Roper Starch Worldwide (2000). *Lessons from the Environment: The Ninth National Report Card on Environmental Attitudes, Knowledge, and Behaviors.* Washington, D.C.: NEETF.

National Environmental Education and Training Foundation & Roper Starch Worldwide (1998). *The National Report Card on Environmental Knowledge, Attitudes and Behaviors (The Seventh Annual Survey of Adult Americans).* Washington, D.C.: NEETF.

Wals, A. E. J. (1994). *Pollution Stinks! Young Adolescents' Perceptions of Nature and Environmental Issues with Implications for Education in Urban Settings.* De Lier, The Netherlands: Academic Book Center.

Watzlawick, P., Weakland, J., & Fisch, R. (1974). *Change: Principles of Problem Formation and Problem Resolution.* New York: W. W. Norton.

Weizsäcker, E. v., Lovins, A. B., & Lovins, L. H. (1998). *Factor Four: Doubling Wealth—Halving Resource Use, The New Report to the Club of Rome.* London: Earthscan.

Wilson, E. O. (1994). *Naturalist.* Washington, D.C.: Island Press.

Wright, T. S. H. (2002). Definitions and Frameworks for Environmental Sustainability in Higher Education. *International Journal of Sustainability in Higher Education,* 3(2), 203-220.

BIOGRAPHY

Harold Glasser is an Associate Professor in Environmental Studies at Western Michigan University. His research explores how societies—from ancient to contemporary—make choices about using and protecting the environment. He also uses this research to create multicriteria decision frameworks and tools, information systems, and design prototypes to help support ecocultural sustainability. He teaches a class on the campus as a living laboratory and is Editor of the Selected Works of Arne Naess, Director of the Campus Sustainability Assessment Project, Chair of Western Michigan University's Sustainability Committee, a member of the

Economicology Group, and a steering committee member of the national Higher Education Network for Sustainability and the Environment. Glasser has written on deep ecology, environmental policy, multicriteria analysis, green accounting, education for ecocultural sustainability, and campus sustainability assessment.

CHAPTER 12

THE CONTRIBUTION OF ECOFEMINIST PERSPECTIVES TO STAINABILITY IN HIGHER EDUCATION

Annette Gough

INTRODUCTION

Higher education institutions are the gatekeepers of knowledge production, accreditation, legitimation and dissemination. What they choose to include, exclude, or denigrate can make all the difference to the cognitive and operational capacities of their students as future citizens (e.g. Maher & Tetreault, 2001; Odora Hoppers, 2001).

In the past, in much of higher education women, as students and academics, have struggled for recognition (Christian-Smith & Kellor, 1999; Currie et al., 2002; Davies et al., 1994; Heilbrun, 2002; Kelly, 1985; Merrill, 1999; Morley, 1999), and women[1] have been overlooked in most sustainability programs through being subsumed into the notion of "universalized people" (Braidotti et al., 1994; Buckingham-Hatfield, 2002; Gough, 1999a; 1999b; Salleh, 1997). However, women have a distinctive contribution to make to sustainability policy, pedagogy and research that needs to be foregrounded. This chapter discusses research into the absences of women's perspectives from sustainability policies, pedagogy and research and argues that ecofeminist pedagogies and research methodologies suggest new possibilities for the development of sustainability in higher education.

The ecofeminist movement has developed in parallel with the environment and environmental education movements since the 1970s, but there has been little dialogue between it and the other two. Chapter 24 of *Agenda 21* (UNCED, 1992) had as its overall goal, achieving active involvement of women in economic and political decision making, with emphasis on women's participation in national and international ecosystem management and control of environmental degradation. This perspective has been overlooked to date in most forms and sectors of education, so in this chapter I argue for the power and the promise of adopting such a perspective for the higher education context. At the technical level actions should include

---

[1] While I recognise that women are one of many marginalised groups in society, and discuss the importance of indigenous knowledge systems later in this chapter, the main emphasis in this chapter is on the contribution of ecofeminist perspectives to sustainability in higher education.

Peter Blaze Corcoran & Arjen E.J. Wals (Editors), Higher Education and the Challenge of Sustainability: Problematics, Promise and Practice, 149-161.

increasing proportions of women as decision makers in implementing policies and programs for sustainable development; recognising women as equal members in workplaces both with respect to workloads and finance; and including women's perspectives in the content of higher education courses. However, it is equally important that ecofeminist perspectives inform research and pedagogies in higher education. In developing my arguments I focus on changing institutional and philosophical actions to make women's and men's lives in higher education more democratic, academic knowledge less partial, and sustainability achievable. As Uma Narayan and Sandra Harding argue:

> It is worth recollecting that the deepest forms of sexism and androcentrism – the ones most difficult even to identify, let alone to eradicate, have not been those visible in the intentional actions of individuals (which is not to excuse such overt or covert sexism and androcentrism). It has not been sexist or androcentric motivations or prejudices of individuals – their false beliefs and bad attitudes – that has given women the most trouble. Rather, it has been the institutional, societal, and civilizational or philosophic forms of sexism and androcentrism that have exerted the most powerful effects on women's and men's lives – the forms least visible to us in our daily lives. (Narayan & Harding, 2000, pp. vii-viii)

GENDER EQUALITY AND SUSTAINABILITY

The Rio Declaration (1992, Principle 21) stated: "Women have a vital role in environmental management and development. Their full participation is therefore essential to achieve sustainable development". The recognition of women in the Rio Declaration was the result of lobbying by groups like the Women's Environment and Development Organisation (WEDO) in New York. As WEDO co-chair and former US Congress member Bella Abzug explains, "although we women are the vast majority of grassroots activists, very few of us are in positions of power, setting the priorities and making decisions on issues to be tackled nationally and internationally" (1991, p. 2). So their goals are political: "to encourage women's global solidarity and empowerment, to expand and deepen women's networks, to educate and inform, and to create a local, national and global capacity to act" (Abzug, 1991, p. 2). The Johannesburg Declaration on Sustainable Development (United Nations, 2002) expanded on strategies to achieve women's full participation in sustainable development and included strong commitments to ensuring women's empowerment, emancipation and gender equality within all activities related to enacting the implementation plan from the Summit.

Many texts in the past decade and more have argued that there is a close relationship between women and sustainable development (e.g. Braidotti et al., 1994; Merchant, 1992; Salleh, 1997; Shiva, 1989), and most recognise that, while gender equality is a prerequisite for sustainable development, the relationship is not a simplistic one. Recently Susan Buckingham-Hatfield (2002) explored ways in which women have, and have not, become more involved in environmental decision making since UNCED. She concluded that health care, education and economic status are prerequisites for women to be meaningfully involved in environmental decision making, and that structural inequalities are currently overwhelming progress towards sustainable development. Eradication of poverty is recognised by the UN as being essential for sustainable development, as is the importance of universal basic education. However, although women's literacy levels have increased in the past decade, "economic inequality has seen little change and gendered health difference remain significant" (Buckingham-Hatfield, 2002, p. 233).

WOMEN'S PLACE IN HIGHER EDUCATION

Historically universities have been uni-vocal places exclusively for white males of the ruling classes. When individual members of minority and marginal groups in society have been able to access the structures of knowledge and power represented by the academy, they have had to defend both their presence and expertise on a regular basis. Many authors have documented the university's long association with the development and legitimation of men's knowledge and scholarship, and the tenuous relationships between women and their legitimate claims to knowledge (e.g. Christian-Smith & Kellor, 1999; Currie et al., 2002; Davies et al., 1994; Gillett, 1981; Heilbrun, 2002; Kearney & Ronning, 1996; Kelly, 1985; Luke & Gore, 1992, Morley, 1999).

Although universities had been in existence for centuries, it was only in the latter half of the nineteenth century that women were able to enter the academy as students. Margaret Gillett (1981) wrote of the experiences of women seeking admission to McGill University in Canada:

> She could not defend herself, she could not argue for nor claim intellectual equality because she could not know what it was that she did not know. She could not even be a party to deciding whether or not she ought to know. Thus, even though she was high on a pedestal, the politics of ignorance kept her in her subordinate place. (in Kelly, 1985, p. 4)

Although women are still confronting issues of the politics of ignorance in many places around the world[2], at the university level in English speaking nations this politics was being substantially challenged in the late nineteenth century. The first woman to graduate from a university in the British "Empire" was in Canada in 1875. In 1877 the first woman graduated in New Zealand, and in 1883 the first woman graduated from the University of Melbourne, the first in Australia. Although women had been permitted to enrol in all courses (except in medicine) at the University of London from 1878, they were not admitted as full members of either Oxford or Cambridge universities until 1920 and 1948 respectively. The universities took the women's fees, allowed them to sit the usual examinations but they would not formally confer degrees upon them: "Conservatives feared that the full admission of women to Oxford and Cambridge would erode the 'distinctively manly spirit' on which their fame rested" (Kelly, 1985, p. 2). Women's experiences at these universities during this period are eloquently portrayed by Virginia Woolf ([1929] 1977) in *A Room of One's Own*, and more recently by Ian McEwan (2002).

The content of university courses studied by women is also an issue:

---

[2] For example, 66% of the world's illiterates are women (Kearney, 1996, p.1), and in some areas, such as Pakistani villages, women are 100% illiterate (del Nevo, 1993).

> In the early days of university education for women, opinions were divided between those who thought women should conform to the same educational patterns as men and those who felt strongly that there should be a differential educational pattern – women should study subjects requiring so-called 'feminine aptitudes'... Had the women not pursued a similar course to that taken by the men, it is doubtful whether their qualifications would have been accepted in the community. From that time forward, the University has maintained a common curriculum in all subjects. (Blackwood, in Kelly, 1985, p. vii)

The early women graduates were proud to have entered the halls of academia on men's terms, but this unquestioning acceptance of men determining what it was worthwhile and necessary to know, and constructing and administering rules which perpetuated their own power, was challenged by feminists of the late twentieth century who drew attention to the socially constructed nature of knowledge and curriculum, and the need to change its form and focus to recognise the "politics of knowledge". For example, Sandra Harding notes that,

> Feminists have argued that traditional epistemologies, whether intentionally or unintentionally, systematically exclude the possibility that women could be 'knowers' or *agents of knowledge*; they claim that the voice of science is a masculine one; that history is written from only the point of view of men (of the dominant race and class); that the subject of a traditional sociological sentence is always assumed to be a man. They have proposed alternative theories that legitimate women as knowers. (Harding 1987, p. 3, emphasis in original)

More recently, Harding (1991; 1998; Narayan & Harding, 2000) has argued that gender is not the only influence on the production of knowledge and that class, race, sexual orientation, culture, ethnicity, age and religion are also significant. These issues are taken up later in this chapter.

Another aspect of women's place in higher education is their rank in the institutions. As Table 1 illustrates, women are still confronted by a "glass ceiling" and are few in number at academic leadership (associate professor and professor) levels (13.8% of female academics compared with 36% of male academics, and only 22% of all Australian academics are female[3]).

Table 1. Distributions of Women and Men Across Ranks: Australian Universities 1996 (from Anderson et al., 1997, Table 6.13).

	Female		Male		Total	
Professor/Prof Fellow	115	4.5	1504	16.6	1619	13.9
Assoc Prof/Rdr/Snr Res Fell	237	9.3	1756	19.4	1993	17.1
Snr Lect/Fellows/Snr Res Fell	687	26.9	3020	33.3	3707	31.9
Lecturer/Research Fellow	1466	57.4	2590	28.6	4056	34.9
Other	47	1.8	201	2.2	248	2.1
Total	2552	100	9071	100	11623	100

If higher education is to contribute effectively to sustainability a dual agenda of advancing women as academic leaders and renewal of higher education institutions

---

[3] Similar figures apply in other English speaking countries – see Kearney & Ronning (1996) and Merrill (1999).

is essential. As Sheryl Bond (1996, p. 51) argues, women are needed in senior decision-making positions, not just for numerical balance but

> to diversify and strengthen the leadership of the academy [because] women... by virtue of their own life experience, embrace the experience and knowledge of women, as well as men... sufficient numbers of women are needed to generate change in institutional policies and practices, in who is admitted to the academy, what is taught, the methodologies of teaching and learning, as well as the definition of questions which excite the mind and warrant investigation. (Bond, 1996, p. 51)

There is a gender dimension in all academic activity – research, teaching and administration – which goes beyond a disciplinary discussion and traverses all fields in academia. There also appears to be a correlation between increasing the number of women in academe and higher education institutions focusing on gender, ethnicity and culture (Maher & Tetreault, 1999). Thus it is vital that these dimensions (and others) are addressed if we are to achieve meaningful social change and achieve sustainability.

AN ECOFEMINIST PERSPECTIVE

The French writer Francoise d'Eaubonne coined the term *ecofeminisme* in her 1974 book *Le féminisme ou la Mort,* in which she "called upon women to lead an ecological revolution to save the planet. Such an ecological revolution would entail new gender relations between women and men and between humans and nature" (Merchant, 1992, p. 184). There are many paths into ecofeminism - from different forms of feminism (radical, Marxist, socialist, liberal), from environmentalism, from the study of political theory and history, and from exposure to nature-based religion. Indeed, just as there is not one feminism, there is not one ecofeminism. However, the origins and trajectories of ecofeminism are contested, and much ecofeminist energy has been consumed in arguing the various ideological positions. What is clear is that ecofeminism is not a monolithic, homogeneous ideology (which causes difficulties for some scholars), but it nevertheless provides a very different notion of what constitutes political change: "while ecofeminism recognises the severity of the crisis, it also recognises that the methods we choose in dealing with the problems must be life affirming, consensual and non-violent" (Diamond & Orenstein, 1990, p. xii). This progressive and critical social theory orientation of ecofeminism[4] (which it shares with the feminist and environment movements) is important in reshaping the basic socio-economic relations and the underlying values of society and institutions.

Since the term ecofeminism was coined in 1974 and named as a grassroots women-initiated environment movement, its meaning has expanded from just being concerned with ecological feminism to now recognising "that there are important connections between how one treats women, people of color, and the underclass on one hand and how one treats the nonhuman natural environment on the other" (Warren, 1997a, p. xi). This enriched vision of ecofeminism is inspiring the research and other work of scholars in a variety of academic and vocational fields including

---

[4] Ariel Salleh (1997, p. 69) notes that a number of prominent ecofeminists, such as Carolyn Merchant (1992) and Vandana Shiva (1989) have been influenced by Marx's work.

anthropology, biology, chemical engineering, communication studies, education, environmental studies, literature, political science, recreation and leisure studies, and sociology.

There are many reasons for taking the philosophical basis of the ecofeminist movement seriously. As Warren (1997b, pp. 13-14) argues, there are significant empirical data which suggest that:

1. the historical and causal significance of ways in which environmental destruction disproportionally affects women and children;
2. the epistemological significance of the "invisibility of women", especially women who know (e.g. about trees), for policies which affect both women's livelihood and ecological sustainability;
3. the methodological significance of omitting, neglecting, or overlooking issues about gender, race, class, and age in framing environmental policies and theories;
4. the conceptual significance of mainstream assumptions, e.g. about rationality and the environment, which may inadvertently, unconsciously, and unintentionally sanction or perpetuate environmental activities, with disproportionately adverse effects on women, children, people of color, and the poor;
5. the political and practical significance of women-initiated protests and grassroots organizing activities for both women and the natural environment;
6. the ethical significance of empirical data for theories and theorizing about women, people of color, children, and nature;
7. the theoretical significance of ecofeminist insights for any politics, policy, or philosophy; and
8. the linguistic and symbolic significance of language used to conceptualise and describe women and nonhuman nature.

These data should form the basis for reforming the policies and practices of teaching, research and administration in higher education if we are to achieve sustainability. However, before moving on, I believe that is important to confront the misreading of ecofeminism as "essentialist", i.e. that women are closer to nonhuman nature[5]. Many ecofeminists (including those mentioned in footnote 5), have argued that ecofeminism is not essentialist. For example, in one of the clearer explanations, Braidotti et al. argue that we must recognise that

> women and nature are simultaneously subjugated, and that this subjugation takes historically and culturally specific forms. If women take themselves seriously as social agents and as constitutive factors in this process, their praxis to end this double subjugation can be rooted not so much in women's equation with nature, but in taking responsibility for their own lives and environment (Braidotti et al., 1994, p. 75).

In this way we pursue a critical social theory within specific cultural, temporal and physical positions as a body of ideas, which avoids essentialism - but we must be vigilant and not reify "women's ways of knowing" (Gilligan, 1982).

---

[5] Accusations of essentialism are an obvious concern of most ecofeminists: the indexes of texts such as Braidotti et al. (1994), Salleh (1997) and Warren (1994, 1997c) all contain entries for 'essentialism'.

HIGHER EDUCATION, SUSTAINABILITY AND FEMINIST RESEARCH AND PEDAGOGIES[6]

Elsewhere I have discussed the significance of the absence of women from the international gatherings that formalised conceptions of environmental education, and the need to recognise them in environmental education research and pedagogy (Gough, 1999a), so I will not repeat those arguments here. Although my personal disposition is towards critical and poststructuralist feminist research and pedagogy, all feminist pedagogies and research methodologies can be applied in the development of sustainability in higher education. We can do many types of research and teaching which puts the social construction of gender at the centre of the inquiry, and we need more stories from women's lives relating to environments that we can use in the development of sustainability in higher education. Like Donna Haraway,

> I want to argue for a doctrine and practice of objectivity that privileges contestation, deconstruction, passionate construction, webbed connections, and hope for transformation of systems of knowledge and ways of seeing. But not just any partial perspective will do; we must be hostile to easy relativisms and holisms built out of summing and subsuming parts. (Haraway, 1991, pp. 191-2)

From an ecofeminist perspective there are many challenges for higher education institutions' research, teaching and administration if they are to have a role in sustainability. For example, Brown and Switzer (1991, p. 16) note that women are less likely to have scientific or economic training than men and, consequently, have less influence in developing curricula which give high priority to issues of importance to women, such as reduction of toxic wastes and information on safety standards. They also note that there is a need to compensate for the effects on research and teaching of the relative absence of women and women's interests from the professions of environmental science and economics: "this absence has meant that many questions on ecologically sustainable development from the fields of health, welfare, household management and social policy have neither been investigated nor included in environmental education" (Brown & Switzer 1991, p. 16).

Feminist research and pedagogy in environmental education is generally still at the stage of attempting to "add women" to traditional analyses rather than engaging in the more distinctive features of feminist research and pedagogy. In many ways environmental education is still mainly at Stage 1 of Kreinberg and Lewis' (1996) model of curriculum change for science education (adapted for environmental education) with only a few moving to Stage 2 and beyond:
Stage 1: Absence of women in environmental education not noticed;

---

[6] In this section I use the terms environmental education and education for sustainability interchangeably. This is because the majority of work to date in this field has been known as environmental education rather than education for sustainability. However, I believe that an important component of the shift from environmental education to education for sustainability as a terminology is the incorporation of ecofeminist perspectives into the latter where they have definitely been absent from the former (see Gough, 1999a, 1999b). I found aspects of Bill Scott's (2002) discussion of sustainability and learning useful in this regard, although I found his silence on the gender dimension alarming.

Stage 2: The search for the missing women in environmental education;
Stage 3: Why are there so few women in environmental education?;
Stage 4: Studying women's experience in environmental education;
Stage 5: Challenging the paradigms of what environmental education research and pedagogy are;
Stage 6: A transformed, gender balanced environmental education curriculum and research agenda.

What makes feminist research distinctive, according to Harding (1987), is that it moves beyond "adding women" and opens up

new empirical and theoretical resources (women's experiences);
new purposes of social science research (for women); and
new subject matter of inquiry (locating the researcher in the same critical plane as the overt subject matter).

This distinctiveness is more than a concern with research methods; it is a concern with epistemologies and ontologies as well as methodologies, and this has significant implications for both research and pedagogy in education for sustainability as we move towards Kreinberg and Lewis' stages 5 and 6.

New empirical and theoretical resources (women's experiences)

Traditionally, environmental education has analysed only male experiences or has constructed universalised subjects that are not distinguished as male or female. Yet there is no universal *man*, only culturally, racially, socio-economically (and so on) different men and women with fragmented identities. Environmental education has not addressed areas of women's experiences and knowledge which are equally resources for environmental education research processes. "One distinctive feature of feminist research is that it generates its problematics from the perspective of women's experiences" and "uses these experiences as a significant indicator of the reality against which hypotheses are tested" (Harding, 1987, p. 7).

> All of us live in social relations that naturalize, or make appear intuitive, social arrangements that are in fact optional; they have been created and made to appear natural by the power of the dominant groups. Thus, it is not necessary to have any *particular* form of human experience in order to learn how to generate less partial and distorted belief from the perspective of women's lives. It is 'only' necessary to learn how to overcome–to get a critical, objective perspective on–the 'spontaneous consciousness' created by thought that begins in one's dominant social location. (Harding, 1991, p. 287, emphases as in original)

Education for sustainability needs to adopt this feature for its research and pedagogy to address the current partial and distorted understandings of ourselves and the world around us that are produced by the traditional approaches to environmental education.

New purposes of social science research (for women)

"In the best of feminist research, the purposes of research and analysis are not separable from the origins of research problems" (Harding, 1987, p. 8). Therefore, if the focus of education for sustainability begins with what is problematic from the perspective of women's experience then it is designed for women (as well as for the environment).

New subject matter of inquiry (locating the researcher in the same critical plane as the overt subject matter)

The best feminist analysis "insists that the inquirer her/himself be placed in the same critical plane as the overt subject matter", i.e. "the class, race, culture, and gender assumptions, beliefs, and behaviors of the researcher her/himself must be placed with the frame of the picture that she/he attempts to paint" (Harding, 1987, p. 9). It is also important to critically examine "the differential ground for the scholar's, the teacher's, and the students' knowledge and authority, in order to put them all into relation with one another" (Maher & Tetreault, 1999, p. 209). Thus, rather than appearing as an anonymous voice, the researcher and teacher is a real individual with specific and locatable interests. In the case of education for sustainability this means that the whole focus for the research and pedagogy in higher education will need to change.

Feminist research provides an opportunity to re-vision the world–"to know it differently than we have ever known it; not to pass on a tradition but to break its hold on us" (Rich, 1990, in Crotty, 1998, p. 182)–by transforming common methodologies and methods and working against the "limiting of human possibility through culturally imposed stereotypes, lifestyles, roles and relationships" (Crotty, 1998, p. 182). The adoption of the three distinctive features of feminist research outlined above will lead to better ("less partial, less distorted") education for sustainability research stories and pedagogical practices that will benefit the field as a whole. Already some academics are moving in this direction. For example, several of the authors in Kearney and Ronning (1996) have adopted some of these approaches, particularly the university curriculum in development studies, which includes a gender dimension in Brazil (D'Avilo-Neto, 1996), and Lotz-Sisitka and Burt (2002) discuss their experiences in writing environmental education research texts. Other examples can be found for environmental studies and science in the work of Whitehouse and Taylor (1996) and Gough (2001).

The features of feminist research and an ecofeminist perspective outlined above resonate particularly well with recent work on indigenous knowledge systems and higher education being conducted in South Africa by, for example, Catherine Odora Hoppers (2001). She discusses the subjugation of indigenous knowledge systems by the past regimes and the need for "the reconstruction of knowledge, the critical scrutiny of existing paradigms and the epistemological foundations of existing academic practice, and the identification of the limitations they impose on creativity" (Odora Hoppers, 2001, p. 73) if South African society is to find sustainable and inclusive ways forward. As I have argued throughout this chapter, an ecofeminist perspective on sustainability includes not only women but also recognises that class, race, sexual orientation, culture, ethnicity, age and religion are significant. We need to support these new perspectives through higher education.

CONCLUSION

Because institutional sexism and androcentrism "have exerted the most powerful effects on women's and men's lives" (Narayan & Harding, 2000, p. viii), gender issues in higher education warrant urgent attention if we are to achieve sustainability. This will involve more than "adding women"; cultural attitudes and practices are key impediments to women's equal access to education and positions in higher education, and to getting sustainability onto the higher education agenda.

Kearney (1996) suggests a strategy for change based on four key premises: the promotion of feminine leadership in the academy, the mainstreaming of the gender dimension in the curriculum, continued research on barriers to women's equality, and acknowledgement of the gender dimension of (sustainable) development. I have discussed each of these issues here to varying degrees.

However, we also need to address the impediments posed by the new managerialism that pervades higher education, with its focus on corporate mission statements, goals, monitoring procedures and performance measures. According to Morley (1999, p. 28), the new managerialism promotes three "Es": economy, efficiency and effectiveness. Two other "Es" are noticeably absent–equity and environment–yet these are the areas of most concern for this chapter (and book). Social justice values are perceived as irrelevant to management theories based on marketisation, consumerism, individual rights and choice, and "equity is off the agenda" (Ball, 1994, p. 125). The challenge for us is to not only get the other "Es"– equity and environment–onto the agenda again, but to have them prioritized as part of corporate mission statements, goals, monitoring procedures and performance measures. This reform will require cultural and ideological change too, not simply structural or technical tweaking.

Finally, it may be useful to think of the reforms in terms of a critical and conserving agenda that reasserts traditional university values from feminist and sustainability perspectives. Currie et al. (2002) include in these values: democratic collegiality, professional autonomy and integrity, critical dissent and academic freedom, and the public interest value of universities, however not all of these are equally relevant in developing ecofeminist perspectives in higher education. Democracy is an active vehicle for social justice, and this is central in our concerns around ecofeminism and sustainability. So too is academic freedom as a key legitimating function of the university. Too often the new managerialist university displays a lack of respect for staff and their intellectual skills. It is the public interest value of higher education institutions that is paramount as part of the university's role "in developing active and critical-thinking citizens... [with] tolerance and compassion... [who] work for social justice... [and] the social transformation of society". (Currie et al., 2002, p. 190) The South Africa government has already recognized that there is a role for universities in their social reconstruction agenda – including "the promotion of a critical citizenry" and addressing "its particular development challenges" (South African Higher Education Task Team 2000, as cited in Currie et al., 2002. p. 190). While this is not as consistent with an ecofeminist and sustainability framework as it could be, it is a step in the right direction – towards developing an ecofeminist perspective to sustainability in higher education.

REFERENCES

Abzug, B. (1991). Women want an equal say in UNCED. *The Network '92,* 9(2).

Anderson, D., Arthur, R. & Stokes, T. (1997). *Qualifications of Australian Academics Sources and Levels 1978–1996.* Canberra, ACT: Evaluations and Investigations Program, Department of Employment, Education, Training and Youth Affairs. Retrieved October 13, 2002 from the World Wide Web: http://www.detya.gov.au/archive/highered/eippubs/eip97-11/front.htm

Ball, S. (1994). *Education Reform–A Critical and Post-Structuralist Approach.* Milton Keynes: Open University Press.

Bond, S.L. (1996). The experience of feminine leadership in the academy. In: M-L. Kearney & A. H. Ronning (Eds.), *Women and the University Curriculum: Towards Equality, Democracy and Peace* (pp. 35-52). London: Jessica Kingsley Publishers & Paris: UNESCO.

Braidotti, R., Charkiewicz, E., Hausler, S. & Wieringa, S. (1994). *Women, the Environment and Sustainable Development: Towards a Theoretical Analysis.* London: Zed Books/INSTRAW.

Brown, V.A. & Switzer, M.A. (1991). *Engendering the Debate: Women and Ecologically Sustainable Development.* Canberra, ACT: Office of the Status of Women, Department of the Prime Minister and Cabinet.

Buckingham-Hatfield, S. (2002). Gender equality: A prerequisite for sustainable development. *Geography, 18* (3), 227-233.

Christian-Smith, L.K., & Kellor, K.S. (Eds.) (1999). *Everyday Knowledge and Uncommon Truths: Women of the Academy.* Boulder, CO: Westview Press.

Crotty, M. (1998). *The Foundations of Social Research: Meaning and Perspective in the Research Process.* St Leonards, NSW: Allen & Unwin.

Currie, J., Thiele, B., & Harris, P. (2002). *Gendered Universities in Globalized Economies: Power, Careers, and Sacrifices.* Lanham: Lexington Books.

Davies, S., Lubelska, C. & Quinn, J. (Eds.) (1994). *Changing the Subject: Women in Higher Education.* London: Taylor & Francis.

D'Avilo-Neto, M.I. with Baptista, C.A. & Calicchio, R. (1996). Women and Development: Perspectives and challenges within the university curriculum. In: M-L. Kearney & A.H. Ronning (Eds.), *Women and the University Curriculum: Towards Equality, Democracy and Peace* (pp. 69-90). London: Jessica Kingsley Publishers & Paris: UNESCO.

del Nevo, M. (1993, May). Learning to heal. *New Internationalist,* 243, 3.

Diamond, I. & Orenstein, G.F. (Eds.) (1990). *Reweaving the World: The Emergence of Ecofeminism.* San Francisco: Sierra Club Books.

Gillett, M. (1981). *We Walked Very Warily: a History of Women at McGill.* Montreal: Goden Press.

Gilligan, C. (1982). *In a Different Voice.* Cambridge, MA: Harvard University Press.

Gough, A. (1999a). Recognising women in environmental education pedagogy and research: Towards an ecofeminist poststructuralist perspective. *Environmental Education Research,* 5 (2), 143-161.

Gough, A. (1999b). The power and the promise of feminist research in environmental education. *Southern African Journal of Environmental Education,* 19, 28-39.

Gough, A. (2001). Pedagogies of science (in)formed by global perspectives: Encouraging strong objectivity in classrooms. In: J.A. Weaver, M. Morris & P. Appelbaum (Eds.), *(Post) Modern Science (Education): Propositions and Alternative Paths* (pp. 275-300). New York: Peter Lang.

Haraway, D.J. (1991). *Simians, Cyborgs, and Women.* London: Free Association Books.

Harding, S. (1987). Is there a feminist method? In S. Harding (Ed.), *Feminism and Methodology* (pp. 1-14). Bloomington: Indiana University Press.

Harding, S. (1991). *Whose Science? Whose Knowledge? Thinking for Women's Lives*. Ithaca, NY: Cornell University Press.

Harding, S. (1998). *Is Science Multicultural? Postcolonialisms, Feminisms, and Epistemologies.* Bloomington: Indiana University Press.

Heilbrun, C.G. (2002). *When men were the only models we had*. Philadelphia, PA: University of Pennsylvania Press.

Kearney, M-L. (1996). Women, higher education and development. In M-L. Kearney & A. H. Ronning (Eds.), *Women and the University Curriculum: Towards Equality, Democracy and Peace* (pp. 1-33). London: Jessica Kingsley Publishers & Paris: UNESCO.

Kearney, M-L. & Ronning, A.H. (Eds.). (1996). *Women and the University Curriculum: Towards Equality, Democracy and Peace*. London: Jessica Kingsley Publishers & Paris: UNESCO.

Kelly, F. (1985). *Degrees of Liberation: A short history of women in the University of Melbourne.* Parkville, Victoria: The Women Graduates Centenary Committee of the University of Melbourne.

Kreinberg, N. & Lewis, S. (1996). The politics and practice of equity: experiences from both sides of the Pacific. In L. H. Parker, L. J. Rennie, & B. J. Fraser (Eds.), *Gender, Science and Mathematics: Shortening the Shadow* (pp. 177-202). Dordrecht, The Netherlands: Kluwer Academic Publishers.

Lotz-Sisitka, H. & Burt, J. (2002) Being brave: Writing environmental education research texts. *Canadian Journal of Environmental Education*, 7(1), 132-151.

Luke, C. & Gore, J. (1992). Women in the academy: Strategy, struggle, survival. In C. Luke & J. Gore (Eds.), *Feminisms and Critical Pedagogy*. (pp. 192-210). New York: Routledge.

Maher, F.A. & Tetreault, M.K.T. (2001). *The Feminist Classroom: Dynamics of Gender, Race, and Privilege.* Lanham: Rowman & Lttlefield.

McEwan, I. (2002). *Atonement*. London: Vintage.

Merchant, C. (1992). *Radical Ecology: the search for a livable world*. New York: Routledge.

Merrill, B. (1999) *Gender, Change and Identity: Mature Women Students in Universities.* Aldershot: Ashgate.

Morley, L. (1999) *Organising Feminisms: The Micropolitics of the Academy.* New York: St Martin's Press.

Narayan, U. & Harding, S. (2000). Introduction. In U. Narayen & S. Harding (Eds.), *Decentering the Center: Philosophy for a Multicultural, Postcolonial, and Feminist World* (pp. vii-xvi). Bloomington: Indiana University Press.

Odora Hoppers, C.A. (2001). Indigenous knowledge systems and academic institutions in South Africa. *Perspectives in Education*, 19(1), 73-85.

The Rio Declaration on Environment and Development. (1992). Reprinted in *Earth Ethics*, 4(1), 9-10.

Salleh, A. (1997). *Ecofeminism as Politics: nature, Marx and the postmodern*. London: Zed Books.

Scott, W.A.H. (2002). Sustainability and learning: What role for the curriculum? University of Bath inaugural lecture, 25 April. Retrieved October 13, 2002 from the World Wide Web: http://www.bath.ac.uk/cree/scott.htm

Shiva, V. (1989). *Staying Alive: Women, Ecology and Development*. London: Zed Books.

United Nations Conference on Environment and Development (UNCED) (1992). *Agenda 21*. Final, advanced version as adopted by the Plenary on 14 June 1992. Rio de Janeiro, Brazil: UNCED.

United Nations. (2002). *Report of the World Summit on Sustainable Development*. Johannesburg South Africa 26 August – 4 September 2002. A/CONF.199/20*. Retrieved April 7, 2003 from the WWW: http://www.johannesburgsummit.org/html/documents/summit_docs/131302_wssd_report_reissued.pdf

Warren, K. J. (Ed.). (1994). *Ecological Feminism*. New York: Routledge.

Warren, K. J. (1997a). Introduction. In K.J. Warren (Ed.), *Ecofeminism: Women, Culture, Nature* (pp. xi-xvi). Bloomington: Indiana University Press.

Warren, K. J. (1997b). Taking empirical data seriously. In: K.J. Warren (Ed.), *Ecofeminism: Women, Culture, Nature* (pp. 3-20). Bloomington: Indiana University Press.

Warren, K. J. (Ed.). (1997c). *Ecofeminism: Women, Culture, Nature* (pp. xi-xvi). Bloomington: Indiana University Press.

Whitehouse, H. & Taylor, S. (1996). A Gender Inclusive Curriculum Model for Environmental Studies. *Australian Journal of Environmental Education*, 12, 77-83.

Woolf, V. ([1929] 1977). *A Room of One's Own*. London: Collins.

BIOGRAPHY

Dr Annette Gough is Associate Professor in Science and Environmental Education in the Faculty of Education at Deakin University, Melbourne, Australia. Here she lectures and researches in environmental education, with a particular research interest in feminist and poststructuralist methodologies. She is also an adjunct professor at the University of Victoria, BC, Canada and a visiting professor at Rhodes University, South Africa. She was the first female president of the Australian Association for Environmental Education (1984-86) and is a life fellow of the Association. She has published extensively on environmental education in many international publications, including her book *Education and the Environment: Policy, Trends and the Problems of Marginalisation* (1997, ACER Press). She was managing editor of the *Australian Journal of Environmental Education* and is an advisory editor for the *Canadian Journal of Environmental Education*, the *Southern African Journal of Environmental Education* and the *Journal of Biological Education.*

CHAPTER 13

SUSTAINABILITY AND TRANSFORMATIVE EDUCATIONAL VISION

Edmund O'Sullivan

INTRODUCTION

The connection that I make in this chapter is the link between my foundational work on transformative learning and its implications for sustainability education in institutions of higher education[1]. From the perspective of 'transformative learning,' developed in *Transformative Learning: Educational Vision for the 21ˢᵗ Century* the fundamental educational task of our times is to make the choice for a sustainable planetary habitat of interdependent life forms over and against the pathos of the global competitive marketplace (O'Sullivan, 1999; O'Sullivan, Morrell & O'Connor, 2002). This perspective shares a point of view of a rising tide of people and communities all over this globe. This emergent vision of life deeply challenges the economic globalization that is moving like a tornado in our world as we begin the new century. A planetary consciousness opens us up into the awesome vision of a world that energizes our imagination well beyond a market-driven consumer industrial military worldview that has the whole world hostage at the present moment.

The idea of transformative vision for higher education starts with the notion of transformation within a broad cultural context. In the larger cultural context, transformation carries the dynamism of cultural change. Let me set out the defining features of what I intend when I use the term transformation. The entry point is best attempted my locating the definition within larger cultural and historical currents and forces When any cultural manifestation is in its florescence, the educational and learning tasks are uncontested and the culture is of one mind about what is ultimately important. There is, during these periods, a kind of optimism and verve that ours is the best of all possible worlds and we should continue what we are doing. It is also usual to have a clear sense of purpose about what education and learning should be. There is also a predominant feeling that we should continue in the same direction that has taken us to this point. Here one can say that a culture is in "full form" and the form of the culture warrants "continuity". We might say that a

---

[1] The first sections of this paper are partly excerpted from my book on Transformative Learning and the Canadian Journal of Environmental Education. Excerpt are made in this work with permission of Zed press and the Canadian Journal of Environmental Education.

Peter Blaze Corcoran & Arjen E.J. Wals (Editors), Higher Education and the Challenge of Sustainability: Problematics, Promise and Practice, 163-180.
© 2004 *Kluwer Academic Publishers. Printed in the Netherlands.*

context that has this clear sense of purpose or direction is "formatively appropriate". A culture is "formatively appropriate", when it attempts to replicate itself within this context and the educational and learning institutions are in synchrony with the dominant cultural themes. There is a clear sense of direction and mission during periods of this kind with little contestation.

Even when a culture is "formatively appropriate", there are times that there seems to be a loss of purpose or a loss of the qualities and features that appeared to have given that particular culture it's florescence. Part of the public discourse, during times such as these, is one of "reform criticism". Reform criticism is a language that calls a culture to task for its loss of purpose. It is a criticism that calls itself back to its original heritage. This is a criticism that accepts the underlying heritage of the culture and seeks to put the culture, as it were, "back on track". When reform criticism is directed toward educational institutions we call this "educational reform".

There is another type of criticism that is radically different from "reform criticism" which calls into question the fundamental direction and values of the dominant cultural form and indicates that the culture can no longer viably maintain it's continuity and vision. This criticism maintains that the culture is no longer "formatively appropriate" and in the application of this criticism there is a questioning of all of the dominant culture's educational visions of continuity. We refer to this type of criticism as "transformative criticism". In contrast to "reformative criticism", this "transformative criticism" suggests a radical restructuring of the dominant culture and a fundamental rupture with the past. Transformative criticism has three simultaneous moments. The first moment is already described as the critique of the dominant culture's "formative appropriateness". The second is a vision of what an alternative might look like to the dominant form. The third moment is some concrete indications of the practical exigencies of how a culture probably could abandon those aspects of its present forms that are "functionally inappropriate" while, at the same time, pointing to some directions of how it can be part of a process of change that will create a new cultural form that is "functionally appropriate".

All of the moments above, in their totality, can be called a "transformative moment". It is a historical moment of moving between visions. It is certainly not the case that historical moments and their labeling go uncontested. Many would say that we are not at a transitional moment in our present historical situation as I am maintaining. Truly, we seem to be living in a time of ferment. The transition period into the new century witness educational systems on a global scale being the object of neo-liberal educational reform that are massively intrusive on the foundational values of educational institutions worldwide. Aronowitz and Giroux in 'Education Under Siege' give us a graphic summation of this penetration and its effects:

> During these years, the meaning and purpose of schooling at all levels of education were refashioned around the principles of the marketplace and the logic of rampant individualism.Ideologically, this meant abstracting schools from the language of democracy and equity while simultaneously organizing educational reform around the discourse of choice, reprivatization, and individual competition (Aronowitz & Giroux, 1993, p. 1).

I would add an ecological dimension to their summation by noting that the neo-liberal program proceeds with a profound autism to almost every aspect of environmental sensitivity that constitutes the fundamental underpinning of the recent concerns of the environmental movement. In this most recent version of "neo-liberal reform", there is no questioning of the "functional appropriateness" of the dominant vision of the global marketplace in virtually any of its aspects. When there is criticism within these quarters, it is a criticism that is completely at home with the dominant cultural form that seeks a further extension of what has been in place since the beginning of the twentieth century; the dominance of the market. The educational reform suggested in this venue continues to encourage us to tool up our educational institutions from the nation state to the transnational marketplace.

The fundamental assumption of this chapter is that we must embark upon a discussion of a transformative vision of education if we are to have an education that holds the core values of sustainability. In attempting to do this, it must be kept clearly in mind that it will involve a diversity of elements and movements in contemporary education. At this point, it is helpful to highlight some of the contemporary educational currents that must be part of an emergent vision of transformative education. Since we are in a transitional period, in which there are many contesting viewpoints, the reader should be appraised of some of the elements that are emerging out of a transformative vision of education. To some extent these trends are operating somewhat separately and independently of one another. It is necessary, at the outset, that some of those elements be named because they will form part of a weave of a new type of integral education that will contest the vision of education for the global market place. It is necessary to couch these elements within a broad cosmological framework which will give the sense of an alternative to our present conventions in higher education that come out of the so called needs of the market.

What we are now coming to understand is that we are living in a period of the Earth's history that is incredibly turbulent and in an epoch in which there are violent processes of change that challenge us at every level imaginable. The pathos of the human today is that humans are totally caught up in this incredible transformation and we have a most significant responsibility for the direction it will take. The terror here is that we have it within our power to make life extinct on this planet. Because of the magnitude of this responsibility for the planet; all our educational ventures must finally be judged within this order of magnitude. This is the challenge for all areas of education.

The mind set of all our current neo-liberal educational ventures serves the needs of our present dysfunctional consumer industrial-military system. Highly significant portions of our modern educational institutions are in aligned with and feeding into industrialism, nationalism, competitive trans nationalism, individualism and patriarchal militarism. All of these must be questioned, at a fundamental level and must be put into question. All of these elements together coalesce into a worldview that exacerbates the global crisis's that we are now facing. There is no creativity here because there is no viewpoint or consciousness which sees the need for new directions. It is a very strong indictment to say that our dominant educational institutions are defunct and bereft of understanding in responding to our present planetary crisis. In addition, a strong case can be made that our received educational vision of our western cultural world view suffers from a deep absence of a "cosmological sense". Somehow this cosmological sense is lost or downgraded in our educational discourse. We are not here talking about shallow changes in fashion. We are talking about a major revolution in our view of the world that came with the paradigm of modernism.

A relevant quote from Susan Griffin will give the reader an anticipation of the overall pathos of our present historical moment:

> The awareness grows that something is terribly wrong with the practices of European culture that have led both to human suffering and environmental disaster. Patterns of destruction which are neither random or accidental have arisen from a consciousness that fragments existence. The problem is philosophical. Not the dry, seemingly irrelevant, obscure or academic subject known by the name of philosophy. But philosophy as a structure of the mind that shapes all our days, all our perceptions. Within this particular culture to which I was born, a European culture transplanted to North America, and which has grown into an oddly ephemeral kind of giant, an electronic behemoth, busily feeding on the world, the prevailing habit of mind for over two thousand years, is to consider human existence and above all human consciousness and spirit as independent from and above nature, still dominates the public imagination, even now withering the very source of our own sustenance. And although the shape of social systems, or the shape of gender, the fear of homosexuality, the argument for abortion, or what Edward Said calls the hierarchies of race, the prevalence of violence, the idea of technological progress, the problem of failing economies have been understood separately from the ecological issues, they are all part of the same philosophical attitude which presently threatens the survival of life on earth (Griffin, 1995, p. 29).

HIGHER EDUCATION FOR COMPREHENSIVE INTEGRITY: THE FOUNDATIONS FOR SUSTAINABILITY

Institutions of higher education, that have a view toward a sustainable world, need to foster a vision of education that has comprehensive integrity. I would say the defining feature of an education with comprehensive integrity is one that has, as its foundational underpinning, a grounding that fosters a 'cosmological sense.' The philosopher Stephen Toulmin in his book *The Return to Cosmology* gives a convenient entry point to the discussion of the term cosmology. He observes that there appears to be a natural attitude taken by humans at all times and in all places when reflecting on the natural world and there appears to be a comprehensive ambition to understand and speak about the Universe as a whole. Toulmin notes that, in practical terms, this desire for a view of the whole has reflected a need to recognize where we stand in the world in which we have been born, to grasp our place in the scheme of things and to feel at home within it (Toulmin, 1985).

It is interesting when one looks at the etymology of words, how certain core concepts interrelate. The etymology of the word ecology "eco" refers to the study of "home". Thus our attempts to situate ourselves as humans in the matrix of the earth and further in the universe is, in essence, an exercise in cosmology. This sense of wholeness is seen in the very breakdown of the word Universe (uni-verse) or one story. Historically, the word university meant an institution where one went to experience one's place in the universe. In the modern university the term cosmology appears, for the most part, to be arcane or obscure. In contemporary philosophy in the twentieth century the study of cosmology is, for all intents and purposes, absent. Nevertheless, the term has been very important in the history of philosophy in the past and will surely be so in the future because of the developments of postmodern science (Berman, 1981; Griffin,1988). Our modern world, in which the university functions, has suffered a loss of the cosmological sense and in losing this we have opened up a mind-set that leads to fragmentary experience.

Stephen Toulmin explores some of the factors that have fostered fragmentation. He maintains that there is a crucial difference between modern scientific worldview and its earlier cosmological predecessors. He contends that traditional cosmology was never preoccupied with any isolated aspect of a phenomena. In contrast, we see in the modernist worldview highly specialized and distinct scientific disciplines which have continued to develop in the 20th and now the 21st Century. Knowledge is bureaucratized with very distinct and clear divisions of labour. Toulmin notes that from the 17th century onward, there would be precious few scholar-scientists who would cross the boundaries of more than one discipline. As a consequence, questions that might have been asked across a whole spectrum of disciplines have "rarely been posed, much less answered"(Toulmin, 1985). Nevertheless, disciplined inquiry in modern science had made impressive achievements. Its achievements overshadowed the fragmentation of thought that would come in its wake. By the end of the 19th century, this disciplined fragmentation would cast a shadow where any attempt at a conception of the whole, as experienced in the organic world view of the pre-moderns, was abandoned. The culture's poets appear always to be the forerunners of a cultural critique. Yeat's poem 'Second Coming' crystallizes this; "things fall apart, the centre cannot hold" (Yeats, 1983) is a poetic indication of the loss of the cosmological sense. Toulmin gives us a feel for how this cosmology is lost in the disciplined inquiry of 19th century science by personifying the banishment of the integrative cosmological task. The bureaucratized disciplinarian says to the cosmologist as natural theologian:

> "You used to run a Department of Coordination and Integration, did you? Well, as you can see, we don't have any such department: all our enterprises run perfectly well without needing to be coordinated or integrated. And now, if you don't mind, would you please go away and let us get on with our work?" In short the disciplinary fragmentation of science during the nineteenth century seemingly made the integrative function of natural theology unnecessary (Toulmin, 1985, p. 235).

COMPREHENSIVE INTEGRITY IN A UNIVERSE CONTEXT

When we speak of education within this larger universe context it must be seen as a pervasive life experience. Formal education programs cannot fulfill all of these requirements. At the same time, formal education must be transformed so that it can provide an integrating context for total life functioning. At the higher levels of formal education, what is needed are processes of reflection on meaning and values, carried out in a comprehensive context. Our universities today flounder for want of a larger and more comprehensive context. Having no adequate larger context in which to function, our higher educational institutions operate in a splintered and fractionated view. One of the most common solutions to this vacuum is in the reinstatement of past forms of humanistic studies in a core curriculum, a curriculum which includes philosophy, ethics, history, literature, religious studies, and some general science. At this point in our own cultural history, these attempts at an integral education do not appear to evoke a sense of committed identification and no unifying paradigm appears on the horizon. As a result, effective education does not take place.

In closely examining this moment of crisis that we are living in, Thomas Berry suggests that we need to return to the story of the universe:

> For the first time the peoples of the entire world, insofar as they are educated in a modern context, are being educated within this origin story. It provides the setting in which children everywhere-whether in Africa or China, in the Soviet Union or South America, in North America, Europe, or India-are given their world and their own personal identity in time and space. While the traditional origin and journey stories are also needed in the educational process, none of them can provide the encompassing for education such as is available in this new story, which is the mythic aspect of our modern account of the world. The story tells us how the universe has emerged into being and of the transformation through which it has passed, especially on the planet Earth until the present phase of development was realized in contemporary human intelligence (Berry, 1988, p. 87).

The plentitude of this universe story is the basis for all educational endeavors and is the proper context for the entire educational process. At the same time, it is understood that the story must also be understood within the limits of personal and social development. Thus it can be understood that the story can be appreciated within a human developmental context. At the university level it is understood that the processes of human maturity allow for a penetration at the most profound levels. What is important to note at this point is that our deepest educational endeavors and commitments would be grounded in a story that would have a cultural, historical and cosmological context of meaning that can be accepted on a broad scale by persons of different ethnic and cultural backgrounds. The universe story is a "grand story" but not a "master-narrative", in post-modern terms.

The University should be a place where the universe story is encountered and engaged. If that be the case, then it would suggest that the universe story can help to guide and direct and guide educational vision. It provides the basis of a functional cosmology for a planetary vision. The Story evokes creative energy. In this context the learner needs, above all, an attraction that entrances and moves them. The basic purpose of the story is to enable us to interact more creatively with the emergent processes of the universe as experienced on the earth in this new century (Berry, 1988). This story potentially provides not only the understanding and the sense of direction that we need, it also evokes an energy needed to create this new situation. It needs to be repeated constantly that we are not now dealing with another historical change or cultural modification such as those we have been experienced in the past. The changes we are dealing with are changes on a geological and biological order of magnitude. The educational vision must be also at this order of magnitude.

THE TRANSFORMATIVE LEARNING CENTRE AND SUSTAINABILITY EDUCATION: THE EARTH CHARTER AS FOUNDATIONAL VISION

Institutions of higher education would be profoundly transformed if core curricula fostered the development of comprehensive integrity. It is not my task, in this chapter, to pursue the implications of what I have just set out in areas of higher education where I have no competence or experience. What I would like to do in ending this piece is to give the reader a sense of what education for comprehensive integrity would mean in the work we are attempting to do at the Transformative Learning Centre in the Ontario Institute for Studies in Education at the University of Toronto (OISE/UT www.tlcentre.org). Our Centre represents a space in the University of Toronto and within the Ontario Institute for Studies in Education that provides a planetary vision of education with a strong social justice orientation. As I have already indicated, the powerful vision of "market driven education" has gained a foothold in institutions of higher education across the world. In the past, education institutions of higher learning, both public and private, depended on the nation state for support. In the 21st Century, through the pernicious influence of the World Bank, the WTO and the IMF, educational institutions are being abandoned by nation states and as a result educational institutions have now become highly dependent on transnational business corporations for financial support. As a result of this marriage of institutions of higher learning with corporate spouses, we are seeing institutions, such as universities compromised, by the market demands and vision of transnational corporate business. Universities today have corporate logo's laced throughout their premises. We have corporations running cafeterias and professorial chairs simultaneously. In the thirty plus years that I have been in the University of Toronto, I have watched this institution slowly but surely becoming increasingly influenced by market forces.

In the beginning of the last decade of the 20th century I was involved founding of the Transformative Learning Centre. The centre emerged as a way of forging an expansive higher educational planetary vision of education with the express purpose of having a 'comprehensive integrity' and that was critical of a market oriented educational values. It was created in September of 1993 by the coming together of several OISE/UT faculty members, some students and interested community partners. What the various faculty members, students and community members were looking for was a way of creating a stronger sense of community and collaboration in broad areas of environmental, feminist, anti-racist, aboriginal, adult and popular education theory and practice. The faculty who came together were basically scholar/activists who from a variety of diverse perspectives were looking at ways of combining inter-disciplinary practices, new knowledges, and alternative strategies for community and global change. From earlier conversations it became clear that we all shared an interest in "transforming" contemporary educational and social paradigms. We were also united by our interest in the role of learning in global and local change and by our preference for university and community partnerships in research and field development. The TLC is redefining transformative learning through its strengths in multi-disciplinary participation and approaches, linking academy and community in many diverse areas of research, practice, and education, ie health, environmental, development, anti-racism, feminism, worker and popular, indigenous, peace, and media education.

The Transformative Learning Centre offers programs at the Doctoral and Masters Level of the University of Toronto. As part of developing foundation courses we have recently endorsed the Earth Charter and I believe it provides a very important foundational ingredient for our vision of education (see Appendix 1). It represents a vision of education that values a planetary context with comprehensive integrity. The Transformative Learning Centre has recently initiated a core course which utilizes the "Earth Charter" as one of the core courses in the program of transformnative learning. A brief background on the "Earth Charter" will help the reader to understand why this initiative was taken by our the centre.

The World Secretariat on the Earth Charter Initiative is presently trying to co-ordinate a world wide program to bring the Earth Charter to the attention of educators world wide. The effort is being mobilized because of the deep commitment to a sustainable world where peace, justice and sustainable communities prevail. There is no question we have reached a critical moment in Earth's history. However, the transition to more sustainable ways of living will only eventuate when people in all cultures and societies understand and support the need for such change. progress will be made when the values that motivate people begin to reflect principles that promote more sustainable ways of living. But, people cannot be forced to change their values. Rather, education is the key to assisting people in the difficult task of re-examining their values systems and encouraging the adoption of more ethically based behaviour by individuals, organizations and governments. Transformative education is needed: education that helps bring about the fundamental changes demanded by the challenges of sustainablility.

Accelerating progress towards sustainablility depends on rekindling more caring relationships between humans and the natural world and facilitating the creative exploration of more environmentally and socially responsible forms of development. In my own work on transformative learning, I use the term development in a very guarded way. I avoid using the term 'sustainable development' and share many of the criticism that Donald Worster makes of it in his excellent article entitled, 'The Shaky Ground of Sustainability'(Worster, 1993). We both are circumspect by the economistic orientation of the term in the famous Bruntland Report. There is ample room for a hermeneutics of suspicion where this term is concerned and in the company it keeps. Our current ideas on development are allied to the notions of growth and development within a market point of view. When development is seen within this light, we have a sole emphasis on development from the point of view of a market economics. This solar emphasis on the market has had an incredible morbid impact on the lives of peoples all over this planet. The tail wags the dog because development is premised on market needs to the exclusion of all other needs on this planet. As part of the transformative learning of resistance, this view of development cued totally to the market must be jettisoned. This does not mean that we can totally abandon all ideas that pertain to this important evolutionary process. We are in need of a new conception of development which will be integral to personal, community and planetary development.

The first educational challenge is to advance understanding of our shared global problems and the need to act with a sense of universal responsibility. The second is to provide people with a framework for critically evaluating their situation and identifying action goals for bringing about positive change. The third educational challenge is to foster a culture of collaboration that facilitates new partnerships between civil society, business and governments.

Many international declarations, such as the Talloires Declaration and Agenda 21, recognise the central role of education in promoting a more just, peaceful and sustainable future. However, serious impediments remain. Educational activities associated with "values" remains a contested field because of concern about "which" values and "whose" values are being promoted. These concerns can be allayed so long as the values being examined represent core values that respect human dignity, are life affirming, and are consistent with those of major cultures around the world. However, at the same time, educators must be aware of the need to avoid proselytising, respect the right of individual learners to independently hold values, and understand that within the search for common ground there remain important values associated with cultural diversity.

The Earth Charter is a document that provides a radical shift in perspective and gives a set of foundational principles that deals with the planetary crisis of our times (www.EarthCharter.org). All of the major concerns and principles that are in the Earth Charter reflect the concerns and visions of the Transformative Learning Centre. The reason why one would frame transformative learning within the Charter is because it casts our local work here in University of Toronto in a larger arc of global significance. In my estimation, its most important and primary feature is that it is a document that speaks first and foremost for the needs of the planet and the conditions of living which will allow us to live and sustain ourselves in a planetary context. It is a document which speaks for the planet as a totality, that is a whole, with all of its variety and diversity. It provides a bio-centric vision of the planet that does not prioritize the human. It is not anthropocentric on that account, so all species in their magnificent varieties and forms are taken into account. This document in its species generosity, departs from all previous documents and proclamations of its kind. As an example, the United Nations document on Human Rights and Responsibilities speaks only to the need of the human species. It is human centric however powerful its vision. The uniqueness of the Earth Charter is clearly set out in its Declaration of Universal Responsibility. As it so eloquently presents itself:

> To realize these aspirations, we must decide to live with a sense of universal responsibility identifying ourselves with the whole Earth community as well as our local community. We are at once citizens of different nations and of one world in which the local and global are linked. Everyone shares responsibility for the present and future well being of the human family and the larger living world.

What is abundantly clear by this declaration is that the Earth Charter does not prioritize or speak for the needs of the "global market vision of educational priorities". Since this market vision has pretensions for the development of educational objectives and priorities all over the world today, the Earth Charter stands as an important counter-position and alternative to market driven priorities for education at all levels. In my estimation, the principles of the Earth Charter should be considered as the starting point for foundational frameworks that foster a true planetary education. This is the first document at such a world level that challenges the "market vision." In assessing the global situation it asserts in the introduction that:

> The dominant patterns of production and consumption are causing environmental devastation, the depletion of resources, and a massive extinction of species. Communities are being undermined. The benefits of development are not share equitably and the gap between the rich and poor is widening. Injustice, poverty, ignorance, and violent conflict are widespread and the cause of great suffering...

This Earth Charter document stands as a testimony that our present systems of development have put us out of phase with the larger evolutionary processes of the earth proper. Our present terminal stage can be characterized as that of high entropy and dissipation. A deep cultural pathology has developed in western society that has now spread throughout the planet. A savage plundering of the entire earth is taking place by industrial exploitation. Thousands of poisons unknown in former times are saturating the air, the water, and the soil. The habitat of a vast number of living species now finds that the harm done to the natural world is returning to threaten the human species itself, and this is on a comprehensive scale. Our earlier discussions of our current planetary crisis indicated that we are the planetary crisis. At this time in the evolution of the planet we may say that the planetary crisis is a consequence of our limited awareness. We seem to be caught up in limited awareness of our actions in relation to the planet and the universe. The identification of the person with the planet comes from an awareness that one's personal identify is intertwined the planet and with the universe as a totality. The foundational experiences of all of the major religions point to an intimate relationship between the person and the cosmos. If this relationship holds true then we are able to say that personal development is integrally related to planetary development. The integral connections that we have had within the Earth matrix in the past appear to have broken and are fragmented. We must understand the precarious nature of the human project at this point and, at the same time, take full responsibility for its ultimate outcome. Our responsibilities must be taken into our conscious awareness and followed through with the resolve that it is the fundamental educational commitment for our time.

REFERENCES

Aronowitz, S. & Giroux, H. (1993). *Education Still Under Siege*. Toronto: OISEPress.
Berman, M. (1981). *The Reenchantment of the World*. Ithaca: Cornell University Press.
Berry, T. (1988). *The Dream of the Earth*. San Francisco: Sierra Club.
Griffin, R. (Ed.). (1988). *Spirituality and Society: Postmodern Visions*. New York: Suny Press.
Griffin, S. (1995). *The Eros of Everyday Life*. New York: Doubleday.
O'Sullivan, E. (1999). *Transformative Learning: Educational Vision for the 21st Century*. London: Zed Books.
O'Sullivan, E., Morrell, A. & O'Connor, M. (Eds.) (2002). *Expanding the Boundaries of Transformative Learning: Essays on Theory and Praxis*. New York: Palgrave Press.
Toulmin, S. (1985). *The Return to Cosmology*. Berkeley: University of Califorma Press.
Worster, D. (1993). *The Shaky Ground of Sustainability*. In: W.Sachs(Ed.) Global Ecology: A New Arena of Political Conflict. London: Zed Books.

Yeats, W.B. (1983). *The Second Coming.* In: Mack, M. (Ed.) Modern Poetry. New York: New American Library.

BIOGRAPHY

Edmund O'Sullivan is professor of Transformative Learning and director of the Transformative Learning Center at the Ontario Institute for Studies in Education at the University of Toronto. He is the author of *Critical Psychology and Pedagogy and Transformative Learning: Educational Vision for the 21st Century.*

APPENDIX 1: THE EARTH CHARTER

PREAMBLE

We stand at a critical moment in Earth's history, a time when humanity must choose its future. As the world becomes increasingly interdependent and fragile, the future at once holds great peril and great promise. To move forward we must recognize that in the midst of a magnificent diversity of cultures and life forms we are one human family and one Earth community with a common destiny. We must join together to bring forth a sustainable global society founded on respect for nature, universal human rights, economic justice, and a culture of peace. Towards this end, it is imperative that we, the peoples of Earth, declare our responsibility to one another, to the greater community of life, and to future generations.

Earth, Our Home

Humanity is part of a vast evolving universe. Earth, our home, is alive with a unique community of life. The forces of nature make existence a demanding and uncertain adventure, but Earth has provided the conditions essential to life's evolution. The resilience of the community of life and the well-being of humanity depend upon preserving a healthy biosphere with all its ecological systems, a rich variety of plants and animals, fertile soils, pure waters, and clean air. The global environment with its finite resources is a common concern of all peoples. The protection of Earth's vitality, diversity, and beauty is a sacred trust.

The Global Solution

The dominant patterns of production and consumption are causing environmental devastation, the depletion of resources, and a massive extinction of species. Communities are being undermined. The benefits of development are not shared equitably and the gap between rich and poor is widening. Injustice, poverty, ignorance, and violent conflict are widespread and the cause of great suffering. An unprecedented rise in human population has overburdened ecological and social systems. The foundations of global security are threatened. These trends are perilous- but not inevitable.

The Challenges Ahead

The choice is ours: form a global partnership to care for Earth and one another or risk the destruction of ourselves and the diversity of life. Fundamental changes are needed in our values, institutions, and ways of living. We must realize that when basic needs have been met, human development is primarily about being more, not having more. We have the knowledge and technology to provide for all and to reduce our impacts on the environment. The emergence of a global civil society is creating new opportunities to build a democratic and humane world. Our environmental, economic, political, social, and spiritual challenges are interconnected, and together we can forge inclusive solutions.

Universal Responsibility

To realize these aspirations, we must decide to live with a sense of universal responsibility, identifying ourselves with the whole Earth community as well as our local communities. We are at once citizens of different nations and of one world in which the local and global are linked .Everyone shares responsibility for the present and future well-being of the human family and the larger living world. The spirit of human solidarity and kinship with all life is strengthened when we live with reverence for the mystery of being, gratitude for the gift of life, and humility regarding the human place in nature.

We urgently need a shared vision of basic values to provide an ethical foundation for the emerging world community. Therefore, together in hope we affirm the following interdependent principles for a sustainable way of life as a common standard by which the conduct of all individuals, organizations, businesses, governments, and transnational institutions is to be guided and assessed.

PRINCIPLES

I. Respect and care for the community of life

1. Respect Earth and life in all its diversity.
    a. Recognize that all beings are interdependent and every form of life has value regardless of its worth to human beings.
    b. Affirm faith in the inherent dignity of all human beings and in the intellectual, artistic, ethical, and spiritual potential of humanity.
2. Care for the community of life with understanding, compassion, and love.
    a. Accept that with the right to own, manage, and use natural resources comes the duty to prevent environmental harm and to protect the rights of people.
    b. Affirm that with increased freedom, knowledge, and power comes increased responsibility to promote the common good.
3. Build democratic societies that are just, participatory, sustainable, and peaceful.
    a. Ensure that communities at all levels guarantee human rights and fundamental freedoms and provide everyone an opportunity to realize his or her full potential.
    b. Promote social and economic justice, enabling all to achieve a secure and meaningful livelihood that is ecologically responsible.
4. Secure Earth's bounty and beauty for present and future generations.
    a. Recognize that the freedom of action of each generation is qualified by the needs of future generations.

b.   Transmit to future generations values, traditions, and institutions that support the long-term flourishing of Earth's human and ecological communities.

In order to fulfill these four broad commitments it is necessary to:

II. Ecological integrity

5.   Protect and restore the integrity of Earth's ecological systems, with special concern for biological diversity and the natural process that sustain life.

a.   Adopt at all levels sustainable development plans and regulations that make environmental conservation and rehabilitation integral to all development initiatives.

b.   Establish and safeguard viable nature and biosphere reserves, including wild lands and marine areas, to protect Earth's life support systems, maintain biodiversity, and preserve our natural heritage.

c.   Promote the recovery of endangered species and ecosystems.

d.   Control and eradicate non-native or genetically modified organisms harmful to native species and the environment, and prevent introduction of such harmful organisms.

e.   Manage the use of renewable resources such as water, soil, forest products, and marine life in ways that do not exceed rates of regeneration and that protect the health of ecosystems.

f.   Manage the extraction and use of non-renewable resources such as minerals and fossil fuels in ways that minimize depletion and cause no serious environmental damage.

6.   Prevent harm as the best method of environmental protection and, when knowledge is limited, apply a precautionary approach.

a.   Take action to avoid the possibility of serious or irreversible environmental harm even when scientific knowledge is incomplete or inconclusive.

b.   Place the burden of proof on those who argue that a proposed activity will not cause significant harm, and make the responsible parties liable for environmental harm.

c.   Ensure that decision making addresses the cumulative, longterm, indirect, long distance, and global consequences of human activities.

d.   Prevent pollution of any part of the environment and allow no build-up of radioactive, toxic, or other hazardous substances.

e.   Avoid military activities damaging to the environment

7.   Adopt patterns of production, consumption, and reproduction that safeguard Earth's regenerative capacities, human rights, and community well-being.

a.   Reduce, reuse, and recycle the materials used in production and consumption systems, and ensure that residual waste can be assimilated by ecological systems.

b.   Act with restraint and efficiency when using energy, and rely increasingly on renewable energy sources such as solar and wind.

   c. Promote the development, adoption, and equitable transfer of environmentally sound technologies.

   d. Internalize the full environmental and social costs of goods and services in the selling price, and enable consumers to identify products that meet the highest social and environmental standards.

   e. Ensure universal access to health care that fosters reproductive health and responsible reproduction.

   f. Adopt lifestyles that emphasize the quality of life and material sufficiency in a finite world

8. Advance the study of ecological sustainability and promote the open exchange and wide application of the knowledge acquired.

   a. Support international scientific and technical cooperation on sustainability, with special attention to the needs of developing nations.

   b. Recognize and preserve the traditional knowledge and spiritual wisdom in all cultures that contribute to environmental protection and human well-being.

   c. Ensure that information of vital importance to human health and environmental protection, including genetic information, remains available in the public domain.

III. Social and economic justice

9. Eradicate poverty as an ethical, social, and environmental imperative.

   a. Guarantee the right to potable water, clean air, food security, uncontaminated soil, shelter, and safe sanitation, allocating the national and international resources required.

   b. Empower every human being with the education and resources to secure a sustainable livelihood, and provide social security and safety nets for those who are unable to support themselves.

   c. Recognize the ignored, protect the vulnerable, serve those who suffer, and enable them to develop their capacities and to pursue their aspirations.

10. Ensure that economic activities and institutions at all levels promote human development in an equitable and sustainable manner.

   a. Promote the equitable distribution of wealth within nations and among nations.

   b. Enhance the intellectual, financial, technical, and social resources of developing nations, and relieve them of onerous international debt.

   c. Ensure that all trade supports sustainable resource use, environmental protection, and progressive labor standards.

   d. Require multinational corporations and international financial organizations to act transparently in the public good, and hold them accountable for the consequences of their activities.

11. Affirm gender equality and equity as prerequisites to sustainable development and ensure universal access to education, health care, and economic opportunity.

   a. Secure the human rights of women and girls and end all violence against them.

   b. Promote the active participation of women in all aspects of economic, political, civil, social, and cultural life as full and equal partners, decision makers, leaders, and beneficiaries.

   c. Strengthen families and ensure the safety and loving nurture of all family members

12. Uphold the right of all, without discrimination, to a natural and social environment supportive of human dignity, bodily health, and spiritual well-being, with special attention to the rights of indigenous peoples and minorities.

   a. Eliminate discrimination in all its forms, such as that based on race, color, sex, sexual orientation, religion, language, and national, ethnic or social origin.

   b. Affirm the right of indigenous peoples to their spirituality, knowledge, lands and resources and to their related practice of sustainable livelihoods.

   c. Honor and support the young people of our communities, enabling them to fulfill their essential role in creating sustainable societies.

   d. Protect and restore outstanding places of cultural and spiritual significance.

IV. Democracy, nonviolence, and peace

13. Strengthen democratic institutions at all levels, and provide transparency and accountability in governance, inclusive participation in decision making, and access to justice.

   a. Uphold the right of everyone to receive clear and timely information on environmental matters and all development plans and activities which are likely to affect them or in which they have an interest.

   b. Support local, regional and global civil society, and promote the meaningful participation of all interested individuals and organizations in decision making.

   c. Protect the rights to freedom of opinion, expression, peaceful assembly, association, and dissent.

   d. Institute effective and efficient access to administrative and independent judicial procedures, including remedies and redress for environmental harm and the threat of such harm.

   e. Eliminate corruption in all public and private institutions.

   f. Strengthen local communities, enabling them to care for their environments, and assign environmental responsibilities to the levels of government where they can be carried out most effectively.

14. Integrate into formal education and life-long learning the knowledge, values, and skills needed for a sustainable way of life.

   a. Provide all, especially children and youth, with educational opportunities that empower them to contribute actively to sustainable development.

   b. Promote the contribution of the arts and humanities as well as the sciences in sustainability education.

    c.   Enhance the role of the mass media in raising awareness of ecological and social challenges.

    d.   Recognize the importance of moral and spiritual education for sustainable living.

15. Treat all living beings with respect and consideration.

    a.   Prevent cruelty to animals kept in human societies and protect them from suffering.

    b.   Protect wild animals from methods of hunting, trapping, and fishing that cause extreme, prolonged, or avoidable suffering.

    c.   Avoid or eliminate to the full extent possible the taking or destruction of non-targeted species.

16. Promote a culture of tolerance, nonviolence, and peace.

    a.   Encourage and support mutual understanding, solidarity, and cooperation among all peoples and within and among nations.

    b.   Implement comprehensive strategies to prevent violent conflict and use collaborative problem solving to manage and resolve environmental conflicts and other disputes.

    c.   Demilitarize national security systems to the level of a nonprovocative defense posture, and convert military resources to peaceful purposes, including ecological restoration.

    d.   Eliminate nuclear, biological, and toxic weapons and other weapons of mass destruction.

    e.   Ensure that the use of orbital and outer space supports environmental protection and peace.

    f.   Recognize that peace is the wholeness created by right relationships with oneself, other persons, other cultures, other life, Earth, and the larger whole of which all are a part.

THE WAY FORWARD

As never before in history, common destiny beckons us to seek a new beginning. Such renewal is the promise of these Earth Charter principles. To fulfill this promise, we must commit ourselves to adopt and promote the values and objectives of the Charter.

This requires a change of mind and heart. It requires a new sense of global interdependence and universal responsibility. We must imaginatively develop and apply the vision of a sustainable way of life locally, nationally, regionally, and globally.

Our cultural diversity is a precious heritage and different cultures will find their own distinctive ways to realize the vision. We must deepen and expand the glob al dialogue that generated the Earth Charter, for we have much to learn from the ongoing collaborative search for truth and wisdom.

Life often involves tensions between important values. This can mean difficult choices. However, we must find ways to harmonize diversity with unity, the exercise of freedom with the common good, short-term objectives with long-term goals. Every individual, family, organization, and community has a vital role to play. The arts, sciences, religions, educational institutions, media, businesses, nongovernmental organizations, and governments are all called to offer creative leadership. The partnership of government, civil society, and business is essential for effective governance.

In order to build a sustainable global community, the nations of the world must renew their commitment to the United Nations, fulfill their obligations under existing international agreements, and support the implementation of Earth Charter principles with an international legally binding instrument on environment and development.

Let ours be a time remembered for the awakening of a new reverence for life, the firm resolve to achieve sustainability, the quickening of the struggle for justice and peace, and the joyful celebration of life.

For more information about The Earth Charter, please contact:
Earth Charter Secretariat
P.O. Box 319-6100
San José, Costa Rica
Tel: (506) 205-1600
Fax: (506) 249-3500
E-mail: info@earthcharter.org
www.earthcharter.org

CHAPTER 14

TEACHING INTERACTIVE APPROACHES TO NATURAL RESOURCE MANAGEMENT: A KEY INGREDIENT IN THE DEVELOPMENT OF SUSTAINABILITY IN HIGHER EDUCATION

Niels Röling

INTRODUCTION

Having worked for almost three decades at the Wageningen University, formerly known as Wageningen Agricultural University, I have experienced first hand how difficult it is to demonstrate the added value of social science. In the early seventies, the university was rapidly expanding. Agriculture in Holland was going through a phenomenal increase in the productivity of labour and land. Agriculture alone ensured the surplus on the balance of trade. The public sector invested huge sums in land consolidation and re-adjudication, research and extension. Agribusiness blossomed. In all, a huge institutional support structure (including the Agricultural University) emerged. Holland, a mouse of a country, developed an elephant of an agricultural industry.

The justification for our Department, then called 'Extension Education', was the idea that innovations, developed by scientists, would need to find their way to farmers (e.g. Rogers, 1961; Van den Ban, 1963). University authorities and faculty were interested in the adoption of technologies by individuals and in their diffusion in communities, and in the attributes of farmers who were likely to be survive the drop-out race. In those days it was easy to defend our contribution.

By the end of the century, that had changed dramatically. To be sure, some still held on to the idea that the university is there to service agriculture and support farmers. But the sharp drop in the number of Dutch secondary school graduates who enter the University showed that that approach could no longer support the social contract of the university. Society is no longer interested in improving agriculture. It is no longer considered a source of pride, but rather seen as a festering societal sore. Agriculture is not a subject school leavers get excited about. At the same time it is the number of students that determines the University's budget.

The university proved resilient. It dropped the word 'agriculture' from its name, and rapidly expanded its initiatives in food science, human nutrition, water purification, biotechnology, nature conservation, spatial planning, the sustainable

Peter Blaze Corcoran & Arjen E.J. Wals (Editors), Higher Education and the Challenge of Sustainability: Problematics, Promise and Practice, 181-197.
© 2004 *Kluwer Academic Publishers. Printed in the Netherlands.*

management of ecological services, etc. It cashed in on its continued popularity among foreign students and is mounting advertising campaigns to promote its new identity among Dutch school leavers.

The University's transformation was partly initiated by social scientists. Van der Ploeg (e.g. Van der Ploeg & Long, 1994) showed that farmers cannot neatly be ordered along a continuum from more to less 'modern' as indeed diffusion of innovations research had assumed. Instead, within the same technological and economic parameters, farmers develop a diversity of 'styles'. In the same vain, Long radically did away with the notion that larger social systems determine people's outcomes and behaviours. Instead, human actors, individually or collectively, have tremendous scope for exerting agency and for choosing their own futures (e.g. Long & Long, 1992).

Our Department embraced the participatory development paradigm (e.g. Chambers & Jiggins, 1987). My own change in perspective is best expressed by the title of an inaugural address: 'Towards an Interactive Agricultural Science' (Röling, 1995). That interest in interaction was inspired by exposure to the IPM Farmer Field Schools in Indonesia and Landcare in Australia. Both these innovative movements were reactions to disasters caused by conventional agriculture. In the Netherlands, the societal problems caused by agriculture made it impossible for the Ministry of Agriculture to continue to support it. Instead, agriculture needed to be controlled and a host of regulatory measures were implemented. Our department had looked at communication as a tool to support compliance with central policies. It soon became evident, however, that the central regulation of the behaviour of antagonistic and recalcitrant farmers was a non-starter. More sustainable forms of land use required 'interactive policy making' (Van Woerkum & Aarts, 1998).

What I want to do in the present chapter is to use this experience to explore the practice of teaching 'interactive approaches to natural resource management' that emerged in response to international student demand for new professional roles and perspectives. I will focus on the theoretical frameworks that we found useful in teaching.

ASSUMPTIONS

My contribution assumes that humans have become a major force of nature (Lubchenco, 1998) and that they use that power to impose linear economic growth scenarios on the basically cyclical dynamics of ecosystems (Holling, 1995). As a result, ours has become a risk society (Beck, 1992). We now face uncertainty with respect to issues for which the stakes are very high. In that situation, puzzle-solving science no longer suffices; we need a 'post-normal' science that is truly democratic and interactive and includes 'extended peers' and 'extended facts' (Funtowicz & Ravetz, 1993).

This chapter is further based on the assumption that we live, not in an epoch of change, but in a change of epoch (Da Souza Silva et al., 2000). We have successfully built a technology and an economy that allowed a sizeable proportion of humanity to escape much of the misery that comes with the proverbial 'vale of tears'. The rest of humanity is bent on making the same 'great escape'. However, in the process of co-evolving our greed and the technologies to satisfy them, we have transformed the surface of the earth. We have taken on the management of the planet and not made a good job of it. We are beginning to realise that we might be getting it wrong, that we are undermining the capacity of the earth to generate human opportunity. Those who have not yet made the great escape are beginning to realise that there might be nothing left for them. The threat of anthropogenic destruction of ecosystems has become a key argument for a continued social contract for public sector science.

The huge issue of humans destroying their own future tends to attract bright, creative and motivated students, many of them women. They want to use their lives to make the world a better place, and are quite distinct from those who choose their careers for selfish reasons. Even if, for the sake of long-term sustainability, it would perhaps be better to attract the latter group to classes on sustainability, I have come to the conclusion that (a) students select you and (b) students cannot be force-fed.

These assumptions are supported by various studies. I provide two examples. A segmentation of the Dutch population in terms of susceptibility to appeals to sustainability showed that more than half (54%) of the population is only interested in environmental issues if they are positioned in terms of individual advantage. Only about 20% of the population considers sustainability an important value (Lampert & Van der Lely, 2000). A comparative study of economics and other students shows that 'economics students are less interested in protecting the environment, or helping the poor than students of other subjects. The future economist is instead mostly interested in gaining power and influence, making money and having a good time' (Gandal & Roccas, 2002; New Scientist, 2002).

The kind of student who is interested in sustainability does not necessarily want answers and solutions. They want challenges, options, perspectives and methodologies with which they can be effective. They want to hear about unfinished business. They know that what you are teaching them has a limited shelf life.

Finally, one needs to walk one's talk. Teaching interactivity without using interactive methods puts the teacher in an untenable position.

FOUNDATIONS OF INTERACTIVE APPROACHES

For some reason, the 'real' social scientists, i.e., anthropologists, sociologists and psychologists, are reluctant to examine the reasons why people would participate, collaborate or engage in concerted action. Whereas economists are always ready to peddle their axiomatic and normative models, social scientists seem deliberately to avoid commitment to a preferred outcome. Instead, they take pleasure in deconstructing and debunking anything that smacks of design or recommendation. Hence there is not much theory that underpins interactive practice.

Given an origin in what was then called extension studies, definitely not a 'real social science', the department with which I am associated has always had to accept working with 'dirty hands'. It is expected to deliver recommendations, indicate preferred professional behaviour and make policy recommendations. Most social scientists who are so inclined end up as business consultants. Our focus has continued to be on training public sector and NGO professionals engaged in pursuing public goods. After embracing the participatory development paradigm, it became increasingly evident that participatory approaches are strong on methods and weak on theory. We needed to develop theoretical perspectives to inform the practice of the future professionals we are responsible for training. In this chapter, I briefly describe the most helpful ones.

The beta/gamma interface

It is customary to define sustainability in terms of hard, objective criteria, such as non-renewable energy consumption, the loss of species, the proportion of the world's fresh water that humans extract, etc. Based on such criteria, indicators such as the 'ecological footprint' have been developed as advocacy tools. Environmental advocates do, of course, prefer such criteria to 'prove' that the world is going to waste. But they assume a science framework within which it is difficult to discuss interactive approaches. Therefore, we had to develop another perspective. In this, we have greatly benefited from the work of e.g. Bawden (e.g. 2000); Checkland (1985); Gunderson, et al. (1995); and, Maturana and Varela (1992).

The key point of the alternative perspective is that sustainability is not an objective property of a given ecosystem, but the emergent property of human interaction. People have become a major force of nature. What happens to ecosystems might be measured in terms of the decline of bio-diversity, holes in the ozone layer, coral bleaching, reduced sperm counts and so on. But the main driving force behind these changes is human activity, or to be more precise, the impact of aggregated individual preferences (Goldblatt, 1996). Any reversal of the increasingly bleak outlook for the 'frayed web of life' (World Resources Institute, 2000) therefore must be based on changing human activity, i.e., on conflict resolution, negotiated agreement, human learning and concerted decision making. It is in that sense that sustainability can be seen as the outcome of human interaction.

Teaching interactive approaches to natural resources management implies operating at the interface of natural and social sciences. In Wageningen, we call this interface 'beta/gamma science', where beta stands for natural science and gamma for social science. In our university, a tendency exists for the applied natural sciences to embrace social sciences when it comes to natural resources management. One example of this development is the Department of Irrigation. Its focus on irrigation schemes, with their designed nature, the necessarily negotiated nature of water distribution, and inescapable power conflicts, made that department the first in the university to embrace a beta/gamma approach. To operate professionally, irrigation engineers have to understand human interaction, in addition to designing culverts, etc.

Many other departments have now drawn the same conclusion for their field. Recently the University and others decided to finance a major inter-disciplinary research programme focusing on developing methods for interactive approaches to technology development[1].

Accepting sustainability as the emergent outcome of human interaction and adopting a beta/gamma approach is but a first step in developing a theory underpinning interactivity. But the implications for teaching are important. In the first place, one is not only dealing with social science students, but with a diverse group of people with hugely different backgrounds who share the insight that interactive approaches are a key to becoming an effective professional. Secondly, one cannot afford to teach only social science. One must be able to explain how social process impinges on ecological process and *vice versa*.

A cognitivist stance

Most of us have been trained to think that the world around us exists, whether we are there to see it or not. We must get to know that world so that we can control it. Luckily, we can build objective knowledge by using scientific methods. Thus we can build a body of true, or, since Popper (1972), potentially falsifiable, knowledge. Every dissertation or other bit of sound research is expected to add to the body of human knowledge. We talk of proof, of validation, of evidence, of cause and effect, and even of the 'end of science' when we know everything there is to know (Horgan, 1998).

Teaching interactive approaches requires a radically different perspective. There is no way, in the sense that it can be explained even by positivist, reductionist science, by which the external environment can be projected on the brain for us to objectively know. We are informationally closed. Even a frog does not bring forth *the* fly, but at best *a* fly, and even then not *any* fly, but a fly that can be eaten (Maturana & Varela, 1992). Given the way brains work, sentient beings are doomed to bring forth realities that allow them to take effective action in their domain of existence. Evolution and learning are two important mechanisms by which organisms construct and grasp livelihood opportunities. In the case of humans, constructing reality, or learning, is a social affair, involving language, the evolution and transfer of culture, and building institutions.

Different people build different realities. Truth is multiple. The US Supreme Court ruled that there is no way by which one can determine which expert witness is right. An expert witness is someone who is believed to be right by certain people. Or

---

[1] This project, called 'The Convergence of Sciences', is funded by Wageningen University (INREF), DGIS, the Dutch bilateral development co-operation agency, and FAO's Global IPM Facility (GIF). The project is a collaboration between the University of Ghana at Legon, the Intitut de Richerche Agronomique du Benin, and Wageningen University. Trans-disciplinary teams of farmers, PhD researchers, and their supervisors from the natural and social sciences experiment with forms of agricultural research and experimentation intended to directly benefit small farmers.

as the Mahayana Buddhism puts it: 'An objective world is a manifestation of the mind itself'.

Doomed to construct we might be, but that does not mean that every construction is equally effective. Cognitivism does not have to be relativistic. We might get it entirely wrong. We can build a cosy coherent reality world, in which our values, theories, perceptions and actions are mutually consistent. But this reality world can become divorced from its domain of existence; for example, it can fail to correspond to ecological imperatives. Then we are in for surprises, for unexpected feedback. Our cosy reality world has become obsolete (e.g. Kuhn, 1970). We might not be aware of it, we might ignore it, and elites might have the power to maintain a life style long after it has become unsustainable. But in the end, coherence will have to give way to the need to rebuild correspondence.

The key to survival is resilience, the ability to note discrepancy, to adapt the reality world to feedback, to relinquish the institutions, organisations and interests built around obsolete reality worlds, and build new ones (e.g. Hurst, 1995). The key asset we have is rapid, deliberate learning. We tend to leave this to science. But when it insists that it builds truth, science forfeits the claim to be a survival mechanism. It becomes a liability. For example, I am convinced that neo-classical economics, with its arrogant reification of the market, is a serious threat to human survival, a blinding insight that reduces resilience. A body of 'true knowledge' is a stumbling block in building the knowledge required for a change of epoch.

In this sense cognitivism is pretty radical. Yet acceptance of the constructed nature of reality is a necessary condition for resilience. It is indispensable in a change of epoch. Survival implies that we deliberately deal with the construction of human knowledge and institutions in the process of building effective and adaptive action.

In my experience, teaching interactive approaches can only be effective if students grasp the difference between the naive realism and positivism that still often accompanies natural science training, on the one hand, and cognitivism, on the other. It is essential that they realise that they will have to deal with multiple equivalent perspectives.

Soft System Methodology

Checkland (e.g. 1981) has made a major contribution by making the distinction between hard and soft systems. Hard system thinking applies to natural systems (e.g. plants) and designed systems (e.g. railroads or computers). Hard systems have given goals, their main interest is goal seeking. Hard systems have given boundaries. Hard systems exist. We can model them. So far this is no news for students with natural science training.

From a constructvist perspective, systems are figments of the imagination, in the sense that a system is a construct imposed on experience for the purpose of gaining understanding. A system therefore is a form of enquiry. Human activity systems are soft systems. Systems do not have goals; people have goals. For humans to become an activity system, in the sense of being able to take concerted action to achieve common purposes, a protracted process of negotiation and goal convergence is required. Hard systems can be seen as a sub-set of soft systems. For example, there is not a tree in Holland that would stand if it did not fit a human purpose. 'Sustainability is the emergent property of a soft system' (Bawden & Packam, 1993).

Checkland (1981) invented soft systems after he, as a trained natural scientist, realised one could not manage a corporation with hard system thinking. 'Goals are the bone of contention'. The goals of soft systems need to be negotiated. Soft systems also do not have given but arbitrary, negotiated boundaries. A soft system therefore implies a process of joint learning by which people come to accept shared goals, agreed boundaries, and scenarios for moving ahead. Checkland developed a Soft Systems Methodology (SSM) which basically describes the steps that people with a shared problem go through to become an effective activity system.

Checkland's thinking, and its application to agriculture at Hawkesbury College[2] was hugely important for us. It stimulated us to improve our earlier rather hard notion of 'agricultural knowledge and information systems' (AKIS), a concept that we had developed to capture the idea that innovation requires synergy among multiple, complementary actors (e.g. Röling, 1988). Especially Engel took the notion of AKIS further and developed a SSM for the 'rapid appraisal of agricultural knowledge systems' (RAAKS) (Engel & Solomon, 1997). SSM was the inspiration for what we now call 'facilitation of social learning' (e.g. Groot, 2002; Maarleveld, 2003).

Platforms

Applying soft systems thinking to natural resource management led to the idea of platforms (Röling, 1994), a notion related to common property resource management (Ostrom, 1992). Natural resource dilemmas require co-ordination at an ecosystem level at which the problems seem to be amenable. Water catchments are a typical example. Very often, at that ecosystem level, a capacity for decision making does not exist. Stakeholders in those ecosystems or their representatives have to meet in some forum, dialogue or platform to negotiate concerted action. Emerging natural resource dilemmas force the increasingly interdependent stakeholders in the resource to drop their pursuit of individual preferences and merge into a human activity system with common purposes. The platform concept makes interactive approaches much more concrete, moves them out of a purely communication discourse, and introduces an institutional element. The notion of platform proved

---

[2] Now University of Western Sidney, New South Wales, Australia. When Richard Bawden became Dean in the seventies, a remarkable experiment started at the College. Instead of being trained as technical scientists, students were provided with experiential understanding of the predicaments affecting Australian agriculture (unemployment, environmental degradation, price squeeze, and suicide that accompany the agricultural treadmill). They were then trained to act as facilitators of the local social processes that could turn around the collapsing rural system. They became soft systems practitioners. Graduates from the college have played a major role in landcare. New leadership has led to a return, on the whole, to a traditional agricultural curriculum. But a growing network of systems practitioners across the globe continues to develop soft systems approaches to natural resource management.

persuasive. For example, it led to a research proposal that was rewarded with major funding from the European Commission[3]. But it also stimulates students and motivates them to analyse and/or design concrete options.

Natural resource management practice all over the world and at different levels of aggregation increasingly applies platform strategies to resolve natural resource dilemmas. A typical example at the global level is a consortium of ten major international agencies called 'The Dialogue on Water, Food and Environment'. Dialogue seems to only way to resolve the fact that, in planning the achievement of their global mandates up to 2050, each agency was assuming a huge increase in the use of fresh water... until they realised they were all talking about the same limited globally available water reserves[4].

Platforms have a number of intriguing aspects. In the first place, they emerge, or are installed, in situations where multiple stakeholders increasingly experience interdependence in their use of a scarce natural resource. This leads to resource conflicts, unsustainable use of the resource, loss of livelihoods, and ultimately, alienation and stress (Van Haaften, 2002). The situation cannot (easily) be resolved by imposing regulation. The only solution seems to create a capacity for concerted action at the resource system level and the level of social aggregation at which the dilemma seems resolvable. This itself is a problem: often appropriate system levels and social aggregation levels do not coincide. Platforms are typically made up of representatives of stakeholder groups. These representatives learn to understand each other and to agree under pressure of skilled chairmen in pursuit of compromise. But this places a heavy burden on representatives to communicate with their constituencies and take them along in the cultural changes they themselves experience. This tends to be a very weak aspect of platforms (Aarts, 1998; Röling, 2002). A very important issue is that platforms cannot be expected to resolve resource dilemmas in the absence of a conducive policy context. Authorities must be prepared to abide by agreements that emerge from the platform. They must be clear and consistent about the resources available to implement platform outcomes. They usually are not. Platforms are expected to miraculously lead to solutions while existing organisations are unwilling to share any of their power (e.g. Dore & Woodhill, 1999). Facilitation of social learning therefore cannot limit itself to the platform itself, but must include the policy makers that provide its policy context (Groot, 2002).

In all, platforms can be seen to manage 'multiple commons' (Steins, 1999). Key issues in successfully managing common property resources apply: drawing boundaries around the resource, limiting access to the resource, effective communication among stakeholders, agreement on the sustainable use of the

---

[3] This project, Social Learning for the Integrated Management and Sustainable Use of Water Resources at the Catchment Scale (SLIM), is one of a number of social science projects that the European Commission has funded to support the introduction of the European Water Framework Directive which became operational in all member states in 2000 and is expected to lead to a 'good status of all waters' by 2015. The achievement of the Framework Directive will be based on regulation, 'right prices' and involvement and information of those involved. SLIM is designed to focus on the third element, the participation and involvement of stakeholders in water catchments.

[4] www.cgiar.org/iwmi/dialogue/dialogue.htm

resource, monitoring that use, and effective sanctions on breaking the agreement (Ostrom, 1992). In addition, such issues as legitimacy of the platform, the integration with existing institutions and processes, incentives for participation, co-ordination with higher levels of decision making, and effective facilitation are crucially important (Röling & Woodhill, 2001).

INTERACTIVE APPROACHES: THE THIRD WAY OF BEING EFFECTIVE

Three dimensions

Table 1 shows three dimensions that recur in the social science literature, be it as organisational co-ordination mechanisms (e.g. Powell, 1997), preferred ways of arranging human affairs (Hood, 1998), rationalities (Habermas, 1994), sources of the wealth of nations (Bowles & Gentils, 2002), etc. In other words, these dimensions recur in widely different disciplinarian perspectives. It is not easy to factor out the key words that capture the diversity and it is not the place here to spend much time on this issue. It seems that the dimensions can most usefully be classified as follows: (1) using instruments, including power or resources, (2) using incentives and relying on market mechanisms, and (3) facilitating voluntary mechanisms such as networks, emergence from interaction, or (social) learning. The first two of these mechanisms have been the subject especially of natural science research and economics. Their tenets are widely known and applied in society. Most elementary school graduates are imbued the logic of using technology, understand the use of power, and are aware of the role of the market and market forces in shaping society. Newspapers, magazines, TV programmes and so forth pay a great deal of attention to technology, regulation and economy.

The third dimension, the use of interactive approaches, is an unknown. There is no widely accepted societal discourse about it, and most people, even if they are well educated, find it very difficult to think in terms of the logic of interactive approaches (see section 3 on foundations). It is my experience that many people in positions of power are so blinded by insights based on legal, technical, or economic theories, and especially rational choice theory, that they find it virtually impossible to accept the 'third dimension'. Yet, there is increasing agreement that technical fixes, regulation, and the market are failing in bringing about a sustainable society, be it from the perspective of reducing the gaps between rich and poor, or in terms of sustaining the ecological services on which human life depends. As Einstein said, one cannot solve problems with the methods that got us into trouble to begin with. It seems, therefore, that interactive approaches deserve considerable attention if only because they hold promise to get us out of predicaments that seem immune to more-of-the-same. Below, I make an effort to provide a basis for teaching interactive approaches. Understanding interactive approaches is a key ingredient in teaching sustainability in higher education and the associated skills are indispensable for professionalism in natural resource management at all levels of aggregation.

Table 1. Three dimensions encountered in different social science discourses.

Dimensions	1	2	3
Rural policy practice (pers. com. Rob Schrauwen)	Regulating	Compensating	Stimulating
Rationalities (Habermas, 1984)	Instrumental	Strategic	Communicative
Bases for Individual Behaviour Change (Kelman, 1969)	Compliance	Identification	Internalisation
Preferred ways of arranging human affairs (Hood, 1998)[5]	Hierarchy	Individualism	Egalitarianism
Organisational co-ordination mechanisms (Powell, 1994)	Hierarchy	Market	Network
Causes of 'wealth of nations' (Bowles and Gintis, 2002)	Resources (such as power or access to natural resources)	Invisible hand of market forces	Social capital, trust, community
Institutionalisation (Giddens, 1984)	Domination, legitimation	-	Signification, communication

Cognition as an overarching theoretical perspective on interactive approaches

Cognition seems to be the overarching theoretical perspective that underpins thinking about interactive approaches. It has the great advantage that it is unfinished business and hence has the capacity to motivate intelligent students. The study of cognition has long been dominated by neuro-psychologists and analysts of formal logical systems who use the computer as the metaphor for the human brain. Increasingly, however, cognition is becoming the field of study of philosophers, biologists, ecologists, anthropologists, multi-agent system modellers, and others[6]. Cognition is emerging as an over-arching and integrating concept that captures the core of what makes sentient beings different from other combinations of matter and energy.

I have earlier referred to the Santiago theory of perception and specifically to its observation that a frog does not bring forth *any* fly (as pure relativists would have us believe), but a fly the frog can eat (Maturana & Varela, 1992). Organisms and their environment are structurally coupled. They maintain this coupling through co-evolution and learning. Knowledge is effective action in the domain of existence.

---

[5] Hood's work is based on Mary Douglas' cultural theory. Of course, Mary Douglas discerns a fourth dimension, fatalism, where the sense of belonging to a group is weak, but the domination by rules is strong.
[6] Some interesting books in this respect: Gilbert and Troitzsch (1999); Gigerenzer and Todd (1999); Holling (1995); Hood (1998); Hutchins (1995, fourth printing 2000).

The cognitive system is a duality of the perceiving organism and its environment. I have settled for Figure 1 as best way to capture the essential elements of cognition.

We observe that cognition includes

1. An *agent* that can perceive the environment or context, has beliefs or theories about it, has emotions that provide criteria for judgement about it, and can take action in it; and
2. The *context*: the environment or domain of existence with which the agent is structurally coupled;
3. An *ecosystem*, i.e., a space in which multiple agents interact and mutually adapt.

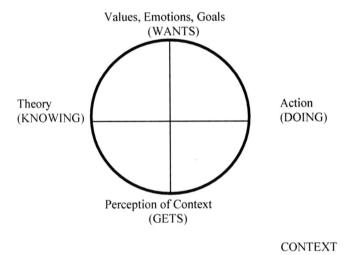

Figure 1. The Elements of Cognition (adapted from Kolb, 1984; Maturana and Varela, 1992; and Bawden, 2000).

Gigerenzer et al. (1999) have contributed two fundamental drivers of the cognitive process:

– *Coherence* (or cognitive consistency);
– *Correspondence* (or structural coupling between agent and domain of existence)

On the one hand, cognition fundamentally assumes a tendency towards coherence among values/emotions, perception, theories and action. But, on the other, it equally requires a tendency towards correspondence between these four elements and the context. Cognitive theory emphasises coherence *and* correspondence. The tendencies toward, and the dilemmas between, coherence and correspondence provide dynamism to cognitive theory. The dilemma between coherence and correspondence is the key to the study of innovation. A typical example is the work of Thomas Kuhn on Scientific Revolutions (1970).

For the purposes of this chapter, I am especially interested in multiple, collective, and distributed cognition because interactive approaches can best be described as a move from multiple to collective and/or distributed cognition. *Multiple* cognition emphasises the existence, in one situation, of totally different cognitive agents with multiple perspectives. *Collective* cognition emphasises shared attributes, i.e., shared myths or theories, shared values, and collective action, e.g. households all engage in waste paper recycling. *Distributed* cognition emphasises different but complementary contributions that allow concerted action, e.g. the operation of a commercial company or the navigation of a battleship (Hutchins, 1995).

Multiple cognitive agents tend to maintain their mutual isolation. But when they become inter-dependent with respect to the use of a resource, they are likely to engage in conflict, work at cross-purposes, and engage in disjoint action. However, multiple agents' perspectives are equally likely to grow into a joint rich picture, they can meet on platforms and decide on collective action. In this way, multiple cognition can grow into collective or distributed cognition. The interest is in how multiple cognitive agents can be facilitated in the direction of collective or distributed cognition. In other words, in addition to coherence and correspondence, we need to understand *convergence* of the elements of cognition.

Using cognitive theory to underpin interactive approaches very much is unfinished business. Table 2 lists a number of research questions suggested by cognitive theory.

Table 2. Eight research questions for studying interactive approaches to solving resource dilemmas.

– What leads multiple stakeholders to perceive mutual interdependence? Can that perception be strengthened?
– Under what conditions does perceived interdependence lead to either conflict or negotiated agreement? Can the choice for the second option be supported?
– By what mechanisms do the elements of cognition (figure 1) converge among stakeholders in a resource? In other words, what social mechanisms lead individuals to become a collective or distributed cognitive agent?
– What mechanisms foster coherence among the converged elements of cognition?
– What mechanisms affect the trade-off between coherence and correspondence?
– What institutional conditions allow effective transformation of multiple into collective or distributed cognition?
– What policy contexts are conducive to such a transformation?
– What is the nature of facilitation required for such a transformation?

In organisational or collective settings, the tendencies towards coherence and correspondence and the dilemmas between them become very interesting social phenomena, including not only power, social pressure, imitation, congruence, convergence, and so forth, but also deviance, social dilemma, conflict, innovation, mutation, evolution and revolution. I emphasise this point because the assumptions underlying the faith in interactive emergence can be quite naïve (Leeuwis, 2000). Relying on interactive emergence to resolve resource dilemmas definitely has a wishful element. Interactive approaches seem the only way out, therefore they must work. The question is whether interactive approaches can be effective in dealing with the rising tide of natural resource dilemmas. If so, under what conditions? Are these conditions likely in our globalising world?

In all, cognitive theory promises to provide a theoretical framework that can be widely shared to deal with the predicament that marks the change of epoch.

Teasing out the 'active ingredients' of interactive approaches

Table 3 cross-tabulates the elements of cognition with the three ways of getting things done. It brings out the totally different and internally consistent nature of the three dimensions. They each provide a coherent logic or blinding insight, if you will. The development of sustainability in higher education requires mastery in all three 'languages' because most situations are characterised by a 'mix' of the three.

The third column begins to establish the active ingredients in interactive approaches. It provides a framework for integrating different social science disciplines that focus on human reasons without assuming motives and therefore becoming axiomatic. These disciplines also allow institutions to affect individuals instead of focusing on methodological individualism. Political science, anthropology, sociology, social psychology and such applied sciences as my own area of communication and innovation studies, provide such perspectives. Many of the scientists involved seem, however, disinclined to apply their science to the development of normative models with respect to 'best practice' for sustainable development. For example, actor-oriented development sociology, as it has been practised for years in Wageningen, seems more interested in the battlefields that result from individual actors exerting agency to realise their 'projects', than in ways of fostering concerted action among actors (Röling & Leeuwis, 2001).

Table 3. Active ingredients in each of the three approaches.

	Using instruments	*Using incentives*	*Interactive appr.*
Predicament, Problem perceived, success perceived ('GETS')	Lack of control over causal factors	Competition, scarcity, poverty	Lack of control over ourselves, disagreement, lack of trust, conflicting interests
Dynamics ('KNOWING')	Causation, use of 'instruments', power, hierarchy	Rational choice in satisfying preferences, struggle for survival, market forces, exchange of values	Interdependence, learning, reciprocity, tendencies toward coherence and correspondence
Values, emotions, goals, purposes ('WANTS')	Control over bio-physical and social resources and processes	Win, gain advantage, satisfaction	Convergence to negotiated agreement, concerted action, synergy
Effect based on ('DOING' 1)	Technology, power differential, use of instruments	Strategy, anticipation, exchange	(Facilitation of) awareness of issues, conflict resolution, agreement, shared learning
Policy focus ('DOING' 2)	Hard systems design, regulation, (social) engineering	Fiscal policy, market stimulation, compensation	Interactive policy making, social process design, facilitation, stimulation

CONCLUSION

A consensus seems to be emerging about a number of crucial statements. However, this consensus has so far not led to serious efforts to redesign society.

– Technical fixes are not sufficient to bring about a sustainable society.
– A sustainable society cannot be regulated on the basis of central power.
– The market fails in bringing about a sustainable society. Sustainability requires more than an aggregation of individual preferences (Goldblatt, 1996).
– There is enough for everyone's need, not for everyone's greed. This famous saying by Mahatma Gandhi reflects the same reality as the famous phrase of George Bush Sr. that the American way of life is not negotiable. A society in which some maintain privileged access to resources through power or market domination cannot be sustainable. Global inequity is a source of friction and political turmoil.

Few will disagree with these statements. Yet, our society is based on quite different premises. In fact, the only agreed way to design future society seems to be the liberation of so-called free market forces. Higher education is one of the few chances we have to bring about change. It is for this reason that the third way of getting things done, through interaction, must be seen as a key ingredient in teaching sustainability. It is only through interaction that we can address the need to take less from the commons, to give more to the public good, and to subject individual greed to collective goals.

REFERENCES

Aarts, M.N.C. (1998). *Een kwestie van Natuur*. Published doctoral dissertation. Wageningen: Agricultural University (in Dutch).

Bawden, R.J., & R. Packam (1993). Systems praxis in the education of the agricultural systems practitioner. *Systems Practice*, 6, 7-19.

Bawden, R. (2000). The Importance of Praxis in Changing Forestry Practice. Invited Keynote Address for 'Changing Learning and Education in Forestry: A Workshop in Educational Reform'. Sa Pa, Vietnam, April 16-19, 2000.

Beck, U. (1992). *Risk Society. Towards a New Modernity*. London: Sage Publications. (First published as Risikogesellschaft: Auf dem Weg in eine andere Moderne. Frankfurt am Main: Suhrkamp Verlag 1986).

Bowles, S. & Gintis, H. (2002). Social Capital and Community Governance. *Economic Journal*, 112 (483) F419-F437.

Chambers, R. & Jiggins, J. (1987). Agricultural research for resource-poor farmers. Part I: Transfer-of-Technology and Farming Systems Research. Part II: A parsimonious paradigm. *Agricultural Administration and Extension*, 27, 35-52 (Part I) and 27, 109-128 (Part II).

Checkland, P. (1981). *Systems Thinking, Systems Practice*. Chichester: John Wiley.

Checkland, P. (1989). Soft Systems Methodology. *Human Systems Management*, 8, 273-289.

De Souza Silva, J., Cheaz, J. & Calderon, J. (2000). Building capacity for strategic management of institutional change in agricultural science organisations in Latin America: A summary of the project and progress to date. San José (Costa Rica): ISNAR at IICA, Proyecto Neuvo Paradigma

Dore, J. & J. Woodhill (1999). *Sustainable Regional Development. An Australia-wide study of regionalism, highlighting efforts to improve the community, economy and environment*. Canberra: Greening Australia.

Engel, P.G.H. & Salomon, M. (1997). *Facilitating Innovation for Development. A RAAKS Resource Box*. Amsterdam: KIT.

Funtowicz, S.O. & Ravetz, J.R. (1993). Science for the post-normal age. *Futures*, 25 (7), 739-755.

Gandal, N. & Roccas, S. (2002). Good Neighbours/Bad Citizens: Personal Value Priorities of Economists. London: Centre for Economic Policy Research, Discussion Paper 3660.

Giddens, A. (1984). *The Constitution of Society: Outline of the Theory of Structuration*. Oxford: Polity Press.

Gigerenzer, G. & Todd, P.M. (1999). Fast and Frugal Heuristics: The Adaptive Toolbox. In: Gigerenzer, G., Todd, P. M., & the ABC Research Group (Eds.). *Simple Heuristics that Make us Smart* (pp. 3-34). New York and Oxford: Oxford University Press.

Gilbert, N. & Troitzsch, K. (1999). *Simulation for the Social Scientist*. Buckingham: Open University Press.

Goldblatt, D. (1996). *Social Theory and the Environment*. Cambridge: Polity Press.

Groot, A. (2002). *Demystifying Facilitation of Multi-Actor Learning Processes*. Published Doctoral Dissertation. Wageningen: University.

Gunderson, L.H., Holling, C.S. & Light S.S. (Eds) (1995.) *Barriers and Bridges to the Renewal of Ecosystems and Institutions*. New York: Colombia Press.

Habermas, J. (1984). The Theory of Communicative Action. Vol. 1: *Reason and the Rationalisation of Society*. Boston: Beacon Press.

Holling, C.S. (1995). What Barriers? What Bridges? In: Gunderson, L.H., Holling, C.S. & Light S.S. (Eds.). *Barriers and Bridges to the Renewal of Ecosystems and Institutions*. New York: Colombia Press: 3-37.

Hood, C. (1998). *The Art of the State. Culture, Rhetoric, and Public Management*. Oxford: Clarendon Press.

Horgan, J. (1996). *The End of Science. Facing the Limits of Knowledge in the Twilight.* London: Abacus
Hurst, D.K. (1995). *Crisis and Renewal. Meeting the crisis of organisational change.* Boston (Mass.): Harvard Business School Press.
Hutchins, E. (1995, fourth printing 2000). *Cognition in the Wild.* Cambridge (Mass.): The MIT Press.
Kelman, H. (1969). Processes of Opinion Change. In Bennis, W., Benne, K. & Chin, R. (Eds.) (1969). *The Planning of Change* (Second Edition) (pp. 222-230). London: Holt, Rinehart and Winston.
Kolb, D. (1984). *Experiential Learning: Experience as a source of learning and development.* New Jersey: Prentice Hall.
Kuhn, T.S. (1970). *The Structure of Scientific Revolutions.* 2nd Ed. Chicago: University of Chicago Press.
Lampert, M. & Van Der Lelij, B. (2000). Milieubeleving Nederlander Magertjes. *Arena,* 2000 (6), 6-7.
Leeuwis, C. (2000). Re-conceptualising participation for sustainable rural development. Towards a negotiation approach. *Development and Change,* 31 (5), 931-959.
Long, N. & Long, A. (Eds.) (1992) *Battlefields of Knowledge: the interlocking of theory and practice in research and development.* London: Routledge.
Lubchenco, J. (1998). Entering the Century of the Environment: A New Social Contract for Science. *Science.* 279: 491-496.
Maarleveld, M. (2003). *Social Environmental Learning for Sustainable in Natural Resource Management, Theory, Practice and Facilitation.* Published doctoral dissertation. Wageningen: Wageningen University.
Maturana, H.R. & Varela, F.J. (1992). *The Tree of Knowledge. The biological roots of human understanding.* Boston (Mass.): Shambala Publications.
Ostrom, E. (1992). *Governing the Commons. The Evolution of Institutions for Collective Action.* New York: Cambridge University Press.
Popper, K.R. (1972). *Objective Knowledge: An Evolutionary Approach.* Oxford: OUP.
Powell, W. (1994). Neither Market nor Hierarchy: Network Forms of Organisation. In: Thompson, G., Frances, J., Levavcic, R., Mitchell, J. (Eds.) *Markets and Hierarchies and Networks: The Co-ordination of Social Life* (pp. 256-277). London: Sage.
Rogers, E.M. (1961). *Diffusion of Innovations.* New York: Free Press, Fourth Edition 1995.
Röling, N. (1988). *Extension Science. Information Systems in Agricultural Development.* Cambridge: CUP.
Röling, N. (1994). Platforms for decision making about eco-systems. In: Fresco, L.O., Stroosnijder, L., Bouma, J. & Van Keulen, H. (Eds.), *Future of the Land: Mobilising and Integrating Knowledge for Land Use Options* (pp. 386-393). Chichester: John Wiley and Sons.
Röling, N. (1995). Naar een Interactieve Landbouwwetenschap, Inaugural Address (in Dutch). Wageningen: Wageningen University.
Röling, N. & Woodhill, J. (2001). From Paradigm to Practice: Foundations, Principles and Elements for Dialogue on Water, Food and Environment. Background Document for National and Basin Dialogue Design Workshop, Bonn, December.
Röling, N. & Leeuwis, C. (2001). Strange bedfellows: How knowledge systems became Longer and why they never will be Long. In: Hebinck, P. & Verschoor, G. (Eds.). *Resonances and Dissonances in Development. Actors, networkls and cultural repertoires* (pp. 47-65). Assen: Royal Van Gorcum.
Röling, N. (2002). Beyond the Aggregation of Individual Preferences. Moving from multiple to distributed cognition in resource dilemmas. In: Leeuwis, C. & Pyburn, R. (Eds.) *Wheelbarrows Full of Frogs: Social Learning in Natural Resource Management* (pp. 25-28). Assen: Royal Van Gorcum.
Steins, N.A. (1999). *All Hands on Deck. An Interactive Perspective on Complex Common-Pool Resource Management Base on Case Studies in Coastal Waters of the Isle of Wight (UK), Connemara (Ireland) and the Dutch Wadden Sea.* Published Doctoral Dissertation. Wageningen: Wageningen University.
Van Den Ban, A.W. (1963). *Boer en Landbouwvoorlichting.* Published doctoral dissertation. Assen: Royal Van Gorcum (in Dutch).
Van Haaften, E.H. (2002). *Linking Ecology and Culture. Towards a Psychology of Environmental Degradation.* Published doctoral dissertation. Tilburg: Tilburg University.
Van der Ploeg, J.D. & Long, A. (Eds.) (1994). *Born from Within: Practice and Perspectives of Endogenous Rural Development.* Assen: Royal Van Gorcum.
Van Woerkum, C.M.J. & Aarts, N. (1998). The Communication between farmers and government over nature: a new approach to policy development. In: Röling, N. & Wagemakers, A. (Eds.). *Facilitating*

Sustainable Agriculture. Participatory Learning and Adaptive Management in Times of Environmental Uncertainty (pp. 272-280). Cambridge: Cambridge University Press.

World Resources Institute (2000). *World Resources 2000-2001. People and Ecosystems. The Fraying Web of Life*. Washington: World Resources Institute with UNDP, UNEP

BIOGRAPHY

Niels Röling (1937) is an emeritus professor of Agricultural Knowledge Systems in the Chair Group Communication and Innovation Studies of Wageningen University, the Netherlands. His current research includes two main areas: (1) the process and facilitation of multi-stakeholder learning towards sustainable use of water at a catchment scale. This research is part of a European-funded research project, SLIM. (2) Interactive agricultural research that involves trans-disciplinary teams of farmers, PhD researchers, and social and natural scientists from Benin, Ghana, and the Netherlands. He is involved in (co) supervision of a substantial number of PhD projects. After an MSc in Rural Sociology and Agricultural Economics at Wageningen and a PhD in Communication at Michigan State University, Röling has spent a life time of research and teaching on the facilitation of voluntary change. It took him from technical innovation to concerted (inter) action as a necessary condition for surviving the impending eco-challenge.

CHAPTER 15

LIVING SUSTAINABLY THROUGH HIGHER EDUCATION: A WHOLE SYSTEMS DESIGN APPROACH TO ORGANIZATIONAL CHANGE

James Pittman

INTRODUCTION

The ongoing health and integrity of our lives are sustained through ecological and social, as well as economic, relationships. Such relationships dynamically join us together in organizations and communities and the ever-changing web of life. In order to sustain the health and integrity, indeed the identity, of our selves, organizations, and communities over time an ongoing focus of our individual as well as collective attention must be reflection on, learning about, and active enhancement of these relationships.

However, our perception of the world perpetuates an individualistic separation between our selves, organizations and communities as well as the natural world, shrouding relationships between these systems in an illusion of static independence. At the behest of the western world, we hold on to a mechanistic Cartesian paradigm that portrays natural and human systems as separate entities subject to hierarchies of human control. Despite this, we must begin to see the true nature of these living systems evolving in relational symbiosis. Without this shift from a dualistic to an interconnected worldview, we will amplify rather than restrain degenerative patterns of oppression, violence, starvation, poverty, disease, toxicity, resource depletion and species extinction.

Among the social institutions that structure human systems, those of higher education are uniquely poised to nurture agents who can design and create such change. Thomas Berry, an elder scholar exploring the wounds in human and Earth relationships, tells us that "the university has a special role to fill as the institution with the critical capacity, the influence over the professions and societal activities, and the contact with the younger generation needed to reorient the human community toward a greater awareness that we exist within a single great interconnected community of the planet Earth" (Berry, 1996). This truth gives institutions of higher education a critical role in catalyzing the heart of a collectively transformative path without any roadmap: learning how we can enhance the sustainability of human systems within our Earth community.

Peter Blaze Corcoran & Arjen E.J. Wals (Editors), Higher Education and the Challenge of Sustainability: Problematics, Promise and Practice, 199-212.

Many students acknowledge this path when they sign the Graduation Pledge of Social and Environmental Responsibility, drafted nearly 15 years ago at Humboldt State University in California and used at hundreds of institutions around the nation. They make a commitment "to explore and take into account the social and environmental consequences of any job [they] consider and...try to improve these aspects of any organizations for which [they] work." (Graduation Pledge Alliance, 1990) Whether or not they ascribe to this pledge, it is imperative that colleges and universities concerned with issues of sustainability embody such values through their own structure and dynamics if student and employees are to learn how to create effective organizational change for sustainability.

This chapter begins with a core assumption of social ecology: that successful sustainability transitions will require fundamental changes in the structure and management of our organizations (Bookchin, 1990). I acknowledge that a number of authors explore how we might successfully create organizational change for sustainability (Anderson, 1999; Capra, 1996; Elkington, 1998; Hawkin, 1999; Nattras & Altomare, 1999) and that others do so with a focus on institutions of higher education (ULSF, 1994; Orr, 1994; Cortese, 1999, Calder, 1999; Creighton, 1998; Collett, 1996; Keniry, 1995; Leal Filho, 1999). While it is not my intention to critique these change strategies, I do contend that such efforts must be ubiquitously amplified. As a student and practitioner of whole systems design, my intention is to place further emphasize on the role of shared vision, sustainability indicators and participatory management for creating meaningful, lasting change within a college, university or other organization.

Moreover, I would contend that collaborative project-based learning among all stakeholders, regardless of their role in a college (or any human system for that matter) is integral to effective sustainability transitions. Just as green cells together in plants absorb the sun's energy to create nourishment at the heart of the web of life, so too students of any age absorb knowledge through experience, cumulatively creating nourishment for our organizations and communities. This makes it imperative for us to design a management structure and organizational culture that helps students and staff learning to cultivate their ability as active agents of collaborative change for sustainability.

Guiding patterns for institutions of higher education so inclined can be found throughout the natural world, helping us to see our organizations, at least metaphorically, as living organisms—dynamic systems adapting in relationship with the larger web of life. Globally, our path is an ongoing journey of design through which we have the opportunity to cultivate our organizations and communities to be in alignment with the regenerative patterns of nature.

ON LANGUAGE, PERCEPTION AND SUSTAINABILITY

It is of primary importance to note that sustainability itself is a concept with meaning that varies according to personal values and perspective. It is perhaps a greater asset than challenge that this rich diversity of meaning transcends objective determination of particular human systems as "sustainable." As such, to label any structure--be it a community, organization, building or other human development--

so definitively is dangerously presumptuous. Alternatively, through inquiry and advocacy we might collaborate on the continual enhancement of systems, policies, behaviors, and practices towards a shared ideal of sustainability

Popular definitions of "sustainable development" as "meet[ing] the needs of the present without compromising the ability of future generations to meet their own needs" (World Commission on Environmental and Development, 1987, p. 43) are mired in anthropocentrism and require an objective generalization of people's current and future needs. In lieu of addressing the true value of diverse needs, "sustainable development" is often used to justify or disguise continuations of rapid and potentially unchecked growth (Elkington, 1998). As such, I feel the word "sustainability," while potentially ambiguous, is a more suitable term for sharing meaning.

In a very basic sense, sustainability means the ability to retain integrity and health over time. For a dynamic vision of sustainability to embody the ideal of change towards being "sustainable," there must be critical assessment guiding continuous improvement. Thus, sustainability is a process characteristic, not a static end state; carrying an implication that health and integrity are retained only through ongoing attention to a dynamic balance of behaviors and conditions. Furthermore, a holistic view of sustainability acknowledges that intra- and inter-systemic relationships influence health and integrity. Thus, as a whole systems design practitioner using the word "sustainability" I mean the dynamic and ongoing enhancement of integrity and health in a given system (individual, organization, ecosystem, community, biome, etc.) through all of its relationships.

Much of the ambiguity about the meaning of such language arises out of our differing personal perspectives and values of what is most important or lacking in a system at any given time. If, for example, the focus of our attention is a financial "bottom line" agenda we limit our attention to economic value in relationships between people, industries and communities. While wee may traditionally view our organizations and communities in this way, human systems are much more complex. We also sustain health and integrity through relationships between individuals and relationships with other organizations and communities as well as relationships with biotic communities comprising natural systems.

Similarly, defining sustainability with an emphasis on ecological integrity without economic stability or social cohesion give rise to equally narrow or intolerant attitudes and policies while hindering collective responsiveness to emerging issues (Wals & Jickling, 2001). Elkington (1998) encourages the use of the "triple bottom line" combining economic, social and ecological aspects of sustainability. Sustainability, from this perspective, depends on the preservation of financial viability and efficiency, minimization of resource throughput and other ecological impacts (if not also restoration of ecological integrity) as well as the active cultivation of trust supporting the ability of people to work together for a common purpose.

It is important to note that strength in one aspect of sustainability can be leveraged to enhance weaker aspects. We tend see this in the context of economic surplus being used to obtain technology for ecological efficiency. However, we often neglect to realize that in an absence of economic abundance, social strengths also contribute to ecological integrity. For example, service-learning initiatives in which students teach family planning in a poverty-stricken community could well embody sustainability more effectively than one million dollars spent on a "green" building. Without attention to comprehensive indicators, decision makers will often overlook potential means for creating change towards sustainability.

In striving towards a comprehensive vision of sustainability, subjective meaning and perception are critical factors. As such, it is important to keep in mind a core principles of Rogers's innovation diffusion theory: if innovations are to be successful, the perceived benefit from change to new ways of functioning must be greater than the perceived cost of the transition itself (AtKisson, 1999). This notion significantly differs from a common assumption that all potential barriers to an organizational or community sustainability transition must first be removed. In using this theory base, universities should ask a number of questions that integrate aspects of the triple bottom line:
- How might individual projects be proposed, designed, implemented or evaluated to promote success?
- How will these change efforts influence overall market visibility, productivity or expenditures?
- How might social dynamics, internally and with external organizations/communities, be affected?
- How do we ensure that the overall ecological impacts of the organization will ultimately be reduced?
- How can change for sustainability best serve students learning in preparation for future life and work?

Just as the definition of "sustainability" is subjective by nature, so to are the answers to these questions. But this level of ambiguity is directly proportionate to the potential benefit we can realize by discussing differences as well as similarities in our personal perceptions of meaning and value. If approached with such comprehensive attention to diverse perspectives, sustainability transitions embody a shift from domination to partnership, from mechanistic thinking to systems thinking and from our current models of management to more holistic and systemic models that emphasize relational interconnectedness (Callenbach, 1993). This gives us a challenge, opportunity and imperative to openly discuss personal perception and diverse values in the context of assessing and redesigning our organizations and communities for sustainability.

WHOLE SYSTEMS DESIGN: SELECTED THEORETICAL ROOTS

Experts have said that the "majority of strategic initiatives driven from the top are marginally effective, at best" (Senge interview), and that the many change efforts fail due to resistance within an organization (Head, 2000). Yet our management traditions perpetuate these ineffective change strategies through mechanistic models of scientific management portraying organizations as rigid, hierarchical machines operating through a combination of command and control. As an alternative, we need to see our organizations as dynamic living organisms that can nurture more decentralized change by embracing comprehensive definitions of sustainability.

Whole Systems Design (WSD) is a collaborative design-based approach to organizational change intended to enhance our collective response to complex problems such as those posed by issues of sustainability. With roots in the social sciences, systems sciences and other disciplines, WSD begins with individuals identifying together a seed of shared vision and organizational ideology. This seed is cultivated through strategic collaboration of a diversity of stakeholders in the design of organizational structures and managerial patterns aligned with shared values. The following section highlights a few of the many individuals and theories that have helped WSD to take root.

Action research, pioneered by Kurt Lewin during the 1950's, is a change management methodology promoting the open flow of information and empowerment to maximize stakeholder involvement in collaborative cycles of planning, implementation and evaluation. It is based on the notion that participation in problem assessment and solution design greatly enhances widespread understanding of and commitment to change (Jessen & Walker, 1994). This decentralized, democratic model of management diverges from earlier, but still pervasive, models of scientific management which sought to optimize efficiency and centralize control by creating a distinct separation between managerial decision-based "thinking" and the "doing" of laborers (Capra, 2002).

In the 1960's, Douglas McGregor further emphasized organizational management promoting stakeholder engagement. He theorized that people have a natural tendency towards achievement and productivity ("Theory X") but are usually hindered by presumptions of irresponsibility and ineffectiveness motivating close managerial supervision and control ("Theory Y") (Weisbord, 1987). Abraham Maslow's "Theory Z" further suggests that people strive not only for responsibility and autonomy but for higher goals such as a life steeped in values as well as a chance to put their ideas into action within a company of which they are proud and in which they can make a difference (Maslow, 1998). Setting the stage for modern management, such theories show value in participatory patterns of management over command and control, acknowledging the subjective importance personal and organizational identity, fulfillment and self-actualization.

Around the same time, the systems sciences influenced a rich diversity of fields with a focus on the relational interconnectedness of systems with various levels of complexity (e.g. individual organisms, organizations, biotic and human communities, etc.). Using systems theory, organizational change theorists critiqued bureaucratic, hierarchical "closed" systems, proposing characteristics of healthy, dynamic "open" human systems: unity centering around compelling purpose, internal responsiveness between component systems which are functionally interdependent and external responsiveness in relationship with other systems in the surrounding environment. (Jessen & Walker, 1994).

Contemporary management research echoes open systems theory, showing that organizations with extended longevity often have a consistent value-based core ideology at the heart of ongoing change in organizational structure and dynamics (Collins & Porras, 1994). Organizational change practitioners highlight the power of shared vision, broad-based dialogue and collaborative learning with concepts such as the "learning organization" (Senge, 1990), the "living company" (De Geus, 1997) and "communities of practice" (Wenger, 1998). With a call to "get the whole system in the room" (Weisbord, 1987), practitioners began to developed whole-scale change events which offer models for optimal stakeholder participation in whole-scale organizational change (Holman & Devane, 1990).

These models and concepts are consistent with cutting-edge efforts to apply living systems theory to organizational change. Humberto Maturana and Francisco Varela suggest that a system is "living" if and only if it demonstrates two integral characteristics: "autopoeisis," the continual, decentralized process of internal self-making through which a system adapts its structure and behavior, as well as "structural coupling," the relationship of information exchange with other systems in the environment, through which a living system guides necessary change (Capra, 1996). Applying living systems theory to organizational change is a capstone to decades of praxis showing that organizational change is less effective when forced through command and control than when triggered by people engaged in intentional exchanges of meaning (Capra, 2002).

In short, these roots of whole systems design urge us to transcend a hierarchical paradigm in which isolated administrative "hero-leaders" struggle to fix isolated parts of a "broken" organizational machine from the top down. It helps us to see that effective organizational change requires the nurturing of "seed carriers" networking and providing leadership in a decentralized manner throughout a living organization (Webber, 2001). In using a WSD approach for creating change, we find that effective change includes, indeed starts with, individuals throughout an organization—every individual. This wisdom has yet to permeate our organizations.

ENVIRONMENTAL RESPONSIBILITY AND SUSTAINABLILITY IN HIGHER EDUCATION

Generally speaking, existing management structures already emphasize sustainability as it pertains to economic dynamics (via finance/accounting, advancement/development offices, etc.) and to some extent social dynamics (via human resources, student services, organizational development offices, etc.). While some institutions do have offices dealing peripherally with ecological aspects of sustainability, there is a general under-representation of environmental issues in traditional management agendas. As such, a college or university must begin to integrate ecological aspects of sustainability into curricular and management structures before they can embrace a comprehensive definition of sustainability in an effective, inclusive and transparent manner.

A recent survey conducted by the National Wildlife Federation (NWF) Campus Ecology Program shows that among responding colleges and university CEOs 64% feel that environmental programs (both curricular and operational) fit the culture and values of their institution, 47% feel such programs are good tools for public relations and 40% feel they are cost effective (McIntosh et al., 2001). Another survey (Velasquez, 2001) found that 47 % of universities have an official document outlining their concept of a "sustainable" university (this could be a master plan, environmental plan, environmental guidelines or environmental statement) and that 59% express their concerns about and profess intentions of accountability for environmental issues as well as the health of their community in the context of a mission statement.

The NWF survey states that only 27% of responding institutions state they have a written declaration that promotes environmental responsible behavior and only 21% have a mission statement that places importance on students learning about environmental responsibility. Interestingly, it states that 48% of institutions consider environmental impact in master planning and a majority of colleges and universities are systematically and regularly setting/reviewing goals for some form of environmentally responsibility—most commonly energy conservation (64%), environmental performance in building design (64%) and solid waste reduction and maximization of recycling (56%). However, only 8% have a system of accountability with incentives or penalties supporting follow through on goals. Additional statistics highlight the following:

- Less than 25% of respondents report that their institution has a council or task force dealing with environmental issues (though of those with such a group, a majority include students).
- Fewer than 14% of surveyed institutions offer any sort of orientation to help students, faculty, and staff learn about campus environmental programs.
- Only 21% or the institutions have an administrator managing general environmental issues beyond regulatory compliance (though 51% do have a recycling coordinator).

Overall, it appears that while change towards sustainability in higher education might be valued at many institutions, effective implementation that is transparent to the student learning process is still at a fledgling stage of development. Where we do see follow-through, there is insufficient recognition of the role of decentralized leadership and behavior throughout an organization. In my own experience, I find this consistent with an unfortunately common and distinct separation between sustainability transitions initiated in curricular contexts of individual programs or departments and change efforts that are initiated in more of a top-down manner.

At several institutions, I have worked with student and faculty leaders dedicated to work on grassroots sustainability transitions in the context of various courses and projects. Unfortunately, traditional hierarchical management structures recognizing only dominant economic indicators often limit the ability of their broad-based efforts to take permanent root in hierarchical or silo-based management structures. Without horizontally cohesive forums for organizational change, these leaders struggle in isolation against the juggernaut of an Ivory Tower and often abandon their efforts as institutional inertia causes the process to crawl slower than student and even faculty turnover rates. Curiously, I have often these patterns coincide with a turn towards "sustainable" development manifesting through initiatives to build a high profile "green" building—ironically, an end not always as cost-effective or "sustainable" as initially proposed.

A number of researchers (Creighton, 1998; Leal Filho, 1999; McIntosh et al., 2001; Pittman, 2001; Velasquez, 2002) concur that lack of sufficient money, time and commitment are most commonly cited in colleges and universities as barriers to sustainability efforts. Such barriers comprise what might be called a "green wall" (Elkington, 1998) preventing the broad-based integration of environmental issues, in particular, into a traditionally economic management agenda. If innovation diffusion theory is correct, though, these barriers may merely be symptoms of a deeper challenge: shifting our perception to realize that sustainability is a priority worthy of our money, time and commitment and empowering those change agents eager to lead action. The participatory integration of diverse sustainability agendas that will enhance our potential to overcome barriers rather than letting them inhibit grassroots efforts to create broad-based change towards sustainability. Such inclusion shifts stakeholder perception to show that a vision of sustainability is successfully saturating organizational culture, nourished by student learning and passion.

It is important to note that the extent to which an institution demonstrates effective change for sustainability will influence the learning experience for students, whether or not they are consciously participating in or studying organizational change. Bloom opines that the "latent curriculum" of lessons taught, or rather modeled, through daily practices and operations is as influential for teaching students as the manifest curriculum in teaching students, perhaps even more powerful unless the two reinforce each other (Rowe, 2002). This suggests that a free flow of information, including specific indicators, showing organizational progress towards a vision of sustainability strengthens student learning. This enhanced flow of information affirms that sustainability efforts are aligned with organizational values and mission, helping champions to develop trust in the system. Thus begins the cycle as change informs learning inspires involvement leads to trust furthers change efforts--ultimately replacing resistance with buy-in and optimizing the potential for continued success.

There is untold potential as we formalize organizational structures around inquiry and action related to sustainability. If we think out of the box, so to speak, and imagine that radical ideas are possible, this potential compels us to ask what patterned model of organizational structure would be best for people and the planet as well as productivity. More specifically, what pattern would help all stakeholders--students of various ages--learn together about living sustainability? It would seem that the current hierarchical pattern of organizational structure and over-representation of economic indicators of progress are not the answer. Potential answers look much like any other living system: individuals organized in nested levels of context and environment that inspire the evolution of ongoing relational improvement. By striving to create such patterns we might recognize the power of organizational structure and culture as students of all ages cultivate their ability as active agents collaborating in living sustainability.

HIGHER EDUCATION, WHOLE SYSTEMS DESIGN AND LIVING SUSTAINABILITY

Regardless of their academic track, students should graduate with a core literacy with which they can be aware of and ideally involved in organizational change for sustainability so as to learn to cultivate similar patterns within organizations wherein the find future employment. Institutions of higher education are then not only imparting knowledge, but also empowering, indeed cultivating, change agents through applied explorations in living sustainability.

Rowe (2002) contends that the spectrum of latent and explicit curricula must include literacy, responsibility and engagement with regard to sustainability issues. She advocates the use of positive vision and experiential project-based learning as tools for students to become active change agents in organizations and communities. This is precisely the pedagogical model that will promote a paradigm of living sustainability: knowledge applied by individuals collaborating on projects creating change for sustainability. These strategies help to create an organizational culture that supports the development of what Calder and Clugston (1999) have called "champions" of sustainability, similar to Senge's "seed carriers" concept (Webber, 2001). Furthermore, project-based experiential learning complements and helps to sustain the managerial application of action research, often missing from current sustainability efforts in higher education (Walker, 2001). Existing management systems, traditional and even environmental, include cycles of strategy design, implementation and evaluation; however, action research through project-based learning opens such processes to involve a broad diversity of stakeholders.

My own research supports these conclusions. While interviewing students, faculty and staff at Prescott College, I asked who should ideally have a hand in organizational sustainability transitions; the vast majority of people felt that "everyone" should be aware of and involved in related projects. Interestingly, when asked what aspects of sustainability they felt were as important as ecological, social and economic more than half added "spiritual." A common theme was that issues of sustainability are often associated with feelings of despair, pain and depression as we face an enormous and overwhelming socio-ecological crisis. From my training in the field of ecopsychology, I am aware that such feelings of despair can often lead to a state of numbing in which individuals feel isolated, disempowered and unable to think about the crisis, much less to create positive change for the future. However, many interviewees also expressed that collaborative work towards a shared vision of sustainability dispels such feelings; they also discuss the role of hope, faith and joy as stakeholders collaborate in response to issues of sustainability.

This further supports whole systems design theory and practice, emphasizing maximization of meaningful stakeholder involvement in management, continual engagement around core values and ongoing assessment towards the design of positive organizational change. Such approaches ultimately extend strategic change efforts beyond formal organizational dynamics, enriching the deeper levels of soil, if you will, comprised by the ad hoc and incidental interpersonal dynamics where many decisions are made within an organization (Capra, 2002). The cumulative effect of WSD for living sustainability has the potential to include increases in the perceived value of and commitment to change efforts as stakeholders, individually and collectively, see their role in the organization as being regeneratively rather than degeneratively aligned with the sustainability of the global community. In accordance with management Theory Y (Weisbord, 1987) and Theory Z (Maslow, 1998) it is likely that organizational change for sustainability will ultimately enhance quality of work life for and productivity of employees, as well as fulfilling needs of students, investors and trustees. (Elkington, 2003)

With this information in mind as well as the work of the aforementioned authors exploring organizational sustainability transitions, I would suggest that several elements are integral to success in organizational change for sustainability:

Institutional Commitment

It is important to make explicit a general statement of how sustainability issues are pertinent to the organization's mission and grounded in a larger contextual relationship. Many institutions are using more universal documents such as the Talloires Declaration (ULSF, 1990) for this purpose; other institutions are have designed their own dynamic document such as Northland College's Sustainability Charter (Northland College, 1996) which broadly outlines commitment as well as documenting how ongoing manifestation through specific projects. This stated commitment will serve as the seed to be cultivated by change agents.

Shared Vision of the Future

The continuity and potential success of change is directly proportional to the number of seed carriers for whom it holds meaning that they can articulate in word and deed. This helps stakeholders to see their own leadership role in living sustainability. Cultivating shared vision enhances organizational memory amidst the turnover of change agents. Shared vision is not just held in documents but comes to life through formal and informal dialogue between individuals engaged in collaborative work. Orientation programs, events and other tools can be used to maximize stakeholder engagement with shared vision.

Sustainability Indicator Reporting

Indicators representing a comprehensive definition of sustainability—a triple bottom line budget, so to speak—present a transparent window into management systems. If stakeholders are to find meaning in the indicator set they must be involved in the indicator selection process and some portion must directly relate to their role in the organization. Accordingly, indicator sets should be unique to or be adapted to fit the culture and context of the organization. Still, inspiration and guidance can be found in existing indicator usage. The Penn State Indicator Report and The Global Reporting Initiative (GRI, 2000) are excellent templates from which to begin.

Participatory Management Structures

Change is most effective when stakeholders take part in iterative cycles of organizational change--design, implementation, assessment and even celebration. Institutions can garner support for and commitment to organizational change by offering of horizontally cohesive forums for stakeholder involvement, such as organization-wide committees, task or project teams, and the aforementioned "whole scale change" events (Holman & Devane, 1999). The Sustainability Assessment Questionaire (ULSF, 1999) and Sustainability Audit (Elkington, 1998) can also be used as a starting point for large group process. Participatory structures will also help change agents to examine, discuss, and collaborate on indicators and projects.

External Partnerships

Change towards sustainability will affect interactions with other organizations, whether in the context of investment, purchasing, service outsourcing, professional association, local community or other intersystemic relationships. Ultimately, organizational culture and values aligned with sustainability should be intentionally used to shape all external relationships. In this manner, institutions of higher education can help to catalyze change for sustainability outside of the walls of the Ivory Tower.

It is important to reiterate that these elements are offered as descriptive suggestions, not prescriptive guidelines. Organizational change efforts must be meaningful and hold value to a diversity of stakeholders as well as grow naturally out of past and current efforts of like intention. As such, there are always aspects of an organizational change that will be emergent and particular to each institution.

LIVING SUSTAINABILITY BEYOND HIGHER EDUCATION

The process characteristic of an organism being living is not ascribed to an individual cell or molecule but of that organism's entire metabolic network (Capra, 2002). Likewise, the ongoing health and integrity or sustainability of our organizations and communities can only be realized in alignment with the larger society (Elkington, 1998) and the interconnected Earth community (Berry, 1996). Thus, we as individuals, organizations and communities will learn to demonstrate living sustainability not in isolation but amidst a dynamic web of relationships. Indeed, at this time the sustainability of our society depends on whether we as individuals, organizations and communities can demonstrate similar patterns of living sustainability in alignment with our shared context.

In this we find an imperative for optimizing financial stability and efficiency, minimizing resource consumption and negative ecological impacts while also maximizing trust and the ability of people to work together for a common purpose. Thus, is important for higher education decision makers to realize that living sustainability through whole systems design--shared vision guiding participatory management using specific indicators for action research creating organizational change towards sustainability--can help us move forward together toward a paradigm of health and integrity in all human systems. Father Berry is correct that institutions of higher education have a special role to catalyze action with an awareness of our interconnected Earth community.

By sharing vision, monitoring sustainability with indicators and designing participatory structures for organizational change we prepare ourselves and each other, as students of a living and learning organization, not just for sustainability in higher education but also in the larger context of our communities and global human systems. Ernst Laszlo eloquently captures this with the following words:

> Systems design advocates anticipatory democracy, where people actively apply their skills to the analysis and design of socially and ecologically sustainable systems by becoming active participants in shaping their future. Groups of people engaged in purposeful systems design form an evolutionary learning community, and such communities make for the emergence of a culture of evolutionary design (Laszlo & Laszlo, 1996, p. 16).

For individuals learning within organizations nested within layers of community all aligned to create a more socially equitable, environmentally sound and economically stable future, it is clear that the university is a microcosm of a societal macrocosm. Indeed, the halls of learning are one of the primary places where students of all ages absorb knowledge so as to nourish our communities. We ought not just seek to develop "sustainable" colleges and universities but rather create a culture in higher education that supports and encourages change agents, champions, or seed carriers of sustainability to continually nourish and cultivate our global society for the better. Design, implementation, and evaluation which are aligned with a vision of relational health and integrity honor a diversity of perspectives, hopes, and concerns. These can move us toward continuous personal, organizational, and cultural healing on a shared path of living sustainability.

REFERENCES

AtKisson, A. (1999). *Believing Cassandra: An Optimist Looks at a Pessimist's World*. White River Junction, VT: Chelsea Green.

Anderson, R. (1998). *Mid-course Correction: Toward a Sustainable Enterprise. Atlanta*: Peregrinzilla.

Berry, T. (1996). Education for the 21st Century [cassette]. Keynote presentation at the Prescott College conference *Sacred Earth Sacred Self^{TM} : Integrating Psychology, Ecology and Spirituality (May, 1996)*. Prescott, AZ.

Calder, W. & Clugston, R. (1999). Critical Dimensions of Sustainability in Higher Education. In Leal Filho, W. (Ed.), *Sustainability and University Life*. (pp. 31-46) Frankfurt: Peter Lang.

Bookchin, M. (1990). *Remaking Society: Pathways to a Green Future*. Boston: South End Press.

Callenbach, E., Capra, F., Goldman, L., Lutz, R., & Marburg, S. (1993). *EcoManagement: The Elmwood Guide to Ecological Auditing and Sustainable Business*. San Francisco: Berrett-Koehler.

Capra, F. (1996). *The Web of Life: A New Understanding of Living Systems*. New York: Anchor Books.

Collett, J. & Karakashian, S. (1996). *Greening the College Curriculum: A Guide to Environmental Teaching in the Liberal Arts*. Washington: Island Press.

Collins, J. & Porras, J. (1994). *Built to Last: Successful Habits of Visionary Companies*. New York: Harper Collins.

Cortese, A. (1999). *Education for Sustainability: The University as a Model of Sustainability*. Boston: Second Nature.

Creighton, S. (1998). *Greening The Ivory Tower: Improving the Environmental Track Record of Universities, Colleges and Other Institutions*. Cambridge: MIT Press.

Elkington, J. (1998). *Cannibals with Forks: The Triple Bottom Line of the 21st Century*. Gabriola Island: New Society.

De Geus, A. (1997). *The Living Company: Habits for Survival in a Turbulent Business Environment*. Boston: Harvard Business School Press.

Global Reporting Initiative (GRI) (2002). *Sustainability Reporting Guidelines*. Amsterdam: Global Reporting Initiative.

Graduation Pledge Alliance (GPA) (1990). *Graduation Pledge of Social and Environmental Responsibility.* Manchester College, Retrieved May 21, 2003, http://www.manchester.edu/academic/programs/departments/Peace_Studies/files/gpa.htm

Hawkin, P., Lovins, A., & Lovins, H. (1999). *Natural Capitalism: Creating the Next Industrial Revolution.* New York: Little, Brown & Company.

Head, T. (2000). Appreciative Inquiry: Debunking the Mythology Behind Resistance to Change. OD Practitioner, 32(1), 27-35.

Holman, P. & Devane, T. (1999). *The Change Handbook: Group Methods for Shaping the Future.* San Francisco: Berrett-Koehler.

Jessen, E. & Walker, M. (1994). *Group Process: Theories and Practice in an Educational Setting.* Seattle: Antioch University.

Keniry, J. (1995). *Ecodemia: Campus Environmental Stewardship at the Turn of the 21st Century.* National Wildlife Federation: Washington D.C.

Laszlo, E. & Laszlo, A. (1996). Systems Science and the Humanities. *General Systems Bulletin,* 25(3). 7-17.

Leal Filho, W. (Ed.) (1999). *Sustainability and University Life.* Frankfurt: Peter Lang.

Mcintosh, M., Cacciola, C., Clermont, S. & Keniry, J. (2001*). State of the Campus Environment: A National Report Card on Environmental Performance and Sustainability in Higher Education.* Reston, VA: National Wildlife Federation.

Maslow, A. (1998). *Maslow on Management.* New York: Wiley and Sons.

Nattras, B. & Altomare, M. (1999). *The Natural Step for Business: Wealth, Ecology and the Evolutionary Corporation.* Gabriola Island: New Society.

Orr, D. (1994). *Earth in Mind: On Education, Environment and the Human Prospect.* Washington: Island Press.

Penn State University (2000). *Penn State Indicators Report: Steps Toward a Sustainable University.* University Park: Penn State University.

Pittman, J. (2001). *Sustainability and the Prescott College Community.* Seattle: Antioch University.

Rogers, E. (1995). *Diffusion of Innovations.* London: Free Press.

Rowe (2002). Environmental Literacy and Sustainability as Core Requirements: Success Stories and Models. In: Leal Filho, W. (Ed.), *Teaching Sustainability: Towards Curriculum Greening.* (79-104) Frankfurt: Peter Lang.

Senge, P. (1990). *The Fifth Discipline.* New York: Bantam Doubleday

University Leaders for a Sustainable Future (ULSF) (1990). *The Talloires Declaration.* Washington: ULSF.

University Leaders for a Sustainable Future (ULSF) (1999). *Sustainability Assessment Questionnaire.* Washington: ULSF.

Velázquez., L. (2002). *Assessing Sustainable University Program Effectiveness.* Lowell, MA: UMass Work and Environment program.

Wals, A. & Jickling, B. (2001). Sustainability in Higher Education: From Doublethink and Newspeak to Critical Thinking and Meaningful Learning. *International Journal of Sustainability in Higher Education,* 3(3), 221-232.

Walker, K.(speaker) (2001). Panel discussion at the University Leaders for a Sustainable Future Consultation Assessing Progress Towards Sustainability in Higher Education. Washington D.C.

Webber, A. (2001). [Interview "Learning for Change" with Peter Senge, author of The Fifth Discipline] Fast Company. Retrieved May 21, 2003, http://www.fastcompany.com/online/24/senge.html

Weisbord, M. (1987). *Productive Workplaces: Organizing and Managing for Dignity, Meaning and Community.* San Francisco: Jossey-Bass.

Wenger, Etienne (1998). *Communities of Practice: Learning, Meaning and Identity.* Cambridge: Cambridge University Press.

World Commission Environment and Development (WCED) (1987). *Our Common Future.* Oxford: Oxford University Press.

BIOGRAPHY

James R. Pittman is a consultant helping groups of people create technological and interpersonal solutions supporting multi-stakeholder collaboration on organizational and community change for sustainability. He has worked with the Association of University Leaders for a Sustainable Future, the President's Council on Sustainable Development, the NASA Sustainability and Global Change Program, the EcoSage Corporation, the City of Washington D.C. and the Prescott Center for Alternative Education, among other clients. Mr. Pittman holds a Master of Arts degree in Whole Systems Design and Certificate in Organization Systems Renewal Consultation from Antioch University Seattle as well as a Bachelor of Arts degree in Ecopsychology and Education for Sustainability from Prescott College. As an undergraduate he founded the North American Alliance for Green Education, a consortium of environmental colleges.

CHAPTER 16

DISCIPLINARY EXPLORATIONS OF SUSTAINABLE DEVELOPMENT IN HIGHER EDUCATION

Geertje Appel, Irene Dankelman & Kirsten Kuipers

INTRODUCTION

Experience in The Netherlands has shown that there is much potential in promoting the disciplinary exploration of sustainable development. This article describes the process of developing such disciplinary reviews and the lessons learned so far.

The reality of life is multidimensional and multi-sectoral, and is a manifestation of both sustainable and unsustainable development. Therefore, in order to create a more ecologically sound, socially just and economically viable world, all sectors of society have to be aware of the specific challenges that sustainable development presents to them, and be able to integrate these into their daily activities. Whatever a person's background – economics, management, education, psychology, engineering, medicine, or agriculture – the assumption is that all professionals in the near future will be required to see, judge, and act in accordance with ecological, social and economic dimensions and criteria of sustainable development.

This means that education – not only primary and secondary education but higher education especially - will be required to offer knowledge and skills to meet future expectations and perspectives, regardless of the discipline involved.

The Dutch universities – which signed the COPERNICUS Charter on Sustainable Development in 1993 - were also facing that challenge. But the question was how to move from signing a charter on sustainable higher education to the real integration of sustainable development into the curricula. Herein lies a specific task for all who are involved in the process of curriculum development, use and organization: university teaching staff and university management.

Having university teachers and managers as a principal target group is no easy task. It is useless simply to organize a few workshops or training sessions for university teachers. In universities especially, teachers have something of an aversion to being trained, and above all on issues they do not necessarily see as being relevant to their own discipline. Nor do they want to be instructed 'top-down' on methodologies and the subject matter of their own lectures. If we really want to be effective in integrating elements of sustainable development in overall higher education, it is necessary for the lecturer to see the specific intellectual challenges that sustainable development poses to his or her discipline.

Peter Blaze Corcoran & Arjen E.J. Wals (Editors), Higher Education and the Challenge of Sustainability: Problematics, Promise and Practice, 213-222.

The Working Group on the Disciplinary Exploration of Sustainable Development (DVDO) of the Dutch Network on Sustainable Higher Education was established in 2000[1] with a view to enhancing the processes involved in seeing the linkages, relevance, urgency and timeliness of sustainable development from the perspective of different disciplines. The network for sustainable higher education was set up in 1998. It is a forum for lecturers, students and researchers representing almost all the Dutch educational initiatives on sustainable development to work together, exchange information, and develop new knowledge for higher education. The network consists of five national working groups: Interdisciplinary Education; Criteria for Sustainable Higher Education; North-South; Future Studies; Disciplinary Exploration of Sustainable Development; and one devoted to developing Masters programmes. The National Committee for Sustainable Higher Education (CDHO) has representatives from higher vocational education, universities, ministries, and student organizations. The Committee initiates new activities and manages the working groups.

In 2000, the working group DVDO – in which various Dutch institutions participate - decided to focus on the intellectual challenges posed by sustainable development for each discipline. Its strategy is to prepare a series of publications to challenge lecturers, start a process of reflection and stimulate internal discussion within each discipline. The 'disciplinary review' was devised as a way of exploring the relationships between various disciplines and sustainable development, and thus to challenge lecturers and university boards to integrate sustainable development into the disciplinary study programmes.

A disciplinary exploration of sustainable development might seem to be a contradiction in terms, because it is commonly accepted that a prerequisite of sustainable development is a multidisciplinary approach and skills. However, meeting the university disciplinary structure, educating students for sustainable development means educating students in disciplinary knowledge. Offering a sustainable development perspective within the disciplinary knowledge base is a first step in understanding the relevance of one's own discipline for sustainable development and of sustainable development for that discipline. And it can help in stimulating willingness (and eagerness) to work in a multidisciplinary way. To complement the training in sustainable development, another step might be to confront students with a multidisciplinary project. The interdisciplinary nature of sustainable development also enables students to explore the areas beyond the boundaries of their disciplines.

UNDERSTANDING UNIVERSITIES

In presenting sustainable development as a challenge for university teachers, it is first necessary to understand the higher education context. During the disciplinary review project (which is described in the following section) the processes that take place in a university organization had to be analyzed and discussed in depth. The working group felt that a particular set of conditions must exist that would provide a

---

[1] The authors are both coordinators and participants of this Working Group.

favourable environment for developing an intellectual challenge of this kind. The first step in identifying these conditions is to gain an insight into various aspects of the higher education system as a whole and into aspects of the university organization.

When we scrutinize the higher education system in this way we are confronted with the internal and external environment of these large and complex organizations, as well as with the rapid changes they have undergone in recent decades, the related communication processes and the way they are adapting. Generally speaking, organizations interact and learn from feedback from their internal and external environment. In the case of universities, the external environment has changed radically from what it was only a generation ago (Maassen & van Vught, 1996), and they have adapted accordingly (Baggen, 1998). The way external development outcomes are experienced depends on interaction processes, systems and instruments, and is often influenced by political and financial considerations and various other unpredictable incentives or factors. When conditions are such that they are seen as new opportunities, this is reflected in the development of curricula.

One of these developments in the external environment is the growing awareness of the importance and relevance of sustainable development, which appeals to the role of universities in society and their social responsibility. Therefore, information and communication with stakeholders (including students, the labour market, the business community, the education ministry, branch associations, etc.) on sustainable development continues to be a stimulus for a sustainable development integration process.

The formal structure of a university generally consists of three organizational levels: the university board, the faculties and the research departments. The university board and the faculties spend much of their time dealing with managerial and administrative aspects, while the departments are concerned with the core business of the university: research and education. Democratic structures, professional autonomy, and a diffuse decision making power, are among the key elements influencing innovation processes in higher education. Mintzberg (1983, p. 210) writes:

> In the Professional Bureaucracy, with operator autonomy and bottom-up decision making, and in the professional association with its own democratic procedures, power for strategic change is diffuse. Everybody, not just a few managers or professional representatives, must agree on the change. So change comes slowly and painfully, after much political intrigue and shrewd maneuvering by the professional and administrative entrepreneurs.

THE NEED FOR AN INTELLECTUAL CHALLENGE

A bottleneck for the integration of sustainable development in curricula is the fact that university professionals are not always convinced of the value of integrating sustainability in their courses, or lack the motivation to do so. (Bras-Klapwijk et al., 2000). University professionals generally operate and communicate in their own division's disciplines, subdisciplines, specializations and, possibly worldwide, professional associations. However, issues of sustainable development need a broader, multidisciplinary approach (Orr, 1994) and cross-disciplinary borders. This is in stark contrast to the existing, research-focused, reward, career and status systems in the academic world. Jaffee (2001, p. 39) pointed out the unintended consequences of these research-focused systems in a teaching versus research organization paradox:

> Higher education has been struggling for many years with a trade-off between teaching and research. Historically, the large comprehensive university has defined quality on the basis of the faculty's research activity and scholarly productivity. To encourage these efforts, the primary academic rewards – such as tenure, promotion, reappointment, and merit pay – were based on the publication of research articles and books. However, it is now widely recognized that the means and rewards employed to achieve this goal have had the unintended consequence of neglect for teaching.

The question is how to gain the disciplinary professionals' attention, how to motivate them (despite the reward systems), and how to start a communication process on the issue of sustainable development within a discipline and with the lecturers. The working group decided to use the development of 'disciplinary reviews' as a catalyst.

The development of a disciplinary review is based on exploring the concept of sustainable development from a disciplinary perspective, and a discipline from a sustainable development perspective. It became clear during the first year of the project that this strong focus on substance is essential in maintaining the involvement of the lecturers, but it was not sufficient on its own.

Paying enough attention to exploring the sustainable development concept and making it explicit in relation to the discipline is one of the main conditions for a successful integration process. The assumption of professional autonomy and a democratic culture of decision-making needs a method of working that fits in with the organizational characteristics and the change processes concerned.

Looking at change processes in general, it is possible to identify four phases (Bullock & Batten, 1985; Burnes, 1992): (1) exploration, (2) planning, (3) action, (4) integration.

The experience of the working group so far has been that the process is non-linear. Some of the institutions started to define a strategy and to integrate sustainable development into curricula while the first phase (exploration) was still in progress. Some of the review publications analyzed the integration of sustainable development in one course. It was common for a strategy for integrating sustainable development to be discussed in meetings at all decision-making levels. This intervention at different levels is in line with the diffuse decision making structure in universities. The working group found that composing a disciplinary review involves collecting information, communicating and analyzing the situation, and is therefore a first step in integrating sustainable development into curricula.

In fact, the insights that were gained in the process of developing a disciplinary review itself became one of the project's main results. This process can be viewed as one of long-term development, with careful procedures for planning, development and implementation and is not unlike other innovation processes in higher education (de Graaff, 1993). It is also continuous in that: it takes several years before the perspectives are integrated into the curricula and before a follow-up starts again with phase one.

A MULTIDIMENSIONAL INTERACTION APPROACH

The process of composing a disciplinary review can never be seen in isolation from other processes and interactions with educational and other developments and internal and external incentives. One of the developments is the Bologna Declaration, a European agreement that affects the curricula of European universities. It changes the Dutch university structure into a bachelor-master system. For some disciplines, this is an opportunity to restructure the whole curriculum, including the content of the courses. Structure changes of this kind might offer opportunities for integrating sustainable development issues into the curricula, especially when the aim is to broaden the education programmes.

Another incentive is formed by the outcomes of quality assessments. In general, outcomes of quality assessments cause a shake-up, especially when they reveal opportunities for improvement (Hulshof & Warps, 1998).

The purpose of a quality assessment is to monitor and improve quality, to enhance accountability and to inform the "outside" world of performance. Such evaluations for study programmes take place once every six years. Review committees screen all such study programmes and report their main findings publicly and make recommendations. A quality assessment set up by the Association of Universities in the Netherlands (VSNU, 2000) was carried out at all Dutch universities for the Chemistry study programmes. One of its findings was that much more attention should be paid to integrating sustainable development issues into the curricula content (focusing on life-cycle analysis, renewable resources, recycling by-products etc.).

Also a quality assessment on environmental sciences was published recently (VSNU, 2002), which again recommends integrating sustainable development into the curriculum of the education programmes as well as structural measures to be taken at the highest organizational level, with reference to the Copernicus Declaration of 1993.

LECTURERS INTEGRATE SUSTAINABLE DEVELOPMENT

As mentioned, the DVDO working group embarked in 2000 on building an intellectual challenge. The project was known as 'Lecturers integrate sustainable development: challenges in the 21st century'. Early experience in the field of integrating sustainable development in the curricula indicates that the concept of sustainable development has to be made explicit within a specific discipline and that lecturers have a strong need for concrete examples and case studies. The project examined the extent to which the theory and practice of a number of disciplines contributed to or obstructed sustainable development.

An initial assumption of the working group was that a top-down approach would be unsuitable in complex organizations such as universities, and that it would be more feasible to start from within a discipline's own existing processes. The group also expected to find different cultures within a university, clustered along disciplinary lines, so that a measure of success would be how closely the work could connect to that diversity. The question that therefore arose was how sustainable development would fit in with the organizational values and subcultures of a university.

The disciplines focused upon in 2000 were: economics, management sciences, physics, and history. The following fifteen disciplinary explorations were published in early 2003 and distributed to universities and polytechnics in the Netherlands: management sciences, economics, physics, history, biology, mathematics, health, civil engineering, mechanical engineering, computer science, law, philosophy, general introduction to sustainable development, and reviews of economics and management sciences for polytechnics. Disciplinary reviews of physical planning, psychology and chemistry explorations will be published in early 2004. Nine universities and polytechnics are involved in the project.

During the exploration phase, the authors used a mix of data gathering techniques to analyze the content of sustainable development in relation to a specific discipline. For example, the physics publication was based on interviews with physicists working in companies and at universities, whereas the economics publication was based on literature and knowledge distilled from within the discipline. The authors clarified concepts in interviews with other players in the process within the subdisciplines and discipline. For instance, the authors of the review of business and management studies screened the 121 largest companies in the Netherlands on sustainable development related issues. After the content of each discipline had been studied in depth, the insights acquired were shared with as many lecturers as possible via seminars. At some of the universities, the first draft of the 'Disciplinary Review on Sustainable Development' was discussed during curriculum committee meetings (physics and economics). The final version was presented and discussed during, nationally organized, disciplinary seminars. The disciplinary review of health science was discussed at an international conference on health and sustainable development. Other contributors to the seminars were a variety of stakeholders including lecturers and students, who made presentations and participated in the discussions. The faculty board and management also participated in several meetings.

The processes that led to the disciplinary reviews and meetings, however, seemed just as important, because of the valuable strategic lessons that can be derived from them with regards to the best way to challenge lecturers and others to integrate sustainable development into their disciplines.

METHODOLOGY

The steps used for generating the disciplinary reviews on sustainable development can be divided as follows:

1.   Appointing a coordinator for each institution. Determining the terms of reference.

2. Selecting the discipline to be reviewed.
3. Selecting the researcher to execute the review.
4. Developing the disciplinary review (through literature study and interviews).
5. Submitting the text of the disciplinary review to professionals for feedback.
6. Publishing the review.
7. Arranging meetings with lecturers (from the institution and outside).
8. PR and follow-up.

Each of the subprojects was coordinated by the sustainable development focal point of one of the participating institutions: the project coordinator. They either executed the research for the disciplinary review themselves or appointed a research consultant. First of all, the project coordinator had to make clear what goals needed to be achieved and who was to be involved. He or she had to formulate a number of benchmarks. What is the target group? How can the interest of the target group be raised? How to elicit a response and engage them in debate? And how can the target group be challenged to make a contribution to integrating sustainable development in education? In his or her plan, the project coordinator also needed to take into account the culture of the discipline concerned. What is the best way to create support for the results of the project? What are the criteria for the success of the project and how should they be measured? The institution carrying out the review established these criteria and also needed to determine what was required for disseminating the knowledge outside the institution. The project coordinator formulated what he or she hoped to achieve and identified the resources that were required (i.e. disciplinary sources, manpower, contacts, media, funds).

TAKING STOCK

The disciplinary reviews executed so far have developed fundamental insights into the content of disciplines in relation to sustainable development and have triggered valuable discussions and experiments in higher education institutions. Although the limited time and funds somewhat restrict the scope of the disciplinary reviews, useful lessons can be learned.

One of the most important results of the project so far has been that the approach of the project through the development and presentation of disciplinary reviews on sustainable development offers important challenges to lecturers in terms of content. Because the reviews start from the lecturer's own discipline, and her or his colleagues will have contributed, the content of the review can easily be recognised by the target group.

The review publication is a reference book that is strongly oriented to content. It provides examples of sustainable development that are intricately linked to a specific discipline. It is from within their own expertise that lecturers are being invited to branch out to the wider world of sustainable development.

The process of developing the disciplinary reviews itself is a learning process from which important lessons can be learnt for future initiatives. The criteria or conditions for the success of such a process, which are within the scope of the

'Methodology Working Group', can be of use in similar processes of exploring the relationship between specific disciplines or sectors and sustainable development.

In the faculties concerned, the development of a disciplinary review turned out to be an instrument in the decision making process on sustainable development at all levels. A strong point in the project and a condition for achieving a real integration process is the presentation of sustainable development as a concept. The disciplinary reviews are therefore not only an instrument for further communication and integration, but also a product of a learning and communication process themselves.

The review process is adapted to the disciplinary culture of the university environment, but also to the dynamics of a university organization. On the other hand, it can challenge such disciplinary approaches and organizations to look beyond the borders of their disciplines. The group found that integrating sustainable development is not an isolated process. It is important that the integration of sustainable development through disciplines review processes interacts with other (external) incentives and developments, such as major system changes and quality assessments.

CONCLUSIONS AND CHALLENGES

Rather than challenging the disciplinary structure of institutions of higher education, this chapter has shown that challenging disciplines to take sustainable development on board is also a valuable approach. In this effort, the development of disciplinary reviews can be very instrumental. The process of integrating sustainable development into the disciplines turns out to be a non-linear, ongoing, cyclic learning process. The changing context in which many universities are currently operating is one of the main sources of new opportunities for learning for change, as sustainable development requires.

Disciplinary reviews can play an important role in promoting the integration of sustainable development in disciplinary education, because of their content, the process of disciplinary analysis, and the emergence of actual and potential disciplinary relationships with sustainable development. They can offer an important stimulus for considering the role of the discipline in the present world and in the future. The reviews not offer only a challenge to lecturers, but they can also be a trigger for other users, including students, to learn how sustainable development is relevant to them and their professional field.

Although the disciplinary explorations have been published and widely distributed in The Netherlands, a detailed analysis of what is involved in moving from an exploration phase to an integration phase has yet to be performed. Early results from the follow-up projects confirm that changing to a bachelor-master system might offer new opportunities for the integration process. Cultural factors, such as the attitude towards change processes and sustainable development as a concept, might also play a role. These preliminary conclusions need to be studied in more detail. Further analysis of the follow-up projects will help distinguish the role of disciplinary reviews in the integration of sustainable development.

REFERENCES

Baggen, P. (1998). *Vorming door wetenschap: universitair onderwijs in Nederland 1815-1960*. Delft: Eburon.

Bras-Klapwijk, R. M. A. de Haan & K.F. Mulder (2000). *Training of Lecturers to Integrate Sustainability in the Engineering Curricula*. In: Van der Bor, W, Holen, P., Wals, A.E.J. & Filho, W. (Eds.) Integrating Concept of Sustainability into Education for Agriculture and Rural Development. Frankfurt am Mein: Peter Lang Scientific Publishers.

Bullock, R.J., and Batten, D. (1985). *It's just a phase we're going through: a review and synthesis of OD phase analysis*. In: Group and Organization Studies, 10, December, 383-412.

Burnes, B. (1992). *Managing Change*. London: Pitman Publishing.

DVDO (2002). *Working Document on Disciplinary Reviews: disciplinary exploration of sustainable development*. Nijmegen: DVDO/UCM.

Graaff, de, E. & P.A.J. Bouhuijs (Eds.) (1993). *Implementation of Problem-based Learning in Higher Education*. Amsterdam: Thesis Publishers.

Hulshof, M.J.F. & J.H.J.M. Warps (1998). *Kritische factoren bij Onderwijsvernieuwing*. Onderzoek in opdracht van het ministerie van OC & W. Den Haag: SDU.

Jaffee, D. (2001). *Organization Theory: Tension and Change*. New York: The McGraw-Hill Companies, Inc.

Maassen, P. A.M. & Van Vught, F.A. (Eds.) (1996). *Inside Academia, The Netherlands: Centre for Higher Education Policy Studies (CHEPS)*. Utrecht: Uitgeverij De Tijdstroom BV.

Mintzberg, H. (1983). *Structure in fives*. New Jersey: Prentice-Hall, Inc., Englewood Cliffs.

Orr, D.W. (1994). *Earth in Mind. On Education, Environment, and the Human Prospect*. Washinghton DC: Island press.

Vereniging van Universiteiten (VSNU) (2000). *Onderwijsvisitatie Scheikunde en Scheikundige Technologie. Utrecht: VSNU*.

Vereniging van Universiteiten (VSNU) (2002). *Onderwijsvisitatie Milieuwetenschappen*. Utrecht: VSNU.

BIOGRAPHIES

Geertje Appel is a programme manager of the sustainable development programme at the Vrije Universiteit (Free university), Amsterdam. The programme aims to integrate sustainable development into the university curriculum with emphasis on content, value education and (philosophical) reflection. The activities resulted in education materials with a strong focus on incorporation of sustainable development issues into the regular course content of disciplines. As a member of the working group Disciplinary Exploration of Sustainable Development she co-ordinated the disciplinary review projects on 'Economics and sustainable development' and 'Law and sustainable development.' After her study on sociology and environmental sciences at the University of Amsterdam and the University of Utrecht Geertje worked as a consultant in the field of management systems and change processes (environmental management and quality management) in complex organisations.

Irene Dankelman is programme coordinator sustainable development at the University of Nijmegen, Netherlands, since 1999. As an ecologist she has specialized in social and policy aspects of sustainable development, and in particular in gender and environment. She has published widely on these issues. Dankelman has worked for national and international NGOs, the Dutch government and with the United Nations, and is presently (board) member of a number of international and national organisations, including the Women's Environment and Development Organisation (WEDO) and the Commission on Sustainable Development in Higher

Education (Netherlands). As a consultant and as co-chair of the National Platform Johannesburg, she was extensively involved in the World Summit on Sustainable Development (2002).

Kirsten Kuipers is project coordinator at the Center for Environmental Studies and Sustainable Development at the University of Nijmegen. Currently she coordinates a interuniversity project on the integration of sustainable development into the curriculum of several academic disciplines. She is editor of a series so called disciplinary reviews on sustainable development, ranging from 'History and Sustainable Development' to 'Civil Engineering and Sustainable Development'. For this project she herself wrote a review called 'Philosophy and Sustainable Development', issued in June of 2003. She organized several courses on topics like sustainable technology development and corporate social responsibility. In earlier years she worked for several environmental organizations and as a freelance (radio)journalist.

CHAPTER 17

THE PROMISE OF SUSTAINABILITY IN HIGHER EDUCATION: A SYNTHESIS

Arjen E.J. Wals & Peter Blaze Corcoran

The promise of various perspectives that have been offered in Part Two each will resonate in varying degrees depending on the reader's own background, biases and preferences. It would be naïve to say that they are all complementary and, when combined in the 'right' way, will lead to a 'best' version of sustainability in higher education. When studying the many perspectives of sustainability and the role of higher education in society, it becomes quite clear that they are not all compatible. Some are even inconsistent and have very different underlying ideological roots and values prohibiting an eclectic approach of integrating them, but instead demanding critical choices. So how do we deal with these incompatibilities? How can we benefit from these multiple perspectives and others not covered by this book? Part of the answer to these questions lies in the facilitation of what has been called social learning.

The role of the facilitator of processes that explore and develop the potential of sustainability in higher education is a crucial one. Exploring sustainability in higher education can be seen as a process of simultaneous individual and institutional confrontation and self-confrontation in order to arrive at a better understanding of both the potential significance of sustainability for both the institution and for oneself. Adopting such a position means putting emphasis on the process and its facilitation. This brings us to the need for facilitated cultivation of pluralism and conflict in order to create space for social learning in moving towards contextual sustainability in higher education. The process of determining how to become sustainable as an institute of higher education as undertaken by a group can be viewed as a particular manifestation of social learning. Social learning here is seen as a collaborative re-framing process involving multiple interest groups or stakeholders (Vandenabeele & Wildemeersch, 1998). Through discursive dialogue and cooperation between people positioned within different configurations or frames with regards to the key issues involved, as identified by the authors in Part Two, such learning can be intensified and lead to change. Hence, social learning can be viewed as an intentionally created purposeful learning process that hinges on the presence of alternative constructions of reality.

If indeed the exploration of sustainability in higher education involves the reconciliation of diverging norms, values, interests, and constructions of reality then the innovation process should be designed in such a way that differences are

Peter Blaze Corcoran & Arjen E.J. Wals (Editors), Higher Education and the Challenge of Sustainability: Problematics, Promise and Practice, 223-225.
© 2004 *Kluwer Academic Publishers. Printed in the Netherlands.*

explicated rather than concealed. By explicating and deconstructing these differences it becomes possible to analyze their nature and persistence. This is an important step since it helps to improve both the dialogue between the stakeholders involved and to identify strategies for utilizing conflict in the social and individual learning process.

The promotion of sustainability in higher education requires more than consensus in the present, but rather requires a dialogue to continuously shape and re-shape ever-changing situations and conditions. A dialogue here requires that stakeholders involved can and want to negotiate as equals in an open communication process which views diversity and conflict as the driving forces for development and social learning (Kunneman, 1996; Wals & Bawden, 2000). As Wals and Heymann (2004) point out elsewhere, such dialogue rarely spontaneously emerges, but requires careful designing and planning. Sustainability can and perhaps should be a highly contested concept and the potential differences in interests and possibilities can be significant, especially when there are significant power imbalances within a university.

Sustainability in higher education can be regarded as both as the collaborative creation of an ever-evolving product and as an engaging creative process involving a variety of different actors. Moving towards sustainability as a social learning process has up until now received less attention than concepts of sustainability as expert (pre)determined and essentially teachable products (Wals & Jickling, 2002). One question to be raised is: How can academia help develop all-round personal capabilities that generate positive but often unanticipated outcomes? This is a question related to determining the kind of competence that is needed to contribute to sustainability and academia's role in developing such competence amongst all its staff and students. With Raven and Stephenson (2001), we agree that competence here does not refer to getting the job done effectively, after all there is know consensus about what 'the' job entails, but rather to making an effective contribution to society by going beyond boundaries and by influencing the systems in which the competence is developed. From this perspective, sustainability can, at the institutional level, be viewed as a catalyst for systemic institutional and organizational change.

Education for sustainability above all means the creation of space for social learning. Such space includes: space for alternative paths of development, space for new ways of thinking, valuing and doing, space for participation minimally distorted by power relations, space for pluralism, diversity and minority perspectives, space for deep consensus, but also for respectful disagreement (Lijmbach et al., 2002) and differences (Olson & Eoyang, 2001), space for autonomous and deviant thinking, space for self-determination, and, finally, space for contextual differences. This observation reminds us of John Dewey's views on education and democracy, almost a century ago, when he argued that education should realize a sense of self, a sense of other, and a sense of community; it should create space for self-determination as individuals and/or members of groups exercise greater degrees of autonomous thinking in a social context (Dewey, 1916). In Part Three a variety of cases illustrate how different institutions balance emancipatory and instrumental practices in sustainability in higher education.

REFERENCES

Dewey, J. (1916). *Education and Democracy*. New York: The Free Press.

Kunneman, H. (1996). *Van Theemutscultuur naar Walkman-ego. Contouren van postmoderne individualiteit*. Amsterdam: Boom.

Lijmbach, S., Margadant-van Arcken, M., Koppen, C.S.A and Wals A.E.J. (2002). 'Your View of Nature is Not Mine!' Learning about Pluralism in the Classroom. *Environmental Education Research*, 8 (2), 121-135.

Olson, E.E. and Eoyang, G.H. (2001). *Facilitating Organization Change: Lessons from complexity science*. Jossey-Bass/Pfeiffer, San Fransisco.

Raven, J. and Stephenson, J. (2001). *Competence in the Learning Society*. New York: Peter Lang Publishers.

Vandenabeele, J. and Wildemeersch, D. (1998). Learning for Sustainable Development: Examining Lifeworld Transformation Among Farmers. In: D. Wildemeersch, M. Finger & T. Jansen (eds.) *Adult Education and Social Responsibility* (pp. 115-132). Frankfurt am Main.: Peter Lang Verlag.

Wals, A.E.J. and Bawden, R. (2000). *Integrating sustainability into agricultural education: dealing with complexity, uncertainty and diverging worldviews*. Gent, Belgium: ICA.

Wals, A.E.J. and Heymann, F.V. (2004). Learning on the edge: exploring the change potential of conflict in social learning for sustainable living. In: A. Wenden (Ed.) *Working toward a Culture of Peace and Social Sustainability*. New York, SUNY Press.

Wals, A.E.J. and Jickling, B. (2002). "Sustainability" in Higher Education from doublethink and newspeak to critical thinking and meaningful learning. *Higher Education Policy*, 15, 121-131.

PART THREE

PRACTICE

CHAPTER 18

THE PRACTICE OF SUSTAINABILITY IN HIGHER EDUCATION: AN INTRODUCTION

Kim E. Walker, Arjen E.J. Wals & Peter Blaze Corcoran

How do we move from promise to practice in the exploration of sustainability in higher education? In Part Three, we bring examplary practices to the forefront. These practices have in common that they have been carefully studied by both insiders and outsiders, but they differ in scale and scope. Some represent intra-institutional practices, in that they describe different ways in which a single university or unit within a university seeks to respond to the challenge of sustainability in higher education. In this category we have included innovative sustainability initiatives at universities in Denmark, United Kingdom, South Africa. and the United States. Other cases represent inter-institutional practice, in that they describe networks or other forms of cooperation between universities that jointly seek ways to explore sustainability within their institutions. In this category Part Three contains a review of a group of universities involved in a sustainability initiative in the USA, and an analysis of an network promoting sustainability in UK universities. Finally, one case from The Netherlands is included highlighting a tool or instrument that has been designed to help institutions systematically move towards sustainability. In this introductory chapter we provide a rationale for using case study methodology to highlight practice. We believe such a rationale is helpful in light of the recent avalanche of case studies reported in educational and other research.

Case study methodology is a common and appropriate research tool used in studies of sustainability in higher education[1]. The decision to publish case studies for a broad audience suggests that others have something to learn from the case study. Research in sustainability in higher education should, ideally, take account of all the complexities of the field. Sadly, this has not always, or frequently, been the case. Fien (2002, p. 244) reports that research in sustainability in higher education remains predominantly a theoretical 'in that few studies have sought to go beyond description to include a critical and theoretical analysis of findings or to ground explanations in social or organizational theory'. Similarly, Corcoran et al. (2004) find that case study research in higher education has been descriptive but not

---

[1] In discussing case study methodology we draw upon an article the three of us wrote for a special issue of Environmental Education Research on the use of case study research in environmental education (see Corcoran et al., 2004).

Peter Blaze Corcoran & Arjen E.J. Wals (Editors), Higher Education and the Challenge of Sustainability: Problematics, Promise and Practice, 229-234.
© 2004 *Kluwer Academic Publishers. Printed in the Netherlands.*

transformative ....the research does not problematise practice, instead it sets up dichotomies of practice'.

Yin (1989, p. 82) explains that case studies allow a researcher to 'reveal the multiplicity of factors [which] have interacted to produce the unique character of the entity that is the subject of study'. It is a method of learning about a complex instance through description and contextual analysis. The result is a description and theorizing about why the instance occurred as it did, and what may be important to explore in similar situations. According to Yin, a case study "... investigates a contemporary phenomenon within its real-life context; when the boundaries between the phenomenon and context are not clearly evident; and in which multiple sources of evidence are used" (Yin, 1989, p. 23).

Case study methodology seems a very appropriate research tool to investigate sustainability in higher education. The case study approach allows the researcher to 'go deep', to learn what works and what does not. However, there is an imprecise understanding of case study and, according to Merriam (1998) it is often misused as a 'catch-all' research category for anything that is not a survey or experiment. Indeed, a case study can accommodate a variety of research designs, data collection techniques, epistemological orientations, and disciplinary perspectives. No matter what the researcher's epistemology, the case study is an appropriate strategy for answering questions about how or why.

Case study research has many differences depending on the purpose of the study, the size of the study, the people involved, the theories developed and the theories tested (Corcoran et al.,2004). Bassey (1999), for example, defines a range of purposes for educational case studies that include theory-seeking and theory-testing case study (Niko Roorda, Chapter 24); story-telling and picture drawing case study (Nan Jenks-Jay, Chapter 21), and evaluative case studies (Wynn Calder & Rick Clugston, Chapter 20). Case studies may involve description, explanation, evaluation and prediction. Many case studies involve people working within their regular environment (Rocky Rohwedder, Chapter 23; Nan Jenks-Jay, Chapter 21; Malcolm Plant, Chapter 22; Heila Lotz-Sisitka, Chapter 25; and Susanne Leth, and Nadarajah Sriskandarajah, Chapter 26).

Case study methods in sustainability in higher education vary according to the researcher's purpose in conducting the case. Frequently, the researcher is an outside evaluator or critical friend who sets out to critique the practices of an institution or series of institutions. William Scott and Stephen Gough in Chapter 19, for instance, provide a critical examination of attempts to introduce sustainability into universities in the UK as do Wynn Calder and Rick Clugston in Chapter 20 when evaluating South Carolina's Sustainable Universities Initiative. The aim for an outside evaluator is multi-purpose and often has internal and external purposes. Internally, the evaluator provides important feedback to the practitioners involved in an innovation and often works with these people to move forward. The evaluator may provide feedback to the institution as a whole in the form of a report on the success or otherwise of the implementation of sustainability in the institution.

Externally, the evaluator may compare institutions in an effort to identify practices that work and those that do not. This work is particularly valuable for those attempting reform in their own institutions. The work also provides important data for funding groups and potential funding groups.

Case studies may be conducted by the practitioners involved in the innovation and examples of such involvement appear in the following chapters: Nan Jenks-Jay's Chapter 21 on sustainability innovations at Middlebury College, Malcolm Plant's Chapter 22 on the development of an MA in environmental education at Nottingham Trent University, Rocky Rohwedder's Chapter 23 on the Environmental Technology Center at Sonoma State University, Niko Roorda's Chapter 24 on the development and application of an auditing tool for sustainability in Higher Education, and Heila Lotz-Sisitka's Chapter 25 on a curriculum innovation at Rhodes University. Where the case study is conducted by an internal practitioner, the aim is usually to engage in a self-study of practice. While this form of case study research has the disadvantage of not benefiting from critical external feedback it can be a valuable tool in improving practices.

In a review of the use of case study methodology in sustainability in higher education (Corcoran, et al., 2004), we wrote that research on innovation in sustainability in higher education has broadly consisted of quantitative studies using a pre-determined set of sustainability indicators or ecological footprints or qualitative studies using, primarily, a case study approach. In some instances the research focuses on a case study approach but uses data from a set of pre-determined sustainability indicators as part of the case study (Niko Roorda, Chapter 24).

The findings from the study showed that case studies in sustainability in higher education rarely included any information on the theoretical approach to the methodology or on the methods used to gather the data. Instead, stories of successes were reported and the data supporting these successes are not readily available for public critique. Although much research in environmental education or on education in general tends to be well theorized, research on sustainability in higher education is not. It was concluded in the earlier study that case study research would be more effective in bringing about change if it was better theorized and documented (Corcoran et al., 2004). Therefore the case studies included in Part Three were selected on the quality of the methodology, diversity of the institutions involved, and potential transferability towards other contexts.

The case studies that appear in the following chapters provide a range of windows through which to view sustainability in higher education.

In Chapter 19, William Scott and Stephen Gough provide a case study of the attempts made by institutions of higher education to introduce considerations of sustainability in the time period after the Earth Summit in Rio. The authors discuss the separation of the work occurring in the 'operations' of the Higher Education Institutions (HEIs) and the 'curriculum' of HEIs and offer some explanations about why the two constituencies are developing separately and the special demands on each. The case study includes an analysis of the work of Higher Education Partnerships for Sustainability (HEPS) and specifically, the curriculum toolkit developed by HEPS. They critically analyse the various theoretical positions underpinning the toolkit and conclude with a review of the toolkit and a critical discussion of its explicit and implicit underlying learning theories. Scott and Gough conclude with a positive assessment of the important function the toolkit potentially makes to implementing sustainability into the curriculum of universities.

The authors then provide a brief review of 18 in the United Kingdom HEIs. The review is based on data drawn from the HEI's websites using the internal search engines of those sites. The outcomes of that research are documented.

The case study provides a glimpse of activity happening in sustainability currently in the UK and draws upon a critical review of the literature in the field, a critical analysis of the Toolkit and a study of HEIs websites to demonstrate the take up of sustainability.

Wynn Calder and Rick Clugston, in Chapter 20, also cover a number of universities and draw conclusions across those universities in their case study on South Carolina's Sustainable Universities Initiative (SUI). The three universities involved in the initiative are the University of South Carolina, Clemson University and Medical University of South Carolina. The case study describes the efforts of the SUI to realize their five year plan in its work with faculty, students and operations staff at each of the three universities.

Calder and Clugston analyze the progress of the SUI toward achieving sustainable universities using the seven factors developed by the authors. They argue that the seven factors predict the likelihood of support for an academic reform initiative from the institution to which it is being advanced. The seven factors include: credibility of the 'champions' of sustainability within an institution; support of key administrators, benefits of the initiative for departments and programs, conformity to institutional identity, engagement of the university community, academic credibility, and, availability of resources.

Nan Jenks-Jay's case study similarly highlights the factors needed for an institution of higher education to become sustainable in her critical study of a remarkable innovation at Middlebury College in the USA. Nan Jenks-Jay is a key player in the innovation and, therefore, reports from that perspective. She specifically discusses the processes involved in achieving a high level of success in implementing sustainability at Middlebury College. In Chapter 21, she writes about the need for commitment and support from the various levels within the college and the need to sustain that commitment in order for the initiative to continue being successful. She looks at the role of a forward thinking Board in bringing about change, the integration of campus sustainability initiatives and what students are learning about environmental issues. Community involvement is very important at Middlebury, and important in the success of the sustainability initiative. Equally, she points out that a college needs to be integrally involved in the environmental, economic and social sustainability of the community. Student involvement is also crucial.

In Chapter 22, Malcolm Plant provides a case study of a Masters of Arts in Environmental Education by distance learning. He argues for a specific theoretical position as a basis for the changes described. Plant seeks to provide an education which is transformative and draws on his preferred theoretical position, 'critical realism'. Critical realism, as explained by Plant, involves specific teaching strategies. Students 'become empowered' through the process of advancing a critical theory of environmental education. The case study consists of two critical encounters of two students during their involvement in the course. The voices of the students are heard through: (1) excerpts of the reflections of a student on her readings of a course text; and, (2) the reflections of a student (via email) on an assessment task. In both cases the students critically reflect on their professional and personal lives. The author concludes with a concern that such courses may not survive the bureaucratic demands of institutions.

Rocky Rohwedder's case study is also based on a specific theoretical position in the framework he has developed to describe an initiative at Sonoma State University. In Chapter 23, he talks about the significance of place and how the creation of an environmental facility on campus provides a vision for a university campus. He emphasizes the civic responsibility of a campus in contributing to environmental improvement and the lessons students learn as a result of the university taking this responsibility. Rohwedder uses a case study of his own institution to exemplify two broad questions: (1) what are the typical landscapes of higher education institutions; and, (2) what ideals and principles do these landscapes teach? From there, the author asks, How are the physical and curricular landscape of institutions congruous and incongruous with the notion of sustainability?

The challenge, according to Rohwedder, is how to reshape the structure and the behavior of learning institutions so as to align them with the lessons learned in the classroom about nature. He also discusses the importance of an environmental facility to act as a model for other institutions, for the community and as a base for interdisciplinary science environmental education.

Niko Roorda in Chapter 24, describes his Auditing Instrument for Sustainability in Higher Education (AISHE), a tool for policy development for sustainable development in universities. He explains how AISHE is used, some general conclusions based on audits completed to date and two cases are discussed. Two main issues emerged from the audits so far: communication about sustainability is the main point for improvement in sustainability initiatives; and, improvements in vision and policy about sustainability have a high priority.

The two cases involve an economics study program and an environmental technology study program. The author explains the process of conducting the AISHE audit with the staff involved in each program (separately) and the outcomes of each study in terms of the findings about the implementation of sustainability and the usefulness of AISHE in bringing about change.

Heila Lotz-Sisitka writes a case study about a sustainability innovation at Rhodes University, South Africa. In 1990 the Murray & Roberts Chair of Environmental Education was established in the Faculty of Education. A key focus of the position has been to respond to socio-economic issues through education and training programs and research. The case study focuses on a community capacity building program which aims to support adult environmental educators working in community based and other settings. The course is participatory, focuses on community issues and fosters links with the community. It is characterized by reflexivity and change. The key features of the course are: responsiveness; flexible course structure; participation; praxis; and, assessment as learning. Chapter 25 describes the ways in which the innovation at Rhodes has impacted on environmental education courses in other parts of the university and on environmental education courses in other universities in South Africa.

Susanne Leth & Nadarajah Sriskandarajah (Chapter 26) provide a case study of a teaching and learning innovation in forestry education at the Royal Veterinary and Agricultural University, Copenhagen, Denmark. The chapter describes the teachers' efforts to integrate a new interpretation of sustainability into forestry education. The case study is about a theory of change and the implications of thinking differently about curriculum and teaching and learning strategies. The authors use data from a qualitative study they themselves conducted. Specifically, the authors studied an innovative workshop aimed at allowing students to experience participatory learning experiences, problem solving and curriculum development. A model of change involving thinking and learning in forestry education was developed as a result of the study.

We hope that Part Three will resonate with the readers. After all, the extent to which such case studies benefit others depends not only on the way this learning process is documented and shared, but also on how others relate to the case. In other words, it depends on what the readers distill from the case and the way they infuse their own learning into their own institutional context. At the same time we hope that the practice highlighted here sheds new light on both the problematics and the promise of sustainability in higher education.

REFERENCES

Fien, J. (2002). Advancing Sustainability in Higher Education: Issues and Opportunities for Research. *International Journal of Sustainability in Higher Education*, 3 (3), 243-253.

Flint, (1999). Institutional Ecological Footprint Analysis: a Case Study of the University of Newcastle, Australia. *International Journal for Sustainability in Higher Education*, 2 (1), 48-63.

Lave, J. & Wanger, E. (1991). *Situated learning. Legitimate peripheral participation.* Cambridge, Cambridge University Press.

Merriam, S.B. (1998). *Case Study Research and case Study Applications in Education.* San Francisco: Jossey-Bass Publishers.

Walker, K., Corcoran, P.B. & Wals, A.E.J. (2004). Case Studies, Make-Your-Case Studies, and Case Stories: A Critique of Case Study Methodology in Sustainability in Higher Education. *Environmental Education Research*, 10(1).

Wals, A.E.J. & A.H. Alblas (1997). School-based research and development of environmental education: a case study. *Environmental Education Research*, 3 (3), 253-269.

Yin, R. (1993). *Applications of case study research.* Beverly Hills, CA: Sage Publishing.

Yin, R. K. (1989). *Case Study Research: Design and Methods.* Beverly Hills, CA: Sage.

BIOGRAPHY

Kim E. Walker is a Visiting Research Fellow at the University of Bath. Previously she was Visiting Research Professor at George Washington University and Academic Manager, Sustainability in Teaching and Learning at the University of Technology, Sydney. Email: Kimewalker@aol.com

For the biographies of Peter Blaze Corcoran and Arjen E.J. Wals please refer to the "About the editors" section at the end of this Volume.

CHAPTER 19

EDUCATION AND SUSTAINABLE DEVELOPMENT IN UNITED KINGDOM UNIVERSITIES: A CRITICAL EXPLORATION

William Scott & Stephen Gough

INTRODUCTION

In this chapter we critically examine attempts to introduce considerations of sustainable development into the UK higher education sector in the 11-year period since the Earth Summit in Rio. We begin by briefly reviewing significant developments in response to *Agenda 21*, Chapter 36 through an examination of government policy interventions, and external cross-institutional support offered to the sector to facilitate the integration of sustainable development issues within curricula. We provide exemplars relating to policy, programmes and strategy, and comment on their effectiveness and strategic value within the sector. We then comment critically on the work of the Higher Education Partnerships for Sustainability (HEPS) programme, in particular through its innovative Learning for Sustainable Development curriculum toolkit, and end by exploring what might well be learned from all this work. We argue that there is now a priority need for integrated and integrative leadership, both within and across sectors, which synthesises existing knowledge and best practice, and makes them available to ongoing initiatives, and a need to encourage innovation without debilitating prescription of what counts as either sustainable development or programme of education appropriate to its achievement.

It is clear that, across sectors, a significant range of activities have been initiated in response to Agenda 21 and carried out by Central Government, Local Authorities, Institutions, NGOs, and others. For a more detailed and far-ranging review than is possible here, see Reid et al. (2002). In this chapter, we focus on work carried out within higher education which is an important subset within the overall context of sustainable development and learning (Scott & Gough, 2003). We set out the agenda established at Rio (see also Hopkins et al., 1996), identify key stakeholders, identify significant initiatives in terms of policy, programmes and strategy, and then comment on the main issues and challenges we now face. The Key Stakeholders in this sector are Universities (higher education institutions – HEIs) national Government and its agencies (most notably university funding councils and research councils), Business, and NGOs.

Peter Blaze Corcoran & Arjen E.J. Wals (Editors), Higher Education and the Challenge of Sustainability: Problematics, Promise and Practice, 235-247.

DEVELOPMENTS POST-RIO

The targets set at Rio for education were ambitious. They included:

> *Develop* networks. *Develop* cross-disciplinary courses. *Build* partnerships with business, with other stakeholders, and internationally. *Establish* national and/or regional centres of excellence in research and education. *Promote* public awareness building. *Identify* workforce training needs and assess measures taken to meet them. *Encourage* professional associations to review codes of practice. *Develop* national and regional environmental labour market information systems.

In terms of what has happened, the following are prominent initiatives and activities:

1991   Government establishes expert advisory committee on environmental education in higher and further education ("the Toyne Committee") to make recommendations for improvement

Government funds a survey to produce a directory of environmental courses in higher and further education

The Economic and Social Science Research Council (ESRC) funds its £15m Global Environmental Change programme (1991-2000)

1992   Second report on This Common Inheritance published in September included a section on knowledge, education and training

1993   The Toyne Report recommends that all institutions sign up to a recognised environmental management standard and develop their students towards responsible global citizenship

1996   A review of Toyne notes only modest progress; Government asks organisations representing higher education to take action to encourage greater awareness of environmental matters

1997   Forum for the Future begins its Higher Education 21 Project (HE21) to promote examples of best sustainable practice in HE

1998   Government establishes the Sustainable Development Education Panel (SDEP) with terms of reference to identify gaps, opportunities, priorities and partnerships for action in providing sustainable development education in England across all sectors

2000   The Higher Education Funding Councils in the UK funds Forum for the Future to carry out the Higher Education Partnerships for Sustainability (HEPS) project to help higher education institutions deliver and share strategic sustainable development objectives (2000-2003). Currently 18 UK Higher Education Institutions (HEI) work within HEPS

2002   ESRC funded its Environment and Human Behaviour research programme (2002-2003)

OVERVIEW

Throughout this period there have been efforts to promote both sustainable development related research *by* the sector (e.g. Research Councils' continuing interests, with multi-disciplinary approaches now being favoured), *and* wide-ranging development work focused on sustainable development *within* HEIs (e.g. HE21 and

HEPs) although there is little evidence of much overlap between these two strands; indeed, understandably only a minority of research has actually focused on work in HEIs themselves. Furthermore, in the sustainable development related development work there has been considerable tension between attempts to influence the curriculum within HEIs, and interest in changing management practices in relation to environmental/sustainability issues, for example, in relation to transport strategy, energy management, effluent and waste disposal, water and other resource use, resource procurement, and the like. Looking dispassionately across the sector, and over time, it seems clear that there has been more interest in (and progress towards) issues of managing the environment than in curriculum change. Whether one reviews the outcomes of the HE21 initiative, the current work of HEPS, contemporary websites of UK universities (on which we comment later), or the research literature, this conclusion seems inescapable. See, for example, Ali Khan, 2002; Johnston and Buckland, 2002.

Of course, it is clear that there are two quite separate constituencies at work here: environmental managers don't deal with curriculum, and curriculum planners and academics don't *have* to think about environmental management. By and large, where there is an academic engagement with sustainable development within HEIs it will arise either from personal professional interest or, because of external accreditation requirements where a curriculum concern with *environmental* issues is now normal and routine (Hedstrom, 1996). However, whilst this seems increasingly the case, the same cannot be said of *social* issues as legislative, regulatory, accreditation or peer pressures do not yet extend here, and thus any consideration of sustainable development is likely to be partial.

All Estates Managers, however, *do* have to deal with such issues, whether they are 'interested' or not because their jobs *require* them to in an era of increasingly demanding legislation and regulation.

The traditions of personal responsibility and academic freedom militate against forced curriculum innovation in higher education. The unsubtle, determinist and somewhat hectoring approaches which have characterised some curriculum interventions have the same effect. In the HE21 initiative (1997-99) universities were presented with a series of documents which specified "what sustainability learning is required by different professions" (Ali Khan, 2002, p. 15) which they were then expected to implement. No matter how wonderful such specifications may well have been, it is unsurprising that little came of them as key stakeholders within higher education have their own (differing) perspectives on what sensibly both needs doing, and can be done.

Although the HEPs project has yet to report formally, there seems little evidence (see later) that the position in relation to curriculum innovation is any better, which makes the development of the HEPS Learning for Sustainable Development Curriculum Toolkit such a potentially interesting and valuable initiative. This is now explored in some detail.

LEARNING FOR SUSTAINABLE DEVELOPMENT: A CURRICULUM TOOLKIT

This toolkit is promoted by *Forum for the Future* in conjunction with its HEPS initiative. A number of theoretical conceptualisations underpin this work. These include:

– HEIs have three different roles in relation to three aspects of sustainable development: they are places of learning and research, businesses, and key community players, in each case in relation to the environment, to society, and to the economy;

– the 'triple bottom line' view of sustainable development (requiring simultaneous improvements in the environment, the economy and society) is seen as useful but over-simplistic. A more complex formulation is adopted with five kinds of capital: natural, human, social, manufactured and financial;

– twelve criteria (linked with the five-fold capital) are adduced to encapsulate a sustainable society. These relate to: non-renewable resource extraction; manufacture and use of artificial substances; the integrity of the ecological system; human health; learning and social skills; employment, creativity and recreation; governance and justice; positive values and social cohesion; positive (for both environment and people) institutional change; safe and supportive living and working environments; resource-use efficiency and the promotion of human innovation; and, accurate valuation of all forms of capital;

– to be compatible with the promotion of sustainable development, learning has to take place in a particular way; it must be learner-focused; 'holistic', draw together economic, environmental and social strands; be compatible both with the physical learning environment and with the socio-economic characteristics of learners; be applicable at a range of degrees of complexity; and be focused on identified learning outcomes.

The toolkit itself has been developed from work done by *Forum for the Future* with the University of Antofagasta in Chile. It provides an ambitious methodology intending to inform learning activities in short courses, in whole degrees and outwith higher education, for example, in business.

There is a seven-stage approach. First, a 'learner profile' is drawn up in order to map the world from the learner's perspective (see Figure 2 in Johnston & Buckland (2002, p. 17) for an example of one of these). The 'learner', e.g. a particular type of professional or graduate, and the organisations or environmental aspects with which the learner interacts most are placed near the centre of the map. Those with whom interaction is infrequent and/or weaker are placed further out. Secondly, prospective course content is identified by listing the knowledge and skills necessary to manage each interaction in a way that is consistent with sustainable development. The categories, 'ecological', 'social', and 'economic' are used to organise these lists, but it is stressed that interesting entries are likely to span categories. Thirdly, identified knowledge and skills are scored in terms of their ability to contribute to the twelve criteria of a sustainable society. This enables, fourthly, the specification of desired learning outcomes and, fifthly, the design of delivery mechanisms. The sixth stage is a 'values audit', designed to check whether the course, as now designed, is compatible with the values of staff and students. Following this, a course guide can finally be prepared.

This approach recognises the significance of the contexts in which learning takes place, and is a determined attempt to generate practical progress towards sustainable development, undaunted by difficulties of definition of terms or institutional inertia. See www.forumforthefuture.org.uk and www.heps.org.uk and, Johnston and Buckland (2002) for further detail.

The toolkit raises a number of issues. Emphasis has been given to thinking originating in economics and in natural science: thus its approach to learning is more managerialist than emancipatory (compare e.g. Huckle, 1993; and, Kemmis & Fitzclarence, 1986). This is evident in:

- the target population for learning, i.e. graduates and professionals generally;
- the emphasis on measurable learning outcomes informed by the 'twelve features of a sustainable society', rather than a pedagogy of individual and collective self-discovery;
- sustainable development is ultimately about what happens to the five kinds of capital. For an alternative approach, however, (UNESCO's 'Teaching and learning for a sustainable future' programme' see:
- http://www.unesco.org/education/tlsf/ this is ultimately "about the process of learning to make decisions that consider the long-term economy, ecology and equity of all communities";
- most fundamentally perhaps, learning is seen as instrumental to the achievement of sustainable development, rather than being, of itself, a vital and substantial *aspect* of any ongoing process of sustainable development.

The toolkit maps the context of the learner, and includes a values audit to keep course development on track. However, at least three important issues remain:

- elaborate (and expensive) pedagogies aren't always necessary. Where people want to act sustainably, but cannot do so because they lack knowledge or skills, simple information provision may well suffice. (The converse holds equally well – simple information provision will often not suffice, since learners may be indifferent or actively disinclined towards sustainable behaviour);
- people do not learn things just because educators think them important. Learners bring important knowledge, values and skills to the learning process, and these are productively supplemented through external inputs. The question is, how far to privilege the prior knowledge of the learner, and how far that of the external expert and/or educator? *The toolkit* sets out a clear 'expert' element which is considered to be beyond the scope of negotiation in that the toolkit ranks knowledge and sets learning objectives in relation to its pre-specified 'twelve features of a sustainable society';
- any strategy for social change needs to take account of learning which happens incidentally, and independently, of teaching programmes – and sometimes despite them.

O'Riordan (1989) makes a distinction between two world views, the one conservative and nurturing, the other radical and manipulative. The toolkit and most other examples we could have chosen for comparison here, predominantly exhibit the former. Cultural theory, however, (James & Thompson, 1989; Schwarz & Thompson, 1990) posits mutually interdependent rationalities: the hierarchical; the egalitarian; the individualistic and the fatalistic. A glance at both the social ambitions and the pedagogies of the toolkit show that they are overwhelmingly inclined towards an egalitarian view, in which things make sense if they are *fair* and *just*. However, in all societies there will continue to be conflicts between private and collective interests, between local and national priorities (Blaikie & Brookfield, 1987, p. 83), and between now and the future (Dobson, 2003; Greenall Gough, 1993; Pearce & Kerry Turner, 1990).

In the toolkit other worldviews and rationalities are insufficiently explicit. A sustainable world will not only be a world of justice and collaboration because no such world is possible. For example, when the eighth 'feature of a sustainable society' requires that: "The structures and institutions of society promote stewardship of natural resources and development of people", this cannot be done without losers being created who will, whatever the curriculum may tell them, be unlikely to be pleased. Such issues cannot be wished, legislated, or educated away. Further, even if the achievement of "trusted and accessible systems of governance and justice" is fundamental to sustainable development, as the toolkit tells us it is, then it is clearly *not* fundamental to being a professional (say, an engineer). Indeed, one might argue that good governance is a prerequisite for socially and environmentally responsible engineering, rather than the other way round. The toolkit recognises this problem, offering a number of processes to prioritise course content in Higher Education so that it reflects the values of students and institutions: however, the more it does this the less it challenges those values and, in particular, the less likely it is to bring about learning.

It is important to remember that many other institutions, both organisational and cultural, have a bearing on the success or failure of initiatives. HEIs have external responsibilities to business organisations and to Research Councils which are to a greater or lesser extent – but increasingly – mediated through market mechanisms. If a University's graduates cannot find work then new students will be less likely to come. If research grants are not won then the research effort will falter and funding for research will fall. Of course it is true that many businesses have some sort of policy relating to sustainable development, and may take account of it when recruiting staff. Similarly, research councils increasingly recognise environmental issues as a theme of research; but in both cases the relationship of sustainable development to learning may be poorly articulated, and at the margin other things are quite likely to be deemed more important. HEIs have no choice but to respond to this ordering of priorities. They must also respond to the interests of internal stakeholder organisations such as their governing councils, which may or may not place sustainable development high on their lists of priorities as far as the academic curriculum is concerned. However, as we noted earlier, the picture is more positive in relation to the management of HEIs' physical resources, where potential savings from sustainable resource procurement or energy saving measures are increasingly unlikely to be ignored, just as conforming to existing legislative frameworks in relation to pollutants will remain a priority. It is fair to say that the toolkit recognises and seeks to address many of these difficulties. The toolkit is a sophisticated attempt to address this problem of specialisation: but the problem still remains that someone can only be an engineer (or a forester, economist, project manager, or any other kind of professional) with a cross-disciplinary concern for sustainable development if they *are first* an engineer, that is, they have met the assessment requirements within those accredited programmes considered fundamental for admission to that profession. Hence, any focus on sustainable development remains essentially secondary.

EVIDENCE FROM PRACTICE

What follows here is a brief survey of 18 UK HEIs. The websites of these institutions were surveyed on 13 January 2003. Searches were made using the sites' internal search engines for the terms 'sustainable development' and 'sustainability'. A search was also made for 'HEPS', and 'Higher Education Partnerships for Sustainability'. All these sites ranked hits in order of relevance. Where there were multiple hits, the first 30 were examined. The results are summarised in Table 1 where institutions that are part of the HEPs programme are marked *.

Table 1. Survey of UK Higher Education Institutions (*HEPS members) conducted on 13 January 2003.

HEI	Comments
1 *	228 hits for 'sustainable development'. 278 hits for 'sustainability'. NB total number of hits not consistent on subsequent tries! 'Environmental Policy Statement' makes no reference to 'curriculum' or 'learning'. 'Centre for Environmental Research and Training' offers much expertise in a wide range of sustainable-development related areas. Much other specialist research relating to sustainable development in progress. Short text describing HEPS.
2	548 hits for 'sustainable development'. 218 hits for 'sustainability'. Wide range of activity in specialist fields at the local, UK and European levels. Some energy and environment and sustainable procurement figures for the university. Detailed information on HEPS even though the University is not one of the partner institutions.
3 *	Number of hits not specified. More than 30 for both sustainable development and sustainability. Many links to papers, presentations, and courses of a specialised nature. Link to the 'Environmental Action Network' which considers both curriculum and management issues and includes senior administrators. No mention of HEPS here.
4	4787 hits, but none of these appear to be for the phrase 'sustainable development'; mostly being references to 'development'. 211 hits for 'sustainability'. These mainly relate to courses which include some relevant aspect. No reference to HEPS.

5 *	528 hits for 'sustainable development', dominated by the 'Centre for Sustainable Development'. 415 hits for 'sustainability'. These relate mainly to courses and to a lesser extent to research. 2 hits for HEPS give details of the project – but are hard to find.
6	624 hits for 'sustainable development'. One of these leads to the 'Environmental Change Institute'. Others link to courses or research and are often marginal. One leads to the 'Environmental Sustainability Research Cluster'. Again, others seem more marginal. Nothing on HEPS.
7 *	10 hits for 'sustainable development' relating to courses, research and a studentship. 8 hits for sustainability, of a similar nature. No reference to HEPS.
8	More than 200 hits for 'sustainable development'. These mostly refer to individual course or research web pages. 176 hits for 'sustainability' of which nothing was more recent than January 2002. No hits for HEPS.
9 *	3 hits for 'sustainable development' relating to professional studies and a master's level programme. No hits for 'sustainability'. No hits for HEPS.
10	101 hits for 'sustainable development'. These are mostly course information, with some research sites and one reference to energy management policy. 56 hits for 'sustainability' of a similar kind. Nothing on HEPS.
11 *	687 hits for 'sustainable development'. These relate to courses and research, particularly in relation to sustainable rural development. 297 hits for 'sustainability', of a broadly similar kind, though more publications are mentioned. Brief mention of HEPS.
12	73 hits for 'sustainable development'. Mostly these are poorly matched to the whole phrase sustainable development. A link to the 'Centre for the Study of Environmental Change and Sustainability' was last updated in September 2000. More recent links the same Centre was found by search for 'sustainability' (162 hits). Nothing found on HEPS.
13 *	306 hits for 'sustainable development'. These are mostly information about courses in environmental science. 85 hits for 'sustainability'. Again, many of these are poorly matched. No hits for HEPS.
14	324 hits for 'sustainable development'. These are mostly papers or details of teaching. There is an interesting link to the 'Institute for Health Research'. 298 hits for 'sustainability'. Several links to the 'Institute for Environment, Philosophy and Public Policy'. Access to papers and research project details. No hits for HEPS.
15 *	44 hits for 'sustainable development', but poorly matched to sustainable development *per se*.. 32 hits for 'sustainability', of broadly the same kind. No hits for HEPS.
16	225 hits for 'sustainable development'. The first of these is a paper on the importance of sustainable development in higher education curriculum development. Also links to the 'Centre for Development Studies', and to papers on 'what is a sustainable university'. 138 hits for 'sustainability', of a similar nature. Two hits giving details of HEPS.
17 *	154 hits for 'sustainable development', including current information on sustainable development in Wales and the University's role in it. 271 hits for 'sustainability, including links to Cardiff Business School, as well as course and research. No hits for HEPS.
18	773 hits for 'sustainable development', including several to the 'Overseas Development Group'. 394 hits for 'sustainability', including those to the 'Centre for Social and Economic Research on the Global Environment'. No hits for HEPS.

DISCUSSION

One needs to be extremely cautious about the degree of significance attached to a limited survey of a small sample of this kind, particularly when the results depend in great degree upon the workings of each particular university's internal search engine. Nevertheless, one may say that if universities' websites represent the view they want the world to have of them, then by and large sustainable development is a minor constituent of that image. That said, there is clearly a great deal of detailed research and teaching taking place which relates to sustainable development, or aspects of it. There is also evidence that some institutions take sustainable development seriously as a driver of their internal management processes. There is no sense, however, that sustainable development has yet become a major strategic parameter of university life which links the business, learning, research and community functions of institutions in the way envisaged by HEPs, UNESCO and other sustainable development initiatives. This is not a case against these initiatives, merely a glimpse of the scale of the challenge they face.

It is important to view developments in higher education in the wider national context, and to acknowledge that sectors are not free-standing; thus what is possible and what is happening in higher education is affected by, and affects, what is happening and is possible across the sector.

In their review of progress to date, Reid et al., (2002) noted:

> A plethora of initiatives, however, does not amount to a national strategy, and many disparate initiatives, valuable in themselves, have not yet been linked to advantage, locally, nationally and/or farther afield. Often this is because teachers, local government officers, NGO employees, and others have lacked the understanding and/or the infrastructural support to realize such integration, particularly at the interface of bottom-up with top-down approaches.

Reid et al. went on to cite as examples of this, the development of unconnected life-long learning and sustainable development initiatives in some local authorities, and the lack of integration between school curriculum development and LA21, and noted that work related to sustainable development continues to be seen as a costly bolt-on to existing programmes, rather than as a means and opportunity better to achieve existing goals.

This last point applies with some force to the higher education sector, and there is a further concern. Sophisticated initiatives such as HEPs stand a chance of effecting integrated curriculum change because of the quality of the argument, buttressed as it is by sturdy (but not impregnable) theoretical foundations. But if you don't buy into the idea, perhaps because you're not persuaded by the seven-stage approach, the twelve criteria or the five-fold capital model, or perhaps because you are pursuing these issues after your own fashion, where does this leave you? Are you to be counted amongst those who are working to help students to be "both capable and willing to accelerate change to a sustainability society" (Johnston & Buckland, 2002, p. 16), or not? The confidence and assertiveness of HEPS, rather like HE21 before it, might lead the unsure to believe that here is a definitive word on sustainable development within the higher education sector.

Nothing, of course, could be farther from the case. What we all come to know about sustainable development can only develop through our practice, our sharing of this, and the learning that we shall do as a result of both of these. It is because of this that what individuals and groups in HEIs do, jointly and severally, remains of utmost importance. Our review suggests that there is now a priority need for, firstly, integrated and integrative leadership, within higher education and between this and other sectors, which synthesises existing knowledge and best practice, and makes them available to ongoing initiatives, and, secondly, a need to encourage innovation and a culture of open sharing of what has been done and learned.

Such leadership and innovation might well include the following among its priorities:
- commissioning research, particularly into [i] the mainstreaming of sustainable development issues into learning and [ii] the actual and potential relationship between sustainable development and [life-long] learning;
- better use of existing research; long-term cross-sector strategic planning; development through education of transferable skills and flexibility;
- cross-sector monitoring and evaluation of initiatives and learning in sustainable development;
- networking of practitioners in order critically to examine effective practice;
- promotion of, and leadership contributions to, international developments.

Such leadership and innovation might also bear in mind:

There are dangers in being over-prescriptive about what counts either as sustainable development, or as learning that contributes to it. As nobody really knows what sustainable development will turn out to entail, there is considerable merit in encouraging institutions, groups and individuals to explore what they are interested in, and then to come together and share and analyse what emerges. Maintaining such collaborative processes, and keeping in touch, are crucial if professional and institutional development is to be optimised.

Sustainable development, if it ever happens, will be a process in which everyone learns all the time. Its cause is unlikely to be advanced by any group which simply asserts its right and authority to teach others without learning itself. Aiding collaborators do what *they* want to do more effectively will be more helpful that telling them they should really be doing something else. Respecting the varied institutional and professional contexts of collaborators not only recognises that they have unique contextual insights and strategic understandings, but also have on-going institutional commitments that also demand much of their attention, and which mean that progress in relation to sustainable development will be heavily contingent.

Encouraging, facilitating and supporting sequences of small steps may well be more productive ways of exploring issues and gaining confidence than being faced with the imperative of taking giant leaps on the grounds for example, that development is urgent. It is particularly important that experience and confidence be achieved at a pace that makes sense to all stakeholders. As problems, failures and disappointments are to be expected, the priority must be for all to share, and learn from them.

CONCLUSIONS

The World Summit on Sustainable Development made only two references to learning: [i] a re-affirmation of the internationally-agreed Millennium Development Goals of universal primary education by 2015 and the elimination of gender disparities in primary and secondary education by 2005, and [ii] confirmation of civil society's crucial role in sustainable development – with the need for the public to be (at the very least) aware of this. None of this is controversial or new. Indeed, from 1998 there has been an emerging focus on learning in the UN when the 6th session of the UN Commission on Sustainable Development (UNCSD) agreed seven priority areas for action which were all confirmed by the Summit:

1. Clarifying and communicating sustainable development concepts and key messages.
2. Reviewing national education policies and reorienting practices, including teacher education and higher education teaching and research.
3. Incorporating education within national sustainable development strategies and planning processes.
4. Promoting sustainable consumption and production through education.
5. Promoting investment in education.
6. Identifying and sharing innovative practices.
7. Raising public awareness.

The need for learning seems implicit in all of these, especially in the second, and the fact that the UN General Assembly has designated 2005-14 as the 'decade of education for sustainable development', further strengthens higher education's role. It is also hard to see any of the UN's goals being achieved without a great deal of learning across all sectors – arising from experience, from work in schools, colleges and universities, from training, and through professional and institutional development. The Summit, however, merely gives learning a supporting role in relation to a wide range of issues relating to development, health, good governance, trade, and environment (ie, to sustainable development). But this supporting role for learning is unhelpful as the approach assumes that learning is only important *after* experts have decided what – in terms of development, health, good governance, trade, environment, and so on – should be done or learned. This recourse to external expertise is unsatisfactory for a number of reasons.

Teachers in universities know that their job is to promote learning *by* their students, rather than to promote sustainable development, and may well resent being told that their priorities ought to be otherwise. Thus, if sustainable development does require learning, then *learning goals* must be a fundamental part of it. Environmental, and other goals to be achieved *through* learning that experts with no contextual authority deem important, will not do by themselves. As it is often not possible to know what needs to be done, and even where it is possible to say what needs to happen from a particular perspective (development say) and/or a particular discipline (economics, say) such perspectives (or disciplines) are not necessarily congruent. Under these circumstances, seeking to promote learning seems the only sensible way forward.

The divisions which we routinely make of knowledge into disciplines, of policy-making into ministries, and of sustainable development into economic, environmental and social components – whilst useful and necessary ways by which complex entities are made manageable – remain simplifications. For example, the economy, the environment and society are *not* separable, and sustainable development cannot arise from the independent insights of economists, environmental scientists and social scientists, working with different assumptions and methodologies. Learning is required across the institutions they represent, the constituencies they serve, and the literacies they employ. Without this, there will be no sustainable development.

Lastly, how people view the world, and what they do in relation to this, matters because it makes a difference to how things turn out; that is, to how the human-environment relationship co-evolves. And what people learn, matters because it informs and enables what we can do next. Reassuringly, what people learn isn't always what others try to teach, which is why people, and what they learn, are crucial to (and for) sustainable development. Such factors are fundamental to HEIs being able to think about sustainable development, and for progress to be made in ways that make contextual and cultural sense to them.

REFERENCES

Ali Khan S. (2002). Sustainable development education in the UK: the challenge for higher education institutions. *Planet.* December 2002 15.

Blaikie, P. and Brookfield, H. (1987). *Land Degradation and Society.* London: Methuen

Dobson, A. (2003). *Economic Behaviour: Value and Values.* In W.A.H. Scott & S.R. Gough (Eds.) *Key issues in sustainable development and learning: a critical review.* London: RoutledgeFalmer.

Greenall Gough, A. (1993). *Founders in Environmental Education.* Geelong: Deakin University Press.

Hedstrom, G.S. (1996). Foreword, In W Wehrmeyer (Ed.) *Greening People: Human Resources and Environmental Management.* Sheffield: Greenleaf, pp. 9-10.

Hopkins, C., Damlamian, J. and López Ospina, G. (1996). Evolving towards education for sustainable development: an international perspective. *Nature and Resources,* 32(3) 36-45.

Huckle, J, (1993). Environmental education and sustainability: a view from critical theory. In J. Fien (Ed.) *Environmental Education: A Pathway to Sustainability.* Geelong: Deakin University Press, pp. 43-68.

James, P. and Thompson, M. (1989). The plural rationality approach. In: J. Brown (Ed.) *Environmental Threats: perception, analysis and management.* London: Belhaven Press, pp. 87-94.

Johnston, A. and Buckland, H. (2002). How can higher education produce graduates with the capacity to accelerate change towards a more sustainable society? *Planet,* December 2002, pp. 16-17.

Kemmis, S. and Fitzclarence, L. (1986). *Curriculum Theorizing: Beyond Reproduction Theory.* Geelong: Deakin University Press.

O'Riordan, T. (1989). The challenge for environmentalism. In: R. Peet and N. Thrift (Eds.) *New Models in Geography.* London: Unwin Hyman, pp. 77-102.

Pearce, D. and Kerry Turner, R. (1990). *Economics of Natural Resources and the Environment.* Hemel Hempstead: Harvester Wheatsheaf.

Reid, A.D., Scott, W.A.H. and Gough, S. (2002). Education and Sustainable Development in the UK: an exploration of progress since Rio. *Geography,* 87(3), 247-255.

Schwarz, M. and Thompson, M. (1990). *Divided We Stand: Redefining politics, technology and social choice.* Philadelphia: University of Pennsylvania Press.

Scott, W.A.H. and Gough, S.R. (2003). *Sustainable development and learning: Framing the issues.* London/New York: Routledge Falmer.

BIOGRAPHIES

William Scott is Professor of Education at the University of Bath where he directs the Centre for Research in Education and the Environment. He edits the international refereed academic journals: Environmental Education Research, and Assessment and Evaluation in Higher Education, is a Fellow of the Royal Society of Arts, and a member of the Research Commission of the North American Association for Environmental Education. He is a Fellow of the Royal Society of Arts, a Trustee of the Living Earth Foundation, and works extensively with local and national NGOs with interests in environmental, conservation and sustainability issues. He is particularly interested in the role of learning within sustainable development, in the contributions that teachers and institutions can make to this, and in the problems of researching (and evaluating the worth of) such activities. For more details about the research Centre please go to: http://www.bath.ac.uk/cree/ (e/mail cree@bath.ac.uk).

Stephen Gough is co-author (with William Scott) of *Sustainable Development and Learning: framing the issues* (published by RoutledgeFalmer in August 2003) and co-editor (also with Professor Scott) of the companion volume *Key Issues in Sustainable Development and Learning: a critical review*, in which leading authorities from across the social sciences comment upon key issues for sustainable development. He is Senior Lecturer and Director of Studies for Advanced Courses in the Department of Education at the University of Bath, England.

CHAPTER 20

LIGHTING MANY FIRES: SOUTH CAROLINA'S SUSTAINABLE UNIVERSITIES INITIATIVE

Wynn Calder & Rick Clugston

INTRODUCTION

Reorienting colleges and universities toward sustainability is difficult, given economic and disciplinary realities. For higher education to make a significant contribution to a major societal challenge, the issue must be clearly recognized, backed by sufficient external funding, and commanding of academic prestige (Bok, 1990). Since the early 1990s, education for sustainable development has been under-funded and under-supported, both within and outside of the academy. Sustainable development is not a recognized societal priority. Traditional disciplines still view sustainability with suspicion and external funding rarely supports related research and teaching. At a small minority of schools across the United States, highly committed presidents, faculty members, staff, and students have made sustainability a priority in major dimensions of campus life. At a larger minority of schools, eco-efficiency in operations[1] or new courses in environmental studies are present. Evidence of an authentic institutional commitment to sustainable development in the USA, however, is rare (Calder & Clugston, 2002).

Public research universities in the USA are particularly driven by the search for major grants and academic prestige. Thus for the three public research universities of the state of South Carolina to embark in 1999 on a five-year course toward sustainability was to swim upstream against powerful currents in the disciplines, the economy, and the conservative culture of South Carolina. The principal investigators and manager of this Sustainable Universities Initiative (SUI) now acknowledge that their expectations for what could be accomplished were unrealistically high. SUI, which set out to accomplish significant reforms in teaching, research, operations, student life, and outreach, adapted its strategy over time in order to "light many fires and see which burned brightest" (Sustainable Universities Initiative, 2002, p. 1).

The Sustainable Universities Initiative, involving the University of South Carolina, Clemson University, and Medical University of South Carolina, was funded by a Danish foundation. The foundation engaged the Association of University Leaders for a Sustainable Future to conduct an on-going formative

---

[1] Refers to physical plant operations, including energy, water, air, transportation, etc.

Peter Blaze Corcoran & Arjen E.J. Wals (Editors), Higher Education and the Challenge of Sustainability: Problematics, Promise and Practice, 249-262.

external evaluation of SUI.[2] In this case study, written by two members of the evaluation team, we explore the SUI experience to illuminate its implications for those higher education institutions striving to make sustainability more central to the academic and operational functioning of a university.

THE SUSTAINABLE UNIVERSITIES INITIATIVE

The Sustainable Universities Initiative emerged out of a dialogue between representatives of the three universities, outside consultants, and the trustees of the foundation. In addition to funding higher education for sustainability initiatives, the foundation has a special commitment to South Carolina,[3] and funding SUI combined these two central commitments.

The University of South Carolina (USC), located in the state's central capital city, Columbia, supports programs of study ranging from liberal arts and sciences to business, law, and medicine. USC enrolls over 37,000 students and employs nearly two thousand faculty. Clemson University is located in northern South Carolina surrounded by 17,000 acres of University farms and woodlands devoted to research in forestry and agriculture. Clemson has 70 fields of undergraduate and graduate study, serves 17,000 students and supports nearly one thousand faculty. Medical University of South Carolina (MUSC), located in Charleston on the southern coast of the state, has a medical center and educates health professionals, biomedical scientists, and other health related personnel. MUSC enrolls 2,300 students, mostly graduate and professional, and supports about five hundred faculty. Together these three universities educate about sixty percent of all those in higher education in the state.

The foundation gave the nascent SUI a planning grant in 1998 to develop a funding proposal, which was reviewed by internal and external stakeholders. To demonstrate high-level support for the project during this planning process, the three university presidents signed a declaration in which they pledged:

> We therefore singly and collectively commit to:
>
> 1. Fostering in our students, faculty and staff an understanding of the relationships among the natural and man-made environment, economics and society as a whole.
> 2. Encouraging students, faculty and staff to accept individual and collective responsibility for the environment in which they live and work.
> 3. Serving as a center of information exchange for other institutions within the state.
>
> 4. Operating existing facilities and constructing new facilities to maximize efficiency and minimize waste, thereby protecting the environment and conserving resources." (SUI, 1998)

---

[2] Based in Washington, DC, the University Leaders for a Sustainable Future (ULSF) external evaluation team has worked closely with SUI representatives since 1999 in a formative role. We have conducted site visits at the three institutions, frequently visiting the University of South Carolina, where the SUI administrative office is located. The single greatest challenge for the evaluation process has been to measure success (using barometers such as the Five Year Plan) while honoring an SUI philosophy of seeking champions and nurturing them, and pursuing opportunities as they emerge.

[3] South Carolina is the location of the U.S. headquarters and manufacturing facilities of the company whose founder created the foundation.

The Sustainable Universities Initiative was officially born in April 1999 with a $4.5 million US dollar grant to fund an ambitious Five Year Plan (1998). The plan and proposal were developed by a principal investigator on each campus and a Steering Committee representing the three campuses and external constituents. A full-time manager was hired to oversee the initiative and carry out the Five Year Plan, which includes the following vision statement:

> The primary focus of our efforts, our strategy, is to change the products of our institutions, and ultimately the state, by working with faculty to expand their teaching and research agendas, and with administrators and operations managers to ensure that our institutions are practicing what the faculty are preaching. The Sustainable Universities Initiative will serve as a catalyst for activities which will make the state's three research universities, other education institutions, and ultimately, the state as a whole, more sustainable. It will also result in a new model for multi-disciplinary and multi-institutional cooperation within South Carolina's higher education community. Finally, it is hoped it will serve as a model for other state assisted colleges and universities nationwide. (Sustainable Universities Initiative, 1998, p. 3)

The Five Year Plan describes a range of activities that the three institutions would conduct in four major areas: 1) Effect Change within Faculty; 2) Student and Community Programs; 3) Campus Operations; and 4) Share Information / Manage Program. Each area was allotted a different percentage of the total funding, but since these areas overlap in practice, it is difficult to separate designated funds with precision. In addition, the universities have contributed both monetary and in-kind support for various programs over the years. The Five Year Plan served initially as a benchmark to measure progress. For each specific area of activity, the Plan laid out quantitative targets listed as "measures of success." As the initiative evolved and took its own course, however, SUI representatives relied on the Plan less and less as a guide to action.

IMPLEMENTING THE FIVE YEAR PLAN

This section describe the efforts of SUI to realize the Five Year Plan's stated objectives in its work with faculty, students, and operations staff. The examples are chosen to highlight both the unanticipated obstacles and opportunities encountered by SUI in its first four years, and to illuminate strategies for advancing sustainability in higher education.[4]

Effect Change within Faculty

The first major area of activity, "Effect Change within Faculty," was to receive about fifty percent of the grant funds. This area concentrates primarily on providing "mini-grants" for new course development and course modules with a focus on sustainability within existing disciplines and programs, and for new and innovative research that is multi-disciplinary and eventually multi-institutional. This area commits to hiring a critical faculty position on each of the three campuses, hosting faculty workshops for infusing sustainability into the curriculum, and sending

---

[4] At the time of this writing, SUI is in the middle of its fifth year.

faculty to relevant professional conferences. This area also supports projects to enhance the first year academic experience, a speaker series to bring outside experts on sustainability related issues to campus, and faculty/student exchanges between participating campuses. Here we describe some examples of the use of mini-grants to foster new research and course development for sustainability.

SUI mini-grants, ranging from $3,000 to $10,000 US dollars, have been a major vehicle for deepening faculty involvement in SUI. A Steering Committee has convened each spring to identify those proposals that, at a modest level of funding, could actually advance sustainability issues. Given the structure of research universities, however, faculty typically cannot respond to initiatives that give so little money and ask them to do so much. Thus the approach of SUI became one of searching for the committed champions and concentrating the funding on their projects, reinforcing what they were already seeking to accomplish but rarely convincing the unconverted to get on board. In rare cases, mini-grants have provided just enough incentive to move an interested but not active faculty member into the ranks of the committed.

The challenge of changing faculty behavior with small financial incentives was highlighted early on at MUSC. At the time that SUI began, each division was increasingly expected to function as a profit center. Thus writing and executing a mini-grant for less than $10,000 US dollars would hardly be justifiable if one could get a National Institute of Health grant for hundreds of thousands of US dollars or earn $30,000 performing and eye operation.

Given MUSC's highly structured degree programs, with tightly controlled exam requirements, a new course in environmental health is very difficult to create. In the first year of the initiative, however, SUI representatives recognized an opportunity to support a young pediatrics faculty member in adding an environmental health component to his introductory course. He is now a nationally recognized expert in children's environmental health. This example highlights SUI's 'lighting many fires' strategy. In a context where faculty members were expected to bring in major research funding, SUI could provide enough resources to strengthen one key faculty member's commitment to environmental issues. With such limited opportunities in research and course development at MUSC, SUI began in the second year of the initiative to consolidate the University's allotment of funds into operations oriented projects such as an experiment with waterless urinals in a few locations, improvements in recycling, and a popular worm composting project.[5]

SUI has had much greater success in effecting faculty development (and students through their class work) at USC and Clemson by influencing established courses and supporting faculty members in creating new courses. The initiative hoped to have a significant effect, for example, by incorporating a sustainability component in all University 101 classes (a general orientation to university life course required for ninety percent of first year students) (Five Year Plan, p. 10). While this goal proved unrealistic, SUI graduate student interns at USC developed and started giving an interactive "Educating for Sustainable Living" presentation in U101 classes that asks students to calculate their own ecological footprints and discusses

---

[5] Information on this and other recycling efforts at MUSC can be seen at http://www.musc.edu/recycle/.

population growth and sustainability. By year 4 (2002) of the initiative, the interns were reaching about 500 out of 3,000 students.

Mini-grants for course development in English at USC and Horticulture at Clemson have been among the most productive of the Sustainable Universities Initiative. In the second year, SUI representatives found an unanticipated opportunity in USC's English 101 course (an introductory literature and writing class required for ninety percent of first year students). Several instructors (mainly PhD students) have incorporated an environmental theme into nine sections of English 101, thereby exposing nearly ten percent of the freshman class to these topics annually and giving these future professors invaluable experience in teaching environmental literacy and sustainability. In addition, all of the environmental sections include a community service component, which requires students to work and learn in the community through affiliation with local organizations. This exposes students to the social and economic dimensions of sustainability.

A horticulture professor at Clemson became an early SUI champion. One of her mini-grant projects in the first year involved designing a "sustainable landscape" at a local Habitat for Humanity community. Building on this work, she began research on primary schoolyard habitats and outdoor classrooms as models of sustainability in conjunction with a professor in the Department of English. In the following years, other Clemson faculty in English became interested in teaching about sustainability through writing. The Horticulture professor, working with various colleagues, has received over $40,000 US dollars in six separate mini-grants for both individual and joint projects over the first four years of the initiative.

The examples in this section illustrate that any sustainability initiative must be adaptive. Significant numbers of students and graduate instructors are learning about sustainability through English at USC. SUI's approach with Clemson's horticulture professor is reaching numerous faculty in English and other departments because this champion is respected and oriented toward interdisciplinary teaching and research.

Student and Community Programs

SUI's second major area of focus, "Student and Community Programs," was to receive about five percent of the grant funds. This modest allotment was decided on the advice of consultants early in the SUI planning process and based on the assumption that impact on student awareness and behaviors is best achieved through faculty and course development. The primary activities in this area include providing information to new students as part of orientation; fostering environmental leadership within student groups; supporting innovative student projects on campus and in the community; and fostering local and regional business, government, and community networks for sustainability. Given that these activities typically involve care for the environment, recycling, and an emphasis on energy savings and water conservation, this area of focus overlaps significantly with that of campus operations (see below). Here we discuss an innovative approach to student attitudes and practices through university housing at USC.

Early efforts were made by SUI to provide information on consumption habits, recycling, and waste reduction to first year students during orientation, but the effects were minimal. SUI representatives quickly concluded that "provision of information about sustainability and environmental action on campus (which we perceived as a service to incoming students) is perceived as an unwelcome distraction by the students themselves... We also have found students to be less interested in sustainability-related projects than we anticipated" (Sustainable Universities Initiative, 2002, p. 16). This disappointment took a positive turn in the third year of the initiative when the Director of Student Development and University Housing, who became a champion for sustainability through his connections with SUI, hired an Environmental Manager for Housing with SUI support. The manager is now addressing the issue of student awareness and responsibility through housing. Instead of simply providing information to incoming students himself, he includes the responsibility for student orientation and environmental stewardship in the contracts of his student Resident Assistants. Thus for academic year 2002/03, he required of RA's that they hold a certain number of meetings on environmental issues with their students, distribute information and hang signs, monitor recycling practices, and so forth. This is proving to be a successful approach to changing student behaviors.

Through additional projects, the Director of Student Development and University Housing and the Environmental Manager are striving to make University Housing a "socially instructive steward of the environment" (Luna and Koman, p. 3). New "green" practices in Housing include high efficiency washers and dryers, electric vehicles for operations staff, flat screen LCD (liquid crystal display) computer monitors for administrative offices and student labs, and a recycle/reuse "move out" program to handle the waste and useable items that 7,000 students discard when they leave campus each spring. The Environmental Manager claims that Housing's sustainability efforts are reaching nearly all of the 35-40% of students who live on campus (ULSF Site Visit Report, February 2003).

The two champions at USC are proving that this area of university life is a critical leverage point for reaching students by making it part of the culture and expectations of dormitory life. Rather than trying to cultivate student interest when they arrive on campus, the officers are embedding sustainable behavior in the structure and practices of the residence halls.

Campus Operations

SUI's third area of focus, "Campus Operations," was earmarked for about twenty percent of the grant funds. This area concentrates on fostering sustainable campus operations through development of a comprehensive Environmental Management System (EMS) on each campus; supporting student interns and special projects on such issues as recycling, energy conservation education, public transportation options, and campus bicycle use; and providing fifty percent matching funds for small scale operations projects, such as recycling at Clemson football games and the installation of energy saving LED (light emitting diode) lights in MUSC buildings.

Here we discuss SUI's attempt to establish an EMS at USC and the unanticipated emergence of sustainable building projects on all three campuses.

Major SUI activity in operations has centered on the development and implementation of an EMS at each of the three research universities (as well as at other public South Carolina colleges and universities after year 3). SUI representatives envisioned that these EMSs would cover all aspects of operations, thus penetrating institutional barriers between operational departments (such as Housing, Building and Construction, and Facilities at USC) and ensuring new levels of efficiency and accountability.

It was not until year 3 of the initiative that an external consultant on EMSs was brought to the USC campus who caught the attention of the official charged with risk management. The risk manager then pushed to develop an EMS for his department, Environment, Health and Safety (EH&S). EH&S is concerned foremost with regulatory compliance and was to some extent motivated by major fines that have been levied at other research universities by the U.S. Environmental Protection Agency since the late 1990s. This EMS received ISO 14001 certification in August 2002 and EH&S staff have reported they are more efficient, accountable, and effective. University Housing is completing its own EMS mid-way through year 5 of the initiative and USC Facilities (which includes general maintenance, indoor and custodial services, energy services, and grounds management) is beginning the process of designing theirs as well.

SUI representatives did not anticipate the technical and logistical difficulties of creating a comprehensive EMS. Numerous European institutions have established EMSs in campus operations, and only one U.S. institution, the University of Missouri-Rolla, claims to have implemented a university wide EMS.[6] Today, a more realistic EMS process is evolving at USC, one department at a time. A major concern, however, is that integrating these now distinct EMSs down the line may prove more difficult than anticipated. For Clemson and MUSC, EMS development is just beginning.

In contrast to the trials of establishing EMSs on SUI campuses, sustainable design initiatives have become some of the brightest fires at all three schools. Ideas gleaned from a SUI-sponsored green building conference in year 3 and a growing trend in sustainable design on campuses in the USA encouraged the Student Development and University Housing Director and Environmental Program Manager to push successfully for the construction of a LEED[7] certified student residence hall. They claim that the hall will "enable students and faculty to work together in residence on issues of sustainability" and constitute "a model for living and learning and for a new paradigm in campus construction" (Luna and Koman, p. 11). Clemson's president has recently stated that all new construction on campus will be LEED certified, and the new master plan makes sustainable design one of

---

[6] In this case, the university is under consent order by the Environmental Protection Agency, making USC the first higher education institution to develop an ISO certified EMS without regulatory encouragement. Missouri-Rolla claims that the EMS covers the educational, research and administrative activities of the campus. See http://campus.umr.edu/ems/.

[7] LEED stands for Leadership in Energy and Environmental Design. See: http://www.usgbc.org/LEED/LEED_main.asp.

three major principles to guide all construction. Two large-scale Clemson projects are underway that will result in seven new LEED buildings. MUSC's president has also shown support for incorporating sustainable design principles in a new campus hospital. This growing enthusiasm has been stimulated by SUI efforts to foster inter-institutional communication and information sharing.

In addition to faculty champions, the operations leaders may have been the best investment for SUI to date. The SUI experience has revealed that facilities staff interest in sustainability is growing. Sustainability is a winning proposal for them because it bestows status and authority through a cutting edge approach to operations. They can save money through eco-efficiency initiatives which is valued by the fiscally minded members of the institution. Students receive training in the facilities offices to learn about energy management, food purchasing and recycling. Operations staff become less custodians and more teachers, reducing their second-class citizenship in academe.

PROGRESS TOWARD SUSTAINABLE UNIVERSITIES

Clugston and Calder (1999) identified seven factors which make an academic reform initiative likely to receive support from the institution to which it is being advanced. Sustainability, like any other reform, will be successful if certain conditions are met. The factors for success quoted below are followed by reflections on the extent to which the Sustainable Universities Initiative achieved them.

1. *How are the "champions" of sustainability initiatives perceived by others in the institution? Do they have the credibility and the personality needed to promote the initiative or are they marginal institutional actors complaining and promoting their narrow self-interest?*

The principle investigators from the three institutions are well established and respected in their institutions: one is the dean of the School of the Environment, one the director of an equivalent school, and the third is very active in cross-disciplinary programs at his institution. Each has been quite successful in their administrative roles and in managing and fundraising for other initiatives. These project managers were also very skilled in keeping the more radical sustainability advocates involved while finding a wide range of effective "champions" to spread the SUI message. In addition, most of the faculty members and facilities staff supported by the initiative are respected on their campuses.

2. *Does the initiative have the endorsement of key administrative leaders at the institution? Is a commitment to sustainability supported by the President or Chancellor and by other high level and influential figures (e.g. senior managers)?*

The declared support of the three presidents was foundational to the SUI initiative. At USC and Clemson, this resulted in the early formation of high level

Environmental Advisory Committees, which bring together senior level faculty and administrators and report directly to the presidents. New presidents came on board during the initiative and they reaffirmed their institutions' support at a mid-course celebration of SUI in early 2002. This event was also attended by the presidents of other state schools, as well as the governor of South Carolina.

3.   *Who benefits from the initiative? Which departments and programs will the faculty and administration perceive the initiative to be strengthening, and which will it threaten? If the initiative promises to empower and strengthen many programs, it will be supported.*

While SUI reached out to faculty in many disciplines, only a small group of champions became the fires that ignited. Even finding qualified applicants to fill critical faculty positions, which is a stated goal of SUI, has been difficult. To date, an Environmental Ethics faculty appointment was supported at USC, and an adjunct instructor in sustainability related issues is supported at Clemson. SUI has made a concerted effort not to threaten academic departments, but rather to let the assistance flow to where the interest lies. The initiative has tried to be broadly representative, and has not been perceived as pursuing the interests of the few. The problem is that many (both faculty and students) are not interested, despite attempts to include them.

4.   *Does the initiative fit with the institution's ethos, its saga, and its organizational culture? Each college and university has a particular story that it tells about itself and a particular "niche" that it fills in the ecology of higher education. How well does the initiative conform to this institutional identity?*

The ethos of these, and almost all research universities, is not sustainability (and the interdisciplinary and "campus as ecosystem" orientation that this requires). As one USC dean said of the university president in the second year of SUI, "He is supportive of SUI, but it's not a priority. His two priorities are obtaining AAU (Association of American Universities)[8] status and meeting the goal of the current 300 million dollar campaign" (ULSF, 2000). A state research university flagship institution is playing at the high level of economic and disciplinary returns, striving for the cutting edge in research for big grants and disciplinary prestige. This usually means narrow, disciplinary research. The case of MUSC dramatically demonstrates the operation of this logic. The fact that the SUI institutions are operating within southern conservative cultures certainly contributes to some of the resistance SUI has experienced over the years and to a perception that sustainability is a left, liberal agenda invading from the north.

The concept of sustainability has been embraced at Clemson in part because its land-grant status (having a strong agricultural orientation and serving the state through teaching, research, extension and outreach) predisposes the community to consider sustainability as part of its story. Indeed, "sustainable environment" was

---

[8] The Association of American Universities was founded to promote the international standing of USA research universities. See http://www.aau.edu/.

recently identified as one of eight emphasis areas in the new Clemson University road map for the future.

5. *Does the initiative elicit the engagement of the college or university community? Is there sufficient publicity (through public awareness events, press releases, articles, etc.) about new policies and initiatives? Is there regular disclosure of progress, successes and failures? Is the process for critique of current sustainability programs and for determining next steps broadly participatory across the academic community?*

Major efforts have been made to engage and inform the broader academic community and the public. SUI sponsored public speaker events have helped raise awareness, especially among students. Outreach has been strong and effective, and SUI conferences and workshops, held in the first three years, were well attended by community members and professionals who cared about sustainability. But again, most faculty members were not interested, and SUI representatives decided that conferences were not the best way to stimulate faculty engagement. Nevertheless the initiative is perceived to be broadly inclusive and participatory.

Furthermore, SUI's program manager and three principal investigators engage in continual reflection and qualitative self-evaluation. This can be seen, for example, in their detailed annual reports (available online). There is, however, a lack of baseline data available to allow SUI to measure its progress with accuracy. University representatives are typically reluctant to collect data and report back, and SUI managers are loath to demand the extra time, energy, and money of a department that might otherwise be moving ahead on new sustainability initiatives. The lack of quantitative data has been an issue of ongoing concern for SUI staff.

Notably, some facilities staff at USC have reported that the extra time they spend on the job learning about sustainability approaches including technical skills, equipment, new vendor contacts, monitoring systems is neither supported nor recognized by their supervisors (Year 4 Annual Report, p. 2). This speaks to the necessity of improving communication and awareness-raising within departments and units in order to ensure greater community awareness and success of the initiative.

6. *Is the initiative academically legitimate? Is it perceived to be grounded in a recognized body of knowledge? Can it claim an academic rigor and validity? If it lacks this basic sine qua non of academic credibility, it will be rejected.*

Except for a few disciplines, such as some environmental sciences and the land use related professions (e.g., Horticulture, Agriculture, Architecture, and Engineering), sustainability is not part of the cutting edge of the disciplines, and thus interdisciplinary research in it receives little support. Most faculty continue to work within their disciplinary silos. Incorporating sustainability into traditional disciplinary education, claimed one USC faculty member, must still be done subversively (Site Visit Report, 2000). Furthermore, sustainability and interdisciplinary work are rarely on the agendas of accrediting boards or funding agencies for research universities.

SUI set out to be an interdisciplinary and inter-institutional initiative, and one of its major challenges was attracting people from different disciplines into this effort. To SUI's credit, a subset of unexpected champions emerged from various disciplines such as English, Horticulture, Philosophy, and Chemistry. SUI has reported great difficulty facilitating inter-institutional communication between faculty regarding research or course development. It has even been difficult, claim SUI staff, to foster communication between colleagues within one institution.

7. *How successful is the initiative in bringing in critical resources (e.g. grants and contracts, state funding, student demand, recognition and support from key stakeholders such as the media or trustees, and state, national and international leaders)? Does the initiative produce cost savings over time (e.g., energy conservation)?*

The $4.5 million US dollar grant to support SUI is not large by research university standards. For these institutions, a grant of several hundred thousand to a million dollars to one faculty member in the sciences is not uncommon. SUI's grant averages out to less than one million per year for all activities at the three universities.[9] In terms of what these university constituencies are accustomed to, SUI has been able to offer only modest financial support for sustainability programs.

By years 4 and 5 of the SUI initiative, the strength of the operations contribution has been modest cost savings (and the hope of much greater cost savings in the future), and this has successfully attracted increasing support from the institutions. While SUI mini-grants have rarely been enough to attract the unconverted to the sustainability paradigm, SUI has started to emphasize using mini-grant money to strengthen faculty who need initial support to develop major research projects funded by big foundations. According to a survey conducted at USC in year 5, this strategy has paid off: a $214,500 US dollar investment in mini-grants has produced $2.27 million US dollars in additional funding for sustainability related research over the course of the SUI initiative. (Year 4 Annual Report, p. 9).

While SUI has achieved several conditions for success, it continues to struggle for others. Its champions and key administrative supporters are in place, though given the large size of these institutions, more are certainly needed to ensure a deeper impact over time. The message that all can benefit from this initiative and from a commitment to sustainability is clearly communicated, but it has not yet reached enough departments and programs. Most departments are not prepared to see sustainability as academically legitimate. Given the southern USA, research university cultures within which SUI is working, the initiative has been remarkably adept at not appearing threatening or ideological. Indeed, SUI has made a point of being highly sensitive to its diverse audiences. SUI's independent funding has been

---

[9] The South Carolina state government gave SUI $300,000 US dollars in 2000 to help expand the initiative to other public colleges and universities in the state. To date, 13 "affiliate" schools are involved, receiving mostly mini-grants to support course development and operations projects.

a key to its success, but in the U.S. research university world, it is still an initiative with very modest resources.

CONCLUSION

While there is considerable experimentation with sustainability in higher education, and some institutions in the USA have made it a central goal of their efforts, it is very difficult for such a focus to gain centrality in a public research university. This case is unique in that it involves all the public research universities of a state and it illustrates the difficulties and possible strategies for engaging such complex institutions.

Given the forces that shape a public university's agenda, the Sustainable Universities Initiative has been quite successful largely by adapting to unanticipated opportunities. Many of the objectives and timelines in the original Five Year Plan proved unrealizable for different reasons ranging from powerful external forces (such as institutional budget constraints) to internal forces such as the sheer technical and logistical difficulty of getting the job done. The most productive individuals and projects could not always be identified. Thus the early defining metaphor for SUI became "light many fires and see which burned brightest" (Sustainable Universities Initiative, 2002, p. 1).

However, as the initiative has evolved and SUI representatives have become more versed in promoting sustainability within their institutions, they have also become better at identifying opportunities. By creating strong administrative legitimization of sustainability (e.g., presidents' statement, Environmental Advisory Committees), accelerating the operational transition to ecoefficiency, and rewarding those few, but significant, faculty champions for their commitment to sustainability, SUI has found the available leverage points for making sustainability a focus of South Carolina's research universities in difficult economic times. The Sustainable Universities Initiative has, to a significant extent, fanned the flames of sustainability projects against the dampening influence of largely unsustainable forces.

Whether these achievements can translate into comprehensive sustainability initiatives in these, and other, research universities will depend on larger forces in the economy and disciplines which no small initiatives can influence. It will require a major reorientation by government and business. Yet initiatives such as SUI are essential for reorienting future leaders and citizens to make the changes in the disciplines and the economy necessary to create a sustainable future.

REFERENCES

Association of University Leaders for a Sustainable Future. (2000, March). *ULSF Site Visit Report, University of South Carolina*. Washington, DC: ULSF.

Association of University Leaders for a Sustainable Future. (2003, February). *ULSF Site Visit Report, University of South Carolina*. Washington, DC: ULSF.

Bok, Derek. (1990). *Universities and the future of America*. Durham, NC: Duke University Press.

Calder, Wynn & Clugston, Richard M. (2002). Higher Education. In John C. Dernbach (Ed.), *Stumbling toward sustainability* (pp. 625-645). Washington, DC: Environmental Law Institute. See http://www.ulsf.org/dernbach/chapter.htm.

Clugston, Richard M. & Calder Wynn. (1999). Critical dimensions of sustainability in higher education. In Walter Leal Filho (Ed.), *Sustainability and university life* (pp. 31-46). Frankfurt am Main: Peter Lang.

Luna, Gene & Koman, Michael. (2002). *Going green saves green – University of South Carolina's green efforts becoming best practices*. Internal USC Housing document, Columbia, SC.

Sustainable Universities Initiative. (1998). *Five Year Plan*. South Carolina, USA. See http://www.sc.edu/sustainableu/.

Sustainable Universities Initiative. (1998). *Statement of University Presidents*. South Carolina, USA.

Sustainable Universities Initiative. (2002). *Year 3 Annual Report (January 1, 2001 – December 31, 2001)*. South Carolina, USA.

Sustainable Universities Initiative. (2003). *Year 4 Annual Report (January 1, 2002 – December 31, 2002)*. South Carolina, USA.

BIOGRAPHIES

Wynn Calder is the associate director of the Association of University Leaders for a Sustainable Future (ULSF), where he has been since 1996. He is editor of the ULSF's biannual report, *The Declaration*. He is news editor for the *International Journal of Sustainability in Higher Education*, and has published recently on both U.S. and international progress in higher education for sustainable development. In 2000 and 2001, Mr. Calder coordinated the North American Higher Education Network for Sustainability and the Environment (HENSE). Prior to this he worked for six years at Harvard University in academic counseling, admissions and administration. In his spare time, he helped improve the freshman recycling operation, which involved over 1600 students. For five summers between 1986 and 1992, Mr. Calder taught Western Philosophy and Biomedical Ethics at the Massachusetts Advanced Studies Program. He has been a mental health worker and researcher in psychology at McLean Hospital outside Boston, where he counseled and studied adolescents diagnosed with character disorders. He received his Master's in Theological Studies from Harvard Divinity School in 1993, where he focused on comparative religion and psychology. He received his undergraduate degree in history from Harvard University in 1984.

Rick Clugston is executive director of the Association of University Leaders for a Sustainable Future (ULSF). ULSF seeks to strengthen commitment in higher education to sustainability in teaching, research, operations, and outreach (see www.ulsf.org). He is publisher and editor of *Earth Ethics*, and deputy editor of the *International Journal of Sustainability in Higher Education*. His recent publications have focused on higher education for sustainable development and the Earth Charter. Prior to coming to Washington, Dr. Clugston worked for the University of Minnesota for 11 years, first as a faculty member in the College of Human Ecology, and later as a strategic planner in Academic Affairs, Continuing Education and the Office of the President. He was a consultant to the State Department of Education, the Minnesota Business Partnership, and various colleges and school systems on educational improvement. Dr. Clugston has taught and published on human development, strategic planning, educational reform, and most recently on environmental ethics, spirituality and sustainability. He received his doctorate in Higher Education from the University of Minnesota (1987), and his masters in Human Development from the University of Chicago (1977).

CHAPTER 21

INTEGRATING EDUCATION FOR THE ENVIRONMENT AND SUSTAINABILITY INTO HIGHER EDUCATION AT MIDDLEBURY COLLEGE

Nan Jenks-Jay

INTRODUCTION

Across the globe from mega cities to small rural villages, few experiences have had as profound an impact on changing individuals and transforming the places where they live, as has education. The Worldwatch Institute's *Vital Signs 2001: The Trends That Are Shaping Our Future,* suggests that education is important in global efforts to achieve sustainable development as a tool to improve health, decrease poverty, control population and to create equity. Author Gary Gardner goes on to say, "And in an increasingly industrialized world, where people are often disconnected from nature, education is indispensable for understanding the vital need to care for the natural world." Therefore, nothing is more critical to creating a more sustainable world than education. *In Earth Rising: American Environmentalism in the 21st Century,* Philip Shabecoff emphasizes that the "shaping of minds begins first with immediate family, but most profoundly with education, with the schools." Until recently, the traditional model of higher education has been limited to bestowing knowledge through our teaching and making new discoveries through our research. Today, a shift is taking place as the "academy" expands the role of education in the area of sustainability. The challenges associated with becoming a more sustainable world are not insurmountable, but they are somewhat daunting. As educators, we need to understand more about the pathways that will lead to this change and how we can accelerate the pace. As an ideal microcosm of society, institutions of higher learning can demonstrate how to achieve goals for sustainability that will have an impact on and be transferable to other sectors of society. Models already exist on many campuses today through teaching, research and practices. Not only are colleges and universities recognizing that they have a responsibility to make this a better world, but many are becoming leaders in the charge. Middlebury College, a liberal arts college in Vermont, is endeavoring to advance education for the environment and sustainability in higher education through an integrated system-wide approach. This case study will describe this college's commitment and some of the far-reaching outcomes that have resulted.

Peter Blaze Corcoran & Arjen E.J. Wals (Editors), Higher Education and the Challenge of Sustainability: Problematics, Promise and Practice, 263-276.

A LONG TRADITION OF ENVIRONMENTAL MINDFULNESS

An early leader in environmental education, Middlebury College established the first program of environmental studies at a liberal arts college in the United States in 1965 shortly after Rachel Carson's book "Silent Spring" was published and five years before Earth Day. On a parallel track in the operational sector of the College, environmental awareness took hold through facilities management's aggressive energy conservation and waste management/recycling programs. Kathryn Hansen suggests in *Environmental Challenges for Higher Education: Integrating Sustainability into Academic Programs*, "if academia responds to society, I believe academia will also do better in its academic endeavors." This was in part the case at Middlebury College where the new ground breaking programs established in the mid 1960's led to stellar academic and college-wide programs forty years later, which are now integral to the College's reputation as a leading liberal arts undergraduate institution.

Both the academic and campus programs came into being as a result of leadership demonstrated by key influential individuals at the College at the time, the chair of the Biology Department and the Vice President and Treasurer. The programs were continued and expanded as a result of the tenacity of a few faculty members, interest of students, ingenuity of facilities management staff and progressively thinking administrators. Although these impressive programs raised awareness and gained attention, these individual efforts did not constitute the long-term institutional commitment we are now striving to achieve in academia.

Programs in the earlier years existed due to the good efforts of dedicated individuals, which led to the senior administration taking serious stock in them and recognizing their value to the College. A new course of action to advance education for the environment and sustainability was charted as part of period of rapid growth and improvement at the College to make Middlebury College a leading national liberal arts college. WorldWatch Institute's *State of the World 2001* report distinguishes human change from natural change – biodiversity, by "its willingness and purposefulness. As the only creatures known to plan change, we boldly dip into our own history and alter the course of our own development."

A focal point of this purposeful change initiative was the designation by the President and Trustees of Middlebury College of exceptional areas of excellence created from existing strengths at the College. The environment was included because the Environmental Studies Program and the campus management programs had already distinguished themselves as being highly successful. President John McCardell's stated, "At Middlebury we are building an academic plan that emphasizes excellence across the curriculum with special attention to the academic "peaks" that are the hallmark of our identity. Environmental Studies and Awareness is one of these six peaks because of the College's long tradition of being on the forefront of environmental education in both the classroom and on campus." This action paved the path for an institutional integration of education for the environment and sustainability that followed.

Middlebury College's sustained commitment has in part been a result of a willingness to change, an endorsement from the highest levels, sufficient funding, guiding principles and standards, strategic planning, rewards and incentives, external influences and an integrated approach.

AN EVOLVING INSTITUTIONAL COMMITMENT

It was a bold and forwardly thinking action by the Trustees and President to designate the environment as a peak based on its historical strengths. More than mere rhetoric, the creation of the Environmental Peak of Excellence linked the academic and organizational sectors through the institution's broadly defined environmental commitment. It brought with it both the philosophical and financial support to facilitate greater advancement and leadership through the Director of Environmental Affairs, a newly established senior administrative position. Efforts and programs that already existed became more prominent and a new infrastructure was being created. Clugston and Calder's *Critical Dimensions of Sustainability in Higher Education* describes conditions for determining success of sustainability initiatives, all which Middlebury College demonstrates including eliciting the engagement of the entire community, bringing critical resources, having academic legitimacy and endorsement of key administrative leaders.

One such step was the Board of Trustees' endorsement of the mission statement for the Environmental Council, a committee that advises and recommends policy to the President of the College. The mission states, "Middlebury College as a Liberal Arts institution is committed to environmental mindfulness and stewardship in all its activities. This committee rises from a sense of concerned citizenship and moral duty and from a desire to teach and lead by example. The College gives a high priority to integrating environmental awareness and responsibility into the daily life of the institution. Respect and care for the environment, sustainable living, and intergenerational responsibility are among the fundamental values that guide planning, decision making, and procedures. All individuals in this academic community have a personal responsibility for the way their actions affect the local and global environment." The Environmental Council, which reports to the President and is chaired by the Director of Environmental Affairs, is comprised of faculty, staff and students. The Council conducts regular campus audits and helps to educate the campus about sustainability. It recommends new policy and college-wide initiatives. Examples include new sustainable design and environmental construction principles and guidelines, the College's first air conditioning policy and a contractor waste management standard. It also administers the sustainable campus grants program, undertook a carbon emissions inventory, initiated a process to become a carbon neutral institution, and assisted in the preparation of a five-year strategic plan for the Environmental Peak in addition to other responsibilities and projects.

New administrative and faculty positions were authorized and staff positions upgraded in support of the Environmental Peak. The Executive Vice President for Academic Affairs and Provost made a new appointment, the Director of Environmental Affairs, a new senior level administrative position with an academic appointment in the Environmental Studies Program. The Director of Environmental

Affairs reports directly to the Executive VP and Provost; chairs the Environmental Council and the Environmental Peak Committee; meets quarterly with a subcommittee of the Board of Trustees; and oversees operating budgets and endowment funds in addition to other responsibilities. As Director of Environmental Affairs, it was my goal being a faculty member and senior administrator to create a vision for the Environmental Peak including its integration into the academic and non-academic sectors of the College. A staff position of Campus Sustainability Coordinator was upgraded to a professional level from that of Recycling Coordinator, which had been previously filled by postgraduates. The Campus Sustainability Coordinator provides internal education, outreach, resources, continuity and links among projects and people. Through this position more can be undertaken and at a greater pace. The Dean of Faculty and the committee on appointments approved five new shared faculty positions with Environmental Studies and the departments of Religion, Geology, Economics, and History, exposing students to environmental and sustainability topics in a wide range of courses that are not even part of the environmental studies curriculum. The environmental studies program is one of the largest majors at the College. Our research studies indicate that 48% of the graduating class has taken at least one environmental course. Forty-three affiliated faculty represent 16 departments. The diverse affiliated faculty, integrated curriculum and team teaching make it a truly interdisciplinary program involving the sciences, humanities and social sciences. A new addition of a distinguished appointment was recommended by the Director of Environmental Affairs and approved by the senior administration. Today, renowned author and internationally noted environmentalist Bill McKibben holds this position as Environmental Scholar in Residence at Middlebury College.

The breadth and depth of this program along with the expertise of its wide-ranging affiliated faculty from all across the disciplines makes integration even more feasible. Courses are taught in Religion involving the Earth Charter; in Physics involving alternative energy; in Geography addressing night light pollution; in Sociology/Anthropology on human ecology; and in Environmental Studies on local foods along with dozens of other courses, theses and special projects exposing students to issues related to sustainability. Thinking well beyond the borders of the immediate campus, environmental and sustainability programming is being discussed with Middlebury's schools in six foreign countries with one program developing a pilot project as a result of a collaboration with the Dean of Languages and Programs Abroad and a grant from the Andrew W. Mellon Foundation.

Through collaboration and innovation much has occurred. A third of the food purchased for Dining Services comes from local sources, thereby supporting Vermont's agricultural community. Dining Services and Facilities Management are cooperating with students who developed a plan and received funding for establishing an organic garden on campus. Dining and Facilities also work together to compost 75% of food waste from the dining halls, 300 tons annually from a student population of approximately 2000. An award winning recycling program diverts, on average, 60% of the College's solid waste for reuse. Taking the concept of reuse to a new level, the College recently recycled 97.6% of the old science center, reusing 1354 tons of material when the six-story building was deconstructed and earned the College the Vermont Governor's Award for Environmental Excellence and Pollution Prevention. College forestlands are managed under Forest Stewardship Council green forestry certification standards in partnership with Vermont Family Forests.

Energy is addressed at various stages and in different sectors through conservation, co-generation, new technology, efficient design and a new thermal comfort policy. Alternative energy vehicles are part of the College's fleet. A meteorological tower will soon be monitoring wind to determine the potential for generating energy atop the College's ski slopes. Through the work of Environmental Affairs and the Alumni and Career Services Offices, we know that today over 600 alumni of the College are working in environmentally related occupations across the globe, and we are making connections with them as useful resources.

Professor emeritus in geography and environmental studies at Northeastern Illinois University, William Howenstine explains that in higher education "integration is threatening", because our academic upbringing has led many to become "disciplinocentric". However, at Middlebury, efforts are transcending boundaries of academic and non-academic programs and between departments to form new collaborations and develop shared goals that result in mutually beneficial outcomes. These cooperative efforts are strengthened as new networks develop, not through compartmentalized sectors or traditional departments, but instead around areas of interest and expertise such as energy, food or land management, for example.

WALKING THE TALK

According to Calder and Clugston in *Stumbling Towards Sustainability* nearly all the major reports and declarations for sustainability stress "higher education outreach and partnerships." A current example of a developing partnership and outreach that demonstrates the College's commitment to campus and local sustainability is being facilitated by Environmental Affairs. It is a proposed expansion of a biodiesel project involving several students' research and a complex arrangement on campus. In order to achieve success, the project requires an agreement from the chemistry department to test fuel, dining services to supply waste cooking oil as a raw product, facilities management and the ski area to use the fuel in running equipment, and a new start-up biodiesel fuel company in Vermont. This project is attracting a lot of interest from faculty, staff and students as well as that of the greater community and the College administration, which has been intrigued by its potential from the beginning. A spin off was an endeavor by several students to retrofit a bus fueled by vegetable oil and drive it cross country during the summer starting from the steps of the College's administration building in Vermont and ending the trip in Washington state.

Another example of Middlebury College's commitment is one that reaches far beyond the campus and involves the College's sustainable design and construction standards recommended by the Environmental Council, endorsed by the Board of Trustees and created by diverse committee of faculty, facilities management staff,

students, administrators, and a consulting architect. So far, these standards have
been applied to five major construction projects and one deconstruction project on
campus, resulting in an increase in energy efficiency, recycling waste materials and
reducing the overall environmental footprint. The College's request for local green
certified wood jump-started a new sustainable forest products industry in Vermont
according to those involved in the forest products industry in the northeast. In a
paper presented at the 2003 AERA annual meeting in Chicago, Walker, Corcoran,
Wals, Scott and Gough discussed the concept of change vs. continuity as it relates to
sustainability in higher education. They described single loop learning as
"improving efficiency" or changes in actions and double loop learning, which
improves effectiveness or brings a change in values. During these sustainable
design and construction projects single learning processes took place by carrying
new knowledge into the next project to improve its efficiency. Double loop learning
took place by significantly altering the way the professionals perceived the
underlying values that the College and its partner Vermont Family Forest introduced
to these projects.

These learning processes are reflected in comments made by Mark McElroy, the
construction manager for Bicentennial Hall, who indicated that in future projects he
and others now knew how to adjust the traditional construction process to
incorporate new thinking and the use of green certified wood with character marks.
He stated afterwards, "people working on the project were initially taken aback, but
by the end realized that it takes a better eye, more creativity and a higher level of
craftsmanship, so they came away with a real sense of pride in what they had done."
The architect wrote about the Ross/LaForce Commons project, "your fundamental
concerns for the managed use and replenishment of our forests is admirable. More
Americans need to be so involved, particularly those of us who've been rather
indiscriminate in our usage. Education is key. You are teaching us that managed
sustainability of forest lands is not only necessary, but economical." Many of the
architects, engineers and other professionals involved in these projects indicated that
they have been so inspired by Middlebury's thoughtful commitment to green design
and local sustainability that they are carrying lessons from these projects to their
clients and projects across the world. For serving as a major catalyst in the green
certified wood industry in the northeast, Middlebury College received the U.S.
Environmental Protection Agency's Environmental Merit Award in 2003.

In describing those who engage in change at different stages, Everett Rogers of
the University of New Mexico lists innovators at the leading edge of change in
society then spreading to early adaptors, the early majority, late majority and lastly
to the traditionalists. As a major sector of society, schools can influence the course
of sustainability by resisting the status quo and being at the forefront of change as
innovators and early adaptors. Recently Middlebury College developed an
innovative strategy that helped to leverage the re-opening of a furniture-making
company that was purchased by its former employees. When the previously
nationally owned company moved the work to China, it left 150 skilled crafts people
unemployed in northern Vermont. In an effort to support the regional economy and
environment, Middlebury College provided a letter of intent committing to purchase
study carrels made from locally harvest green certified wood and shipped with no

packing materials if the company was up and running in time. The College's letter helped attract lenders, leveraged federal grants and gained the confidence of other prospective customers. Currently, green certified wood from the College's own forestland is now being crafted into furniture for the new Middlebury College Library from the employee owned Island Pond Furniture Company in Vermont. Because all of the College's work is essentially about education, students have been involved by documenting the wood's journey from forest to installation with interviews of over 30 individuals in the local forest industry who participated in the process. One student's work was published in a regional magazine and another's led to a photo exhibit featuring the story of the building's green certified wood at the opening of a new student resident hall. An Environmental Studies Senior seminar project conducted an economic analysis comparing the local, green certified wood to traditionally used woods for Vermont Family Forests, a nonprofit organization composed of small wood lot owners to manage, harvest and market their certified forests. Many of the environmental senior seminars involve a service learning component with a local partner. The College supports real life experiences for students as another of the "peaks of excellence" and has appointed a director of Service Learning. In a related project, students in another environmental senior seminar are working on improving the College's ecologically based land management practices with a goal towards achieving environmental and economic sustainability involving its forest and agriculture lands.

TO STAY THE COURSE: GUIDING PRINCIPLES, STANDARDS, PLANS AND ASSESSMENT

To guide the course of becoming a more sustainably aware and committed institution, the College developed guiding principles such as the resolutions approved by the Board of Trustees for the Environmental Council's mission statement mentioned previously and to employ sustainable design standards to College's building projects. They state in part, "The Guiding Principles are the embodiment of the philosophy and spirit of the college community as set forth and adopted by the Trustees. The Principles outline in general terms the College's environmental goals pertaining to construction, renovation, operation and maintenance of campus facilities. The Principles are a statement of purpose that defines how the College and its appointees will make decisions pertaining to the relationship of this built environment and the natural environment. By integrating a long-range environmental consideration into the planning process, Middlebury College will be better prepared to assess risks, identify opportunities and make more informed decisions about the College's future. ...the Guiding Principles provide direction by describing the overarching community, academic and environmental values of the institution along with general guidelines that respect these values." Following the adoption of the Guiding Principles, the College developed specific standards that actually achieve sustainable design and construction practices. Each project is evaluated to determine how to make the process more efficient and effective. In addition, some projects are being rated for proposed certification with the US Green Building Council's Leadership in Energy and Environmental Design Standards (LEEDS). Through LEEDS these projects will receive a third party evaluation. The same is true for the College's green certified forest management and timber harvest practices, which are monitored by an outside third party. Several incoming first year students have indicated that they selected Middlebury College for not only Environmental Studies academic program, but because they were aware of its commitments like the ones above. Alumni interest in the programs is high following reunion visits, articles in the College's alumni magazine, national media coverage, and from presentations the Provost and I have teamed up to give for alumni gatherings in major cities. Generous alumni contributions as major gifts to the College have dramatically expanded the resources available for the Environmental Peak.

To create an integrated institutional vision, the Environmental Peak committee undertook a highly participatory strategic planning process that culminated in a five-year plan of action chaired by the Director of Environmental Affairs. The resulting report and recommendations target all sectors on campus as having specific responsibilities to fulfill the objectives. The report's overarching goals are "to develop leaders, new knowledge and sustainable practices to meet the complex local and global challenges of the future." Having a plan that guides the process is useful in identifying the participants, allocating resources and setting priorities. In this rapidly changing field, recommendations can become quickly dated, so this working document is evaluated and revised regularly to remain current and flexible. In this way it responds to progressive thinking and to new opportunities that arise unexpectedly and are not included in the original plan.

As the process of institutionalizing sustainability evolves, I have become more deliberate about studying the progress, and using assessment and reflection to improve future projects and to set new goals. It became obvious that to change and improve, we needed to create baselines and take time for reflection and assessment. Middlebury is currently considering a set of college-wide sustainability indicators. However, extremely good use is made of a combination of more specific evaluation techniques designed for individual programs to judge success, change actions and improve outcomes. For example, sustainable design and construction standards are being employed in all new construction and renovation with an ongoing evaluation of each project to assure that new knowledge will be carried forward and that the process will be adjusted. In addition, new buildings are commissioned to determine how efficiently the systems are working and make necessary adjustments. To make progress with goals that have been set forth, the College conducts regular sustainability audits with measured results. These campus audits create a long-term comparable database that aids the College in tracking progress and identifying areas requiring improvement. Another level of audit, voluntary environmental audits assure that the College is in compliance with state or federal regulations. Debriefings are scheduled after collaborative projects to consider what was learned and how the process could be improve. Students or outside consultants are enlisted to study various projects, for example, the deconstruction of the old science building, to track the exact weight and destination of all materials that were recycled or reused, to better understand the economics or local impact of such undertakings.

Surveys are conducted to gauge attitudes before and following sustainability discussion courses. The environmental grant program is an experiment in integrating knowledge and action that receives regular assessment. In a report "Our Common Journey: A Transition towards Sustainability" by the National Research Council's Board on Sustainable Development to present a scientific exploration of the transitions toward sustainability, trends, goals and actions are discussed. The report states a need to design strategies and institutions that can better integrate incomplete knowledge with experimental action into programs of adaptive management and social learning "because the pathway to sustainability cannot be charted in advance and will have to be navigated through trial and error and conscious experimentation". An annual evaluation of the campus environmental grant program has enabled the Environmental Council to see what was working and not and then make changes that have expanded the programs' reach and success. For example the process has been made less intimidating to students and staff, who are less unfamiliar with proposal writing and creating budgets. The revised process added checkpoints to keep the grants on track or make adjustments if necessary. Advisors from the Environmental Council were assigned to each grant to assist with any perceived institutional roadblocks that arose. Assessment is an often over looked part of moving the change process forward. Through the different above mentioned evaluation techniques, we are able to regularly identify greater potential, recognize failure, carry information forward, to compare changes over time and to celebrate the successful efforts of many individuals throughout Middlebury College.

INSTITUTIONAL REWARDS AND INCENTIVES

As Middlebury College charts this new course it requires a different way of thinking and operating. To change systems and transform institutions, we are asking individuals to initiate change in their lives and work. In doing so, Middlebury is cognizant of the fact that this effort requires not only the interest of individuals, but also some positive return. Therefore, it has been vital to consider what institutional rewards and incentives are valued and to implement them at the College as part of this effort to advance the environment and sustainability. Rewards and incentives can come in all shapes and sizes.

Innovators are rewarded for having new ideas through the environmental grant program, which supports proposals from fifty dollars to recycle water in Dining Services to thousands of dollars for a wind-monitoring tower. The Environmental Council administers the grants, which are enthusiastically funded by the President. These programs provide a safe place for staff, faculty and students not to be viewed as troublemakers, but instead as entrepreneurs assisting the College in achieving its sustainable goals. The projects have a high profile, and individuals remark that they have felt a great sense of pride and excitement about working for a College that supports such endeavors.

Resources follow good efforts where possible. A new state of the art recycling facility was just completed for Facilities Management's award winning recycling program and a significantly upgraded composting facility. A greenhouse for Dining

Services now supports student projects and is used to grow greens and herbs for campus dining. A consulting architect was hired to assist Facilities Management and the Project Review Committee in developing sustainable design standards for the College.

A deliberate focus of the campus audits has been not only to identify problem areas, but also to identify areas of improvement and to credit those who have made a small or large difference along the way. The acknowledgement of Middlebury College's achievements has not been without the important recognition of the individuals who are making it possible. All of Middlebury's success in the area of education for environment and sustainability can be attributed to the people whose core values are reinforced by positive institutional feedback and an awareness of what is required to do good work. It is not just about doing more with existing time, resources and knowledge. Staff, faculty and students have been sent to conferences and meetings to become more knowledgeable and inspired. Excused time away from the office is provided for staff members serving on the Environmental Council and they receive a letter of gratitude in their personnel file at the end of their term with a copy sent to their supervisor and the President recognizing their contribution to Middlebury College. Because the Environmental Council is a "standing committee," faculty members assigned to it do not have to serve on other committees, an incentive that significantly lightens their committee load, if so desired. In order to take some burden off enthusiastic junior faculty members who teach the environmental senior seminar, a course that undertakes complex sustainable campus and regional projects once a term, a teaching assistant has been added. This supporting position is now viewed as invaluable by all faculty who teach this course. Environmental Studies Program is involved in the review process for the three joint and four shared faculty appointments and can contribute letters for all ES affiliated faculty. Articles profiling people's involvement have appeared in everything from the campus newspaper to national magazines.

UNPRECEDENTED EXTERNAL FORCES

For many years programs addressing the environment forged ahead on their individual campuses without much external input. Characteristic of parochial New England colleges, Middlebury College's environmental initiatives proceeded in isolation with knowing little about what was occurring elsewhere. As societal interest in education for the environment and sustainability efforts has grown in the recent past, so have the connections among those working to achieve similar goals. Once these external links began to develop, Middlebury made good use of them by sharing its experiences at conferences and learning from others. This influence also gave rise to a new level of awareness and responsibility on campuses. External forces helped to reinforce Middlebury College's agenda, thereby giving education for the environment and sustainability added credibility and greater momentum.

External organizations have had a direct effect on the behavior and norms in higher education moving them towards becoming more committed. The list of organizations far exceeds those mentioned below. Outside influence from groups dedicated to the environment and sustainability began having an influence nationwide. At Middlebury, such influences inspired administrators and energized faculty, staff and students. Middlebury became an early signatory of the Talloires Declaration. It was a founding member of the Northeastern Environmental Studies Programs (NEES), a group that exchanges information and provides collegial support from courses to campus sustainability projects. It is a member of the National Council for Sciences and the Environment. Federal agency programs like the Environmental Protection Agency's Green Lights program targeted colleges and universities as participants.   Other organizations dedicated to disseminating sustainability information and best practices in higher education used Middlebury's composting, recycling and local food procurement as model examples. The College gained valuable insight when it hosted a northeast regional workshop for the National Wildlife Federation's (NWF) Campus Ecology Program realizing that by comparison it was well ahead of other schools in many areas, but that it also had room for improvement. This partnership with the NWF led to increased resources for our students when two projects received funding through the student grants programs, one for a wetland restoration and another for biodiesel research. The College's membership in University Leaders for a Sustainable Future provides essential connections to like-minded individuals and institutions around the globe directly at meetings it has hosted and through use of an active listserve. Through Middlebury's involvement with Project Kaleidoscope in Washington, D.C., the director of Environmental Affairs has led workshops in locations across the country for other environmental programs. These stimulating forums not only provide a rich source of new ideas, but also create opportunities to form new alliances with other programs.   The US Green Building Council's sustainable design standards are having an influence on many capital projects within higher education with over 78 colleges and universities currently registered to certify their building projects under one of the Leadership in Energy and Environmental Design (LEED) rankings including Middlebury's new student residence.

New expectations from others have had a strong influence on institutions embracing their responsibility to creating a sustainable future. The Association of Governing Boards of Universities and Colleges (AGBUC), that recommends policy for institution of higher education in the United States, listed "Creating a Sustainable Society and Future" as one of ten policy goals in 1999 - 2000 stating, "Higher education will be expected to play a stronger role." Currently, the New England Governors and Eastern Canadian Premiers have clearly stated that they are looking to the institutions of higher education to play a major role and partner with them to achieve a goal of carbon reduction in the northeast and Middlebury has endorsed their goals.

When Middlebury College underwent reaccreditation recently, the College did not expect the unprecedented recognition that it received from the external review committee, which highlighted the institution's integral environmental goals. The external committee was comprised of presidents and senior administrators from highly respected institutions of higher education. The committee reported its findings after reading the college's internal assessment and shelves of related documents and following 48 hours of on-campus interviews.   The report stated,

"Middlebury College has a leadership position in higher education for its academic programs in environmental studies. It also takes the environment very seriously in siting and constructing buildings, recycling, and forest management. The College has adopted guiding principles for environmental quality, and has a campus sustainability coordinator to facilitate programming. What is most striking is the College's approach to how carefully both the cost and benefits of environmental proposals are considered without the rancor encountered on most campuses." The summary of the report states, "Middlebury has created a beautiful campus with outstanding facilities, excellent maintenance and admirable environmental policies. The curriculum is academically strong, pedagogically forward-looking, consistently student-centered and rigorously interdisciplinary." Middlebury's environmental commitments were categorized as "successes" in six of the ten categories that were evaluated in the overall report. No one anticipated this outcome, but accepted it with a great sense of honor and new realization of its significance.

Nothing is quite as powerful as is the positive feedback from highly regarded external sources including the media. According to Michael Nitz's work at the University of Idaho about the communication of environmental information, "the mass media can distribute and disseminate opinions, while at the same time conveying the impression that these issues are important." Pride and a realization of significance was the internal response when Middlebury's sustainable efforts appeared in articles in *The New York Times, Chronicle of Higher Education*, the Association of Governing Boards of Universities and Colleges publication *Priorities*, and *WorldWatch Magazine* to name a few.[1]

Expectations and legitimatization from highly regarded external entities such as those described above are occurring more often and having a positive influence on colleges and universities including Middlebury.

CONCLUSION

Throughout this chapter, I have described ways that education for the environment and sustainability have been integrated throughout the system while in fact it is quite the reverse. These elements are in essence the integral thread within a higher education learning community that bring together traditional courses, research, operations and management, and the greater society.

In broadening its definition of education for the environment and sustainability, Middlebury College took a critical first step towards acknowledging its value to the College's educational mission, management goals and responsibility to a much greater community. By doing so, the College has increased its level of commitment and ultimately expanded its influence. We learn much from our peer institutions and share our own successes and failures more openly as part of a greater network of colleges and universities striving to commit to something far greater than we have in

---

[1] The first page of *The New York Times* education section featured a story about the deconstruction and recycling of the College's old science center. The College's sustainable campus initiatives appeared in the Association of Governing Boards of Universities and College's (AGBUC) publication *Priorities*, which is received by over 20,000 trustees and senior administrators in the United States.

the past. Individual colleges and universities will undoubtedly have significant impacts on various regions of the world through their own models and on leadership as our students enter the work force and decision making positions. However, institutions of higher education may play a far greater role as a collective force to bring about pivotal change towards becoming a more sustainable world.

REFERENCES

Association of Governing Boards of Universities and Colleges. (2000, Spring). Middlebury College's "Peak of Excellence." *Priorities,* 14, pp. 4.

Association of Governing Boards of Universities and Colleges. (1999). Ten Public Policy Issues for Higher Education in 1999 and 2000. *ABG Public Policy Paper Series.* No. 99-1.

Clugston, R.M. & Calder, W. (1999). Critical Dimensions of Sustainability in Higher Education. In: W.L. Filho (Ed.), *Sustainability and University Life* (pp. 31-46). Frankfurt: Peter Lang Publishers.

Clugston, R.M. & Calder, W. (2002). Higher Education. In: J.C. Dernbach (Ed.), *Stumbling Towards Sustainability* (pp. 631). Washington DC: Environmental Law Institute.

Dobelle, E S., Thomas, R.R., Chabotar, K.J., Fisher, L.G., Gordon, J.W., Jedrey, M., Knable, B., Maisel, L.S., Kenan, W.R. Jr., Putnam, M. (1999, October). Report to the Faculty, Administration, Trustees and Students of Middlebury College.

Flaherty, J. (2001, November 11). Bein' Green. *The New York Times: Educational Life,* pp. 7.

Gardner, G. (2001). Education Still Falling Short of Goals. WorldWatch Institute's *Vital Signs 2001: The Trends That Are Shaping our Future* (pp. 148). New York: WW Norton and Company.

Gardner, G. (2001). Accelerating the Shift to Sustainability. WorldWatch Institute's *State of the World 2001* (pp. 190). New York: WW Norton and Company.

Hansen, K. (1996). Global Changes from the Grassroots Up. In R. Wixom, L. Gould, S. Schmidt, & L. Cox (Eds.), *Environmental Challenges for Higher Education: Integrating Sustainability into Academic Programs* (pp. 141). Burlington, VT: Friends Committee on Unity with Nature.

Howenstine, W.L. (1996) A Resources Approach to Sustainability. In R. Wixom, L. Gould, S. Schmidt, & L. Cox (Eds.), *Environmental Challenges for Higher Education: Integrating Sustainability into Academic Programs* (pp. 68). Burlington, VT: Friends Committee on Unity with Nature.

Mansfield, W.H. (1998, May/June). Taking the University to Task. *WorldWatch.,* 11(3), 24-30.

National Research Council. (1999). *Our Common Journey: A Transition Toward Sustainability.* Washington DC: National Academy Press.

Nitz, M. (2000). The Media as a tool for communication on the environment and sustainability. In: Filho, W.L. (Ed.), *Communicating Sustainability* (pp. 47). Frankfurt: Peter Lang Publishers.

Perrin, N. (2001, April 6). The Greenest Campuses: An Idiosyncratic Guide. *The Chronicle of Higher Education,* Section 2.

Rogers, E. (1995). *Diffusion of Innovations.* New York: The Free Press.

Shabecoff, P. (2000). *Earth Rising: American Environmentalist in the 21st Century.* Washington DC: Island Press.

Walker, K.E., Corcoran, P.B., Wals, A.E.J., Scott, W.A.H., & Gough, S.R. (2003). *Case Study Methodology in Sustainability in Higher Education: Advancing a Critical Research Model,* presented at AERA annual meeting, Chicago.

BIOGRAPHY

Nan Jenks-Jay is director of Environmental Affairs and holds an appontment in the Environmental Studies Program at Middlebury College where the president and trustees designated the environment as a peak of excellence. For two decades, she has been actively involved in the advancement of environmental programs, their transformation of higher education and impact on regional sustainability. She has authored several book chapters and numerous articles and is regularly invited to speak on this topic throughout the country and abroad. She was quoted in an article entitled "Taking the University to Task" as saying "...environmental programs are leading a shift in the very fundamentals of education... We are gradually seeing one of the most traditional and rigid of all institutions, the 'academy,' accept the challenge to educate and prepare students to live and work in a world in which they must individually and collectively effect change and also to recognize its role as a large business and influential entity in acting responsibly with regard to decisions that impact the environment".

CHAPTER 22

SUSTAINABILITY IN HIGHER EDUCATION THROUGH DISTANCE LEARNING: THE MASTER OF ARTS IN ENVIRONMENTAL EDUCATION AT NOTTINGHAM TRENT UNIVERSITY

Malcolm Plant

INTRODUCTION

This case study draws on my experience of developing, tutoring and evaluating an *MA in Environmental Education* by distance education in the Faculty of Education at my university. I aim to show how this course shifts attention away from value-free and instrumental models of environmental studies and environmental education courses towards value-laden and dialectical processes of teaching and learning. To this end, the MA course is underpinned by a socially critical pedagogy and a critical realist philosophy. A socially critical pedagogy is directed towards encouraging students to reflect on their professional roles critically in order to gain insight into their own and other people's understanding of the socio-political origins of environmental issues so enabling them to devise and evaluate educational frameworks that are effective in their different professional fields. A critical realist philosophy inspires a form of tutor-student interaction that is dialectical with the aim of reciprocally disclosing knowledge about environmental issues at deeper levels of reality. In order to put these theoretical issues into context, I begin by arguing that environmental education and environmental studies courses promoting the view that the environment is 'out there' and deserving of our protection overlook the obvious, that human society and the rest of the biophysical world are interconnected intimately through dialectical processes. Moreover, so that I can illustrate how this theoretical foundation is operationalised from a distance, I examine extracts from two (of many) 'critical encounters' taking place between the students and me. I conclude this case study with the observation that pioneering programmes such as the *MA in Environmental Education* course that stress socially critical pedagogy along with a critical realist philosophy of knowledge are faced with an uncertain future given the ascendancy of market-led and discipline-based curricula currently serving the cultural and intellectual needs of students in higher education (see also Chapter 4 by John Huckle).

Peter Blaze Corcoran & Arjen E.J. Wals (Editors), Higher Education and the Challenge of Sustainability: Problematics, Promise and Practice, 277-292.

HIGHER EDUCATION AND VALUE-FREE KNOWLEDGE

Critically reflective learning that penetrates the social and political conditions responsible for degraded environments and the people that suffer them seems far removed from the assortment of environmental science, environmental health, environmental law, and some environmental education courses that infuse the environmental curriculum in higher education. These courses tend to promote the view that technological fixes such as waste management, recycling and reuse, public awareness campaigns and ecological auditing are the preferred remedies for environmental problems. As Luke (2001, p. 187) argues:

> This reactive approach to environmental destruction has, created, in effect, a conceptual zoning code that keeps most environmentalists from investigating how society is organised, how industrial metabolisms are fabricated and where ecological efficiencies might be realised before the end of the pipe disasters occur.

Instead of 'travelling back' up the pipeline into the realm of society and examining the fundamental social and political reasons why, for example, toxic wastes are allowed to pollute the environment, courses that focus on resolving environmental problems tend to regard environmental problems as a technical issue requiring better management within the prevailing economic order. Regrettably, this ethos denies students perspectives on alternative ways of dealing with growing environmental and social problems, trapping them in an industrialist mind-set that desensitises them to the ecosocial crisis and fragments their existence. This is a state of mind that some psychotherapists call 'ontological insecurity' (Kidner, 2001, p. 4). As Jones et al. (1999) argue, few higher education courses foster students' critical awareness of epistemological and value-related questions as a prerequisite for critical thinking about environment and development issues. Consequently, all that such courses seem able to do is to "replay the scenarios" (Baudrillard, 1992, p. 22) so that industrial society appears unable to escape from the trap of the forces of production assuming, naively, that ecological and social sustainability is achievable only within the existing paradigm of unrestrained economic growth.

It is essential to examine the philosophical questions at large here. If environmental education and environmental studies are to break free from their technocentric mindset, they have to overturn the idea that society and Nature are two entirely different realms. Calls such as "we should protect Nature for its intrinsic worth" (Pepper, 1993, p. xi) and slogans such as "saving the wilderness" (Payne, 1999, p. 23) sustain this viewpoint. Since humans (and organisms in general) are engaged endlessly in constructing and consuming their surroundings, it is meaningless to adopt the catchphrase 'Save the Environment' as if it were 'out there', unless it is understood that humanity and Nature are one. I believe this separation works against understanding and resolving the ecosocial crisis since that requires a commitment to accepting that there is no organism without an environment, and no environment without an organism. As Luke (2001, p. 193) sees it:

This reductionist separation of organisms from their environment is the key to the primary conceptual chasm cutting through the ragged reality of the Earth by most rhetoric of ecology.

Therefore, where higher education courses stress the idea of 'saving the environment', students are likely to remain oblivious of what seems to me obvious: human society and non-human Nature do not have independent existence or, as Martell (1994, p. 178) points out succinctly, "their fortunes are locked together in a mutually constitutive dialectical relationship". The term 'dialectical' is used here in the sense of 'interacting forces or elements'. That is, in terms of environment-society relations, humans interact with Nature thereby changing both Nature and society, a continuous process of consuming and being consumed, just as Marx (1875, p. 327) had realized: "Man lives from nature, i.e. nature is his body, and he must maintain a continuing dialogue with it if he is not to die". This insight led to the idea of 'dialectical materialism' that I believe is a crucial concept when searching for ways to sustain life on Earth. It implies that, instead of trying to understand entities in terms of the detailed arrangement of the particulars comprising them (mechanical materialism), we should try to do so in terms of the relational properties between them since it holds that the innermost nature of things is dynamic and conflictual rather than inert and static. Critical realism is a philosophy rooted in this idea of dialectical materialism that I believe has great significance for environmental education, not only because it offers a radical revision of how we see the human-Nature relationship, but also for its value in shedding light on the significance of tutor-student relations in the context of the *MA in Environmental Education* course.

CRITICAL REALISM AND ENVIRONMENTAL EDUCATION

When considering Nature as a particular facet of reality, students ought to be persuaded to examine fundamental ontological questions about what they understand by reality, and epistemological questions about how to know this reality. Indeed, the debate about the nature of reality and its relevance to a philosophical grounding for environmental education has been vigorous for some time (Robottom & Hart, 1993; Mrazek, 1994; Williams, 1996; Huckle & Martin, 2001; Plant, 2001) and is essentially concerned with whether humans 'construct' reality (constructivism), or whether reality exists independently of our knowledge of it (realism). Soper (1995) refers to these two opposing positions as 'Nature sceptical' (of the constructivists) and Nature endorsing' (of the realists). These contradictory positions moved Dickens (1996, p. 2) to write: "our knowledge of the environment and of our relations to Nature are characterised by considerable ignorance". As commonly presented, the realist perspective holds that there is an external reality existing independently of the historical and cultural conditions that produce knowledge. In contrast, the 'strong' constructivist tendency (commonly associated with postmodern philosophy) claims that knowledge is entirely determined by social processes and can therefore tell us nothing about external reality, while 'mediated constructivism', or 'weak constructivism', allows for both greater or lesser degrees of subjectivity. .

Some environmental philosophers argue for a fresh philosophical basis for environmental education that transcends the dichotomy of the realist/constructivist debate (Dickens, 1996; 2000; Huckle, Chapter 4 in this book). Essentially, their claim is what I call a 'common sense' view of reality, that humanity is rooted in the natural world and that people live in their relationship to Nature. That is, not only do humans fashion Nature materially by social practices but also that Nature is experienced and given meaning through the mediation of cultural discourses and representations. This common sense view of reality is the basis of a socially sensitive realist philosophy called critical realism developed by Bhaskar (1978, 1989). It has an ontology claiming the social construction of reality (that is in the 'weak' constructivist sense - see above) whilst safeguarding a conviction that underlying structures and mechanisms of the real world determine social arrangements and understandings (Hughes & Sharrock 1997, p. 164; Davies, 1999, p. 17; Blaikie, 1993, p. 59). From this philosophical perspective, critical realism rejects strong forms of constructivism that assume the biophysical world is purely a human construct, as well as the narrowly calculative rationality of positivism that treats knowledge as simply the accumulation of sense-experiences. That is, critical realism acknowledges that the biophysical world is a concrete reality, *of which humans are a part*, and that we can come to know our relations to it, *and thus to ourselves*, through dialectical processes.

If critical realism is to be a philosophical grounding for environmental education, students and their tutor need to engage with each other dialectically in a collaborative search for deeper examination of the nature of reality *at increasingly deeper levels of understanding*. That is, instead of simply assuming that reality is what experiment and experience tells us what it is (the so-called 'epistemic fallacy'), educational processes should encourage students to probe experience of deeper realities. In fact, Bhaskar (1978, p. 56) uses the term 'ontological stratification' to describe three overlapping domains of reality: the domains of the *real*, the *actual* and the *empirical*. For example, in the domain of the *real*, gravity is the mechanism that governs the Moon's orbit round the Earth since gravity is a fundamental property of all matter that scientists seek to comprehend. The Moon's gravitational pull is a causal power that cannot be observed directly but becomes manifest, for example, in the domain of the *actual* as a 'bulge' in the Earth's oceans. Experience of this bulge occurs in the domain of the *empirical* reality of everyday life as people experience the ebb and flow of the tide, ships leaving and entering a port, and turtles laying eggs on a beach. These three domains are dependent on each other but it is possible for events to occur without being experienced; and for mechanisms to be possessed, say at the atomic level, without being exercised in the actual domain as a sense experience. Incidentally, although the word 'real' is used to denote one level of reality, each of the levels is 'real' so there is some terminological confusion in describing this model of depth reality.

In presenting this stratified view of reality, Bhaskar realised that human experience and consciousness are at the tip of an 'ontological iceberg', reflecting his contention that the domain of the real is richer and more extensive than is generally accepted. While it is easy to study the order of things, it is very difficult to grasp their innermost essence. Indeed, Bhaskar saw the goal of science as a process of trying to capture ever deeper and more basic strata of a reality at any moment of time unknown to us and perhaps not even empirically manifest. It is clear, then, that positivism is unmoved by this notion of 'depth reality' since it is committed to the view that knowledge of reality is simply the way our experiments and sense experiences tell us what it is. Ontological stratification clarifies one's understanding of what learning should be about: that it is a process of seeking a vertical explanation of reality, of deepening knowledge of reality/Nature in an attempt to trace the origins of experience through the level of events to the level of structures and processes - that is, to "penetrate behind or below the surface appearances of things to uncover their generative causes" (Benton & Craib, 2001, p. 125). In the Critical Encounters below, I illustrate how two of my MA students grappled with these theoretical issues as I engaged with them in this search for 'depth reality' served by the dialectical exchanges between us.

MA IN ENVIRONMENTAL EDUCATION COURSE AND SOCIALLY CRITICAL PEDAGOGY

The MA course recruits environmental educators from schools, colleges, universities, conservation organisations, NGOs, wildlife trusts, field studies centres and community organisations (Plant, 1998; 2001). With reference to Table 1, the award of a Postgraduate Certificate in Higher Education is given after completing the first three 'single' modules, an Advanced Graduate Diploma after completing the first six 'single' modules, and of an MA is awarded after completing a further six modules comprising a 'double' module followed by a 'quadruple' module representing the research dissertation.

Table 1. Module titles for the "MA in Environmental Education" course.

Module AN1	Introducing Environmental Education: Impediments and Possibilities
Module AN2:	Perspectives on the Environment: Differing Ideologies and Utopias
Module AN3:	Enquiring into the Environment: What Knowledge? For What Purposes?
Module AN4:	Realising the Potential of Environmental Education
Module AN5:	Environmental Education in Action: Exploring Local Community Contexts
Module AN6:	Review of Professional Progress in Environmental Education
Module AN7/AN8:	World Politics and the Global Environment/Educational Implications
Module AN9/AN12:	Research Dissertation

The course has two main structural features. Firstly, it offers a widening perspective on environmental issues as the programme unfolds. It begins with a focus on the individual student's professional needs and aspirations before moving on to community environmental issues and then to global environment and development issues (Module AN7/8). However, there is overlap of all three perspectives throughout the programme as students engage with learning processes designed to develop their roles and community responsibilities against their 'home' background of ecological, development and educational issues. Secondly, the students are required to address the three strands of 'environment', 'inquiry' and 'education' throughout their learning. Specific reference to these strands arises in modules AN2 (environment/concept of Nature), AN3 (inquiry/nature of knowledge) and AN4 (action research/methodology). The three integrating strands are simultaneously addressed in a double unit, AN7/AN8, which deepens the students' critical reflections with respect to global issues and this is followed by a four-module block earmarked for the dissertation which requires all three strands to be addressed. The principal aim of the MA course includes emphasis on the extent to which it empowers students to adopt a critical stance:

> To facilitate the critical practice of environmental educators so that they become empowered to further the social conditions necessary for realising an ecologically and socially sustainable, democratic and just society.

By 'critical practice' I mean that students reflect on their professional roles critically in order to gain insight into their own and other people's understanding of the nature of reality and the socio-political origins of environmental issues with the aim of refining and evaluating educational frameworks that are effective in their different professional fields. To 'become empowered' requires that the MA students develop the confidence and means to move from insight to action in advancing a critical theory of environmental education. To realise 'an ecologically and socially sustainable, democratic and just society' is the overriding goal of a critical theory of environmental education as espoused by the MA course rationale. Critical theory draws on Habermas (1972; 1974; 1979) who claims that it is cognitive interests - the strategies for interpreting experiences - that determine the objects of reality; theoretical statements do not describe reality, they depend on assumptions embedded in theoretical constructs and common sense thinking (Blaikie, 1993, p. 97). That is, conceptions of reality depend on dialectical engagement that assumes a material reality is apprehendable through social processes, a position held by critical realists and at the heart of tutor-student interaction on the MA course.

The monitoring of students through tutorial exchanges ensures that their responses to the course materials are meeting the aim of the course and that they are fulfilling the requirements of the formal assessment system. The examples shown in Table 2 give some indication of how the assessment criteria associated with each module meet the course objectives. The bold statements are general objectives of the course, coupled with a selected assessment criterion.

In summary, I argue that a socially critical pedagogy of environmental education underpinned by a critical realist philosophy should recognise the following five precepts if higher education environmental education and environmental studies courses are to make an effective response to environmental issues.

1. That it is humanity's use and misuse of Nature that gives rise to environmental problems. (They are ecosocial problems.)
2. That human society and non-human Nature are not completely independent entities. (They are tied together, dialectically.)

Table 2. Example assessment criteria.

Develop students' powers of investigation and critical reflection.
Example assessment criterion: You are asked to be critically reflective in evaluating the views and opinions that you have recorded from your reading and any insights you have gained. (Notes made in students' Research Diaries, following specified reading for Module AN1).
Develop students' action-oriented skills that enable them to shape the social uses of Nature in ways that prefigure a future sustainable society.
Example assessment criterion: You are asked to reflect on your own values and understandings of nature. (Contribution to an essay on how, if at all, a student's environmental ethic has been influenced by the reading of selected texts, Module AN2).
Enhance the student's self-confidence in promoting well-considered arguments for implementing environmental education.
Example assessment criterion: You are asked to show the extent to which the report promotes a clearly argued strategy for EE in the organisation (In relation to a SWOT analysis and action plan for the student's organisation, Module AN4).
Assess the significance of a student's cognitive and cultural development in relation to the environment and environmental education, and to share these perspectives with others.
Example assessment criterion: You are asked to demonstrate how you establish and develop a dialogue related to an environmental issue with a person or persons outside of your professional life and responsibility (Related to a student's involvement in a small scale action research project, Module AN5).
Students to become aware of the significance of environmental education in the wider community and global context, especially with regard to moral dilemmas and other contentious issues arising from the impact of human activities.
Example assessment criterion: You are asked to produce a portfolio including appropriate ideas and extracts as evidence of your critical involvement with the MA programme and the learning experiences it has provided to date (Contribution to the student's overview of how the learning processes have influenced a student's worldview Module AN6).
Improve the student's capacity to express their understanding of environmental issues through writing.
Example assessment criterion: You are asked to demonstrate to your peers and tutor your ability to communicate effectively and show personal engagement and commitment to the debate about environment and development issues. [This is a part requirement for an extended essay on global environment-development issues, Module AN7].
Develop a student's capacity for action inquiry through an extended dissertation that illuminates their professional ability to be proactive in fostering environmental education.
Example assessment criterion: You are asked to reflect critically on how personal experience is related to broader principles, to practice and to literature. [A demonstration of a student's capacity to base an extended piece of research following prior learning on the course and related to their professional context, Module AN9-12.]

3.   That environmental education should involve students and their tutors in dialectical interaction in order for them to disclose knowledge about environmental issues at deeper levels of reality. (Their learning rejects the epistemic fallacy.)

4. That such knowledge enables students to reveal the power structures and ideology that authorizes the unsustainable use of Nature. (Their learning underlines the need for value-laden and emancipatory knowledge enabling them to identify alternative social structures, learning processes and forms of knowledge in support of social and ecological sustainability.)

5. That value-laden and emancipatory knowledge draws on appropriate forms of local knowledge to complement expert knowledge in examining the origins and possible resolution of ecosocial problems. (For example, to value indigenous, lay and tacit knowledge as possible cultural sources for learning our way out of the ecosocial crisis.)

By the end of the first year of the course, it became clear that it would be difficult to tutor students as a coherent group for not only are they widely located geographically, but also they have different professional needs, and contend with varied local ecological and cultural issues. Being remotely situated and largely independent learners, they opted for a postgraduate course offering a flexible approach to their professional needs. For example, in response to the question whether the course texts encouraged her to reflect on the possibilities of social change in her professional context, Sarita Kendall replied: *"Very much so - and this is easily adapted to/acted on in my independent context, where there is constant room for change"*. Moreover, those students who work in conservation, may be 'out in the field' and unable to contact me for long periods. Again, Sarita Kendall (Colombia) notes that the course is *"extremely flexible, allowing me to fulfil a multitude of other commitments involving travelling, etc"*. In my experience, most distance education courses seem more concerned with delivery mechanisms, materials production, hardware and software, student-tutor contact procedures, and so on rather than with consideration of the ways their students go about their learning. To ignore the diversity of cultural and social identities of the students is to risk losing important learning outcomes for the sake of assessing easily measured and manipulated content. If students' readiness to engage with the tutor in dialogue is foreclosed by the course structures and content, they may have little option but to conform to the tutor's norms and practices, and thereby be denied any consideration of their professional needs and cultural contexts. Hence, a fundamental principle underpinning the *MA in Environmental Education* course is the recognition that students should be collaborative developers of their own learning as reflected in the above assessment criteria. In this way, individualised, reflective and dialectical learning counters any tendency for technology to drive developments in distance education through rational and instrumental approaches that are liable to ignore the cultural context, learning styles and needs of students; and it allows the learners to generate their own questions and goals. To bring about a course that is based on dialogue and critical reflection requires materials that are written as 'open text'. By 'open text', I mean that students are encouraged to participate in the transformation of the study materials in ways that are meaningful in the particular socio-political contexts in which they live and work. As Charles Paxton (Japan) writes: *"It [the open text] legitimises and redeems environmental education in my teaching context. It has forced an engagement with my contextual realities"*. The following two

Critical Encounters illustrate how tutor-student dialogue coupled with the open text work in practice.

CRITICAL ENCOUNTER 1

By 'critical encounter', I refer to a situation where a matter of interpretation or understanding with regard to a student's coursework has arisen, or where the student asks for guidance about issues that impinge on their praxis. As an education officer with the Derbyshire Wildlife Trust in the UK, Helen Perkins is responsible for meeting schools' needs for environmental education, and for developing educational programmes that link wildlife concerns with the wider issues raised by the Local Agenda 21 process. In her essay for Module AN2, *Perspectives on the Environment: differing ideologies and utopias*, she reflected on how her reading of the course texts began to shape her environmental ethic. Whilst regarding *"experiences of Nature"* as an important part of this ethic, she wrote:

> [I have] embarked on a mission to understand Nature, to learn about all the things 'out there', and their interrelationships; to understand something of ecological processes. I have seen this scientific mission as a positive thing, a move on from passive and indulgent awe, but hopefully without losing my sense of wonder.

In pursuing her "scientific mission", she examined *"the philosophical question of whether Nature can be seen to exist independent of its observation"*. However, she admitted that she could still have experiences of Nature: *"the stoat running up Long Clough with a rabbit in its mouth, ... I believe there are material objects 'out there' which are not of our making.* Clearly, ontologically, a realist philosophy shapes her ethic but she admits to being confused when trying to unravel the different environmentalists' perspectives on Nature, of *"not being sure of exactly what is 'natural' and what is determined by human domination and influence"*. She sees that part of this problem of interpretation is people's differing life experiences shaping the way they perceive and respond to Nature, that Nature as 'social construction' presents problems if one is searching for a shared worldview. She asks: *"How do we decide which worldview is best?"*

Evidently, Helen was motivated to grapple with some of the difficult conceptual issues regarding human relationship to Nature so I asked her to deepen her examination of the nature of reality in her assignment for *Module AN3: What Knowledge? For What Purposes?* and to reflect on the implications for environmental education of the postmodern tendency to 'retreat from the real'. This module has three purposes: firstly, to examine whether the societal changes taking place today have implications for responding to the ecosocial crisis; secondly, to consider the relevance to our understanding of the ecosocial crisis of the natural and social sciences; and, thirdly, to reflect on the implications of postmodern thinking for environmental education. Significantly, her essay *Modernism/Postmodernism: is the future hybrid?* relates to some of the key conservation and education challenges facing the Wildlife Trusts in attempting to redefine their role in times of social change. In the following extract, she questions whether (natural) science should continue to be the main intellectual resource for solving the ecosocial crisis.

> If, as in the views of postmodernists, science is a fiction and only a fiction, what of the
> nonhuman world out there that is 'active, alive and above all real?' (Merchant, 1994, p.
> 139). If there are no truths in Nature beyond the semiotic, then Nature will 'become
> little more than a private vision and lose all claim to serving as a norm or guide in any
> degree for humanity' (Worster, 1994, p. 68). Here is justification for environmental
> destruction. Developers and even conservationists are able to argue that since Nature is
> entirely a human creation and has many manifestations, then we can do with it as we
> wish: dismantle it and re-fashion it somewhere else or in another form.

Here she sides with the critical realist ontology in opposing a strong social
constructivist perspective that denies "there are features of the world which exist
independent of discourse and social construction" (Dickens, 1996, p. 7). In
supporting a realist vision of Nature conservation in the UK, she poses a dilemma
for science. In her field of work, it is science that provides the basis for making
decisions concerning what is to be conserved: "*The rush to take action [about what
is to be conserved] sometimes appears to be based on a short-term scientific
snapshot which is presented as an unqualified statement*". She takes the example of
the American mink to elaborate this point:

> There are good reasons for the trapping of the American mink to ensure the survival of
> the water vole but who can say what the long-term effects of this decision will be? The
> subject is fraught with ethical and ecological controversy; it is also bound up with the
> relationships between wildlife organisations and funding organisations and competition
> between wildlife NGO's. [Consequently she asks:] "Is it possible to (re) create Nature
> from the privileged position of the science of ecology alone?" ... Conservation in such
> uncertain times can mean fragmentation. In many UK conservation organisations, the
> challenge is finding a way of bringing together the knowledge and experience of
> scientists and social scientists, with more everyday experiences of Nature. This requires
> an acknowledgement of the value of values, the significance of the environmental
> history of individuals and cultures, and the importance of local place.

Thus, in drawing attention to such uncertainties in Nature conservation, Helen
argues that science can become a means for domination of one NGO over another,
conservation officers' views over environmental educators, or one species over
another. With reference to her educational responsibilities for meeting UK National
Curriculum targets, she recognises that this role

> ... hangs on to its ... modernist principles of teaching science of the environment:
> stream dipping, deconstructing owl pellets, practical conservation work; tasks the
> children thoroughly enjoy and which enable teachers to achieve attainment targets.

In order to explore the possibilities for bridging the gap between the scientific
rationale for conservation practices and the culturally-modulated interpretations of
Nature such as the "*environmental history of individuals and cultures*" to which she
refers above, Helen reconsiders the potential of postmodern ideas for re-orienting
educational approaches in the context of conservation. Her suggestion is "*perhaps it
is time to listen to different and more diverse voices*".

Hence, after Helen's initial scepticism that postmodern ideas had little relevance
for redefining the role of conservation, the above extracts reveal her readiness to
appeal to postmodern insights in challenging the tendency for conservation practices
to celebrate scientific rationality and certainty. Her response represents a
reconstructive view of postmodernism in that it opens up the possibility for her

praxis that accepts its transformative features and the valuing of dialogue among different forms of knowledge. Her argument represents not only a commendable response to the conceptual issues raised in the MA course materials but also shows her willingness to reflect critically on the contribution theoretical ideas can make to the challenges facing Nature conservation. This comes across compellingly in the following extract from her Module AN3 essay:

> Postmodernism helps us to realise that there are not necessarily always right answers; it can encourage us to celebrate diversity. It can also encourage us to recognise that what the individual student brings to the learning process is a vital part of that learning. ... Environmental education in Nature conservation organisations needs to draw on both knowledge and values, and postmodernism can help us to articulate the latter. Whilst most of us, however, would wish, despite postmodernism, to continue to search for unity and totality, perhaps our focus should be upon what that searching reveals. The tensions between the old fashioned dichotomies of Nature/culture, and reason/emotion and the routes we might take to try and resolve these are an important focus for environmental education for our hybrid future.

Helen's critical reflections on the implications of postmodern ideas for environmental education in the context of Nature conservation shows the qualities I expect of an *MA in Environmental Education* student. Firstly, she draws on postmodernism to argue that educators within the Wildlife Trusts need to challenge the idea that knowledge based on one form of reasoning (i.e. science) can restrict other forms of knowledge that might have potential for seeing the way out of the ecosocial crisis. This reconstructive postmodern perspective allows conservation practices to move towards a '*hybrid future*' in which the scientists' inclination to describe a concrete reality yields to alternative socially constructed and pluralistic perspective on the biophysical world. Her critical examination of these philosophical issues calls attention to the need to explore the dialectical nature of the human-Nature relationship in making decisions about priorities in conservation practices. Moreover, this Critical Encounter draws attention to the significance of my role as her tutor in participating in a dialectical exchange of ideas enabling us both to engage with fundamental philosophical questions about the goals of environmental education.

CRITICAL ENCOUNTER 2

Sarita Kendall is a conservation officer/environmental educator working for an NGO in Colombia, and her projects involve her in the conservation of aquatic fauna such as freshwater dolphins through partnerships with indigenous communities. In our discussions about her empowerment as an environmental educator related to her ecological and professional context, we established a candid academic relationship that was as important to me as to her in drawing attention to valuing different worldviews. The following extracts from these discussions arose out of an assessment task that asked her to investigate whether post-industrial society has an 'hegemonic hold' on current social thought, and to explore the implications of this claim in the development of environmental education programmes in her local communities. After some thought, she made contact by email:

Dear Malcolm

I'm about halfway through [this module] and, yes, I can see that I shall have problems with the assessment theme:

I don't think it's intellectually trendy to' retreat from the real' in Colombia - it's more a question of how to deal with very real violence.

I don't think postmodernism has anywhere near a hegemonic hold on current social thought here, though there are obviously many manifestations linked to the media, the economy etc.

That indigenous Indian communities have little experience of what an industrial society is let alone a post-industrial one so what relevance has the question to me?

Accepting her difficulty in responding honestly to the task set and avoiding any reference to post-industrialism, I negotiated an assignment with her that asked her to reflect on the socio-cultural events and activities in the local communities in order to explore how these happenings may have given rise to the every day concerns she expressed above. In responding to each other in this way, we agreed on the following title for her assessment: *Development and Knowledge, with Reference to Indigenous Communities in the Colombian Amazon* and I asked that she consider the following questions:

a) What are the cultural issues you have to face in your work with local communities?
b) In what ways did the Rio Conference fail to address the needs of indigenous people? What arguments might the Third World put forward for dismantling the idea of development?
c) How does 'imported knowledge' through the concept of development conflict with traditional ways of seeing the environment?
d) If language and different ways of knowing are issues for you, what problems do these present for environmental educators in the non-formal educational context in which you work?

In the subsequent essay, she wrote:

Words like 'ecology', 'science', 'development', and 'project' are not part of the traditional vocabulary of indigenous peoples in the Amazon region of Colombia. Yet such words are rapidly making their way into the everyday Spanish that is now used by the majority of mestizos and Indians living along the banks of the River Amazon. Of all these words, the one with most power and weight, the one most frequently spoken, is 'development'. Most mixed Amazon communities have been seduced by the idea of development, starting with the arrival of missionaries, doctors and teachers and, more recently, competing for international funding for the sustainable use of local resources.

Clearly, she is aware of the deeper level of the politics of knowledge that 'exports' to non-Western communities 'universal' knowledge aimed at guiding the socio-economic transformation of less developed societies such as those in the Amazon. At the deeper ontological level, such a politics of knowledge is influenced by mechanisms deriving from both global and local interests and are essentially to do with issues of power. For example, she notes a renewed interest in the potential of local knowledge for resource conservation that opposes these global pressures on local communities to change their lifestyles. For Sarita's conservation organisation, a balance needs striking between her need to carry out scientific work regarding the conservation of local resources and the knowledge held by local people. She explains how to achieve this. The knowledge that is gained from her work not only serves the needs of international agencies bent on auditing natural resources in the event of their use in their global economy, but also serve to inform local people how their livelihoods can remain sustainable. Sarita considers the balance required:

> Our main projects concern river dolphins, otters, limnology, aquatic vegetation and fishing. The last is especially important because fishermen see dolphins and otters as competition and we work to persuade people not to harm aquatic mammals. Dolphins in particular have traditionally been protected by the belief that they are dangerous animals with 'spirits' and can transform into human beings, but conservationist stories and myths are being lost. One of our main aims is to build up a picture of the aquatic resources in Amazon areas with a view to advising people on sustainable use and conservation. We use non-scientific terminology and adopt local expressions and names as far as possible. We also rely to a great extent on local knowledge when locating animals, trying to establish the history of fish migrations and so on.

Nevertheless, although there is a commitment to exploring how local ways of knowing can help Sarita in her conservation work, she emphasises the difficulties in interpreting local knowledge within communities that have a strong oral tradition and a cosmology that is very different from the Western one:

> When checking the sources of information we usually ask people how they know such things. The most common answers are 'my father showed me' or 'I saw it'. There is a strong similarity with the ways of knowing documented for the Inuit by Bielawski (1996, p. 222): 'knowing through doing and experience, knowing through direct instruction, and knowing through stories told in order to convey knowledge'. Oral traditions and stories are now much less important in many of the Amazon communities than they used to be and only a few of the older people retain enough Ticuna and Cocama culture to be able to place oral traditions in the overall cosmology; in more isolated Amazon communities there are still shamans who have this knowledge. Our interest in recovering indigenous knowledge has received mixed responses: many think we are doing something useful, some think we are interfering. None of the reactions has been comparable to that of a Pacific coast Colombian Indian who said to me "First they took our gold, then they took our trees, now they're taking our knowledge" with reference to a GEF biodiversity project.

The dilemmas that Sarita faces in her educational work with Indian communities in the Amazon arise from the coexistence of particularism and universalism in development practices. While, on the one hand global capitalism intensifies the demand for assimilation of communities into the universal economic and social systems as demanded by the younger members of Sarita's communities, on the other hand globalisation leads some members of these communities to resist interference from outside so that they can retain their traditional ways. Sarita is aware that in any wholehearted commitment to Western modernisation, local knowledge may be lost but, as she explains, the irony is, that

> ... sometimes slow scientific methods simply confirm information that was readily available from local people [but the entities] that finance projects or enforce regulations are not easily convinced that indigenous knowledge is sufficient justification in itself and demand scientific evidence.

Sarita continues by reflecting on the potential of indigenous knowledge for informing her conservation work:

> There are many examples of how indigenous knowledge has taught scientists new ways of seeing: for example, the case of the Kayapo Amazon Indians whose forest fields and trail-side gardens went unnoticed by Western eyes (Posey 1994, p. 282). The idea that scientific and indigenous knowledge are complementary, and that there should be a two way process with feedback, is attractive. According to Escobar (1995, p. 215), the Third World should not be seen as a "reservoir of traditions". For genuine changes, he says, a move away from development science "to make room for other types of knowledge and experience" is needed. It is paradoxical to expect an appropriation and application of local knowledge by the very modernist framework that constantly eclipses such knowledge. Indeed, there are very few examples of two way processes that work and it is all too easy for local knowledge to become yet another 'resource' in a long history of exploitation unless the emphasis is put on the socio-cultural context, rather than the extractive process. When we first arrived in one community, the Indians said we did not need to worry about protecting dolphins because they looked after them. But gradually they have agreed that we too have something to offer when it comes to knowledge and conservation.

It is clear that Sarita gave considerable critical reflective thought to educational programmes that could best balance the need to practice Western conservation practices while accepting that local knowledge should be incorporated in any plans for the reserve. She shows empathy for the Ticuna community by her efforts to facilitate this merging of interests in sustainability, human rights and equity. Moreover, she was able to illustrate the contradictions and politics of the concept of sustainable development stemming from the dominance of the Western worldview. The global power relations that impinge on the activities of indigenous communities in the Amazon, and the potential of local knowledge for sustaining natural resources, represent causal mechanisms influencing the events in the actual domain and, in turn, in the empirical domain where the student practices conservation. Thus, we both, tutor and student, began to understand the stratification of knowledge underlying everyday knowledge through our dialectical engagement stimulated by the course processes.

CONCLUSION

The foregoing examples of Critical Encounters illustrate how a commitment to probing the realities of students' professional lives can liberate deeper levels of knowledge that counter the positivist tendency to assume reality is what their everyday experiences tell them it is. The dialectical processes of learning that I have described foster fundamental ways of thinking about motivations and worldviews, and help in examining critically those underlying political and economic forces that are responsible for ecological and social harms. By adopting a socially critical stance to learning, I believe that we, the students and I, continue to work actively against what Greene (1973, cited in Smyth, 1989, p. 211) calls "an unthinking submergence in the social reality that prevails". If education is to reflect "praxis-like" terms (Lather, 1986), and if it is to assent to dialectical processes of learning, it follows that it has to reject the hierarchical and instrumental approaches to environmental studies/education to which I drew attention at the beginning of this essay. Significantly, this rejection must be more inclusive of "oppositional viewpoints" about what constitutes teaching and learning (Smyth, 1989, p. 212). If I am to be 'oppositional' and, in turn, to propose that students confront rationalist trends in education, then I am opposed to an input-output model of education. Unfortunately, such ideals have to compete with government-inspired supervision in the form of performance indicators and appraisal schemes that currently proliferate in higher education in the UK, to contend with the enticement of consumer-oriented lifestyles, and the fancied rewards of global capitalism. It follows that higher education courses such as the *MA in Environmental Education* are likely to be subjected to increasing pressure to standardize so that they fit the mould of what is decreed as being efficient, effective and acquiescent, and, as a consequence, to deny students the rich and relevant teaching and learning opportunities that environmental education and environmental studies can offer.

REFERENCES

Baudrillard, J. (1992). Transpolitics, Transexuality, Transaesthetics, in: W. Sterns and W. Chaloupka (Eds) *Jean Baudrillard: the disappearance of art and politics*, London: Verso.
Benton, T. & Craib, I. (2001). *Philosophy of Social Science: the philosophical foundations of social thought*, Basingstoke: Palgrave .
Bhaskar, R. (1978). *The Realist Theory of Science*, 2nd edition, Brighton: The Harvester Press.
Bhaskar, R. (1989). *Reclaiming Reality: a critical introduction to contemporary philosophy*, London: Verso.
Bielawski, E. (1996). Inuit Indigenous Knowledge and Science in the Arctic. In: L. Nader (Ed.) *Naked Science*, London & New York: Routledge.
Blaikie, N. (1993). *Approaches to Social Enquiry*, Cambridge: Polity Press.
Davies, C. A. (1999). *Reflexive Ethnography: a guide to researching selves and others*, London & New York: Routledge.
Dickens, P. (1996). *Reconstructing Nature: alienation, emancipation and the division of labour*, London & New York: Routledge.
Escobar, A. (1995). *Encountering development: the making and unmaking of the anthropology of modernity*, Princeton: Princeton University Press.
Evans, T. & Nation, D. (1989). *Critical Reflections on Distance Education*, London: The Falmer Press
Greene, M. (1973). *Teacher as Stranger*, Belmont: Wadsworth.
Habermas, J. (1972). *Knowledge and Human Interests*, London: Heinemann.
Habermas, J. (1974). *Theory and Practice*, London: Heinemann.
Habermas, J. (1979). *Communication and the Evolution of Society*, London: Heinemann.
Huckle, J. & Martin, A. (2001). *Environments in a Changing World*, Harlow: Pearson Education .
Hughes, J. & Sharrock, W. (1997). *The Philosophy of Social Research*, London & New York: Longman.
Jones, C. P., Merritt, J. Q. & Palmer, C. (1999). Critical Thinking and Interdisciplinarity in Environmental Higher Education: a case for epistemological and values awareness, *Journal of Geography in Higher Education*, 23(3), 349-357.
Kidner, D. (2001). *Nature and Psyche: Radical environmentalism and the politics of subjectivity*, New York: State University of New York Press.
Lather, P. (1986). Research as Praxis, *Harvard Educational Review*, 56(3), 257-277.
Luke, T. W. (2001). Education, Environment and Sustainability: what are the issues, where to intervene, what must be done? *Journal of the Philosophy of Education*, 33(2), 187-202.
Martell, L. (1994). *Ecology and Society: an introduction*, Oxford: Polity Press.
Merchant, C. (1994). "William Cronon's Nature Metropolis", *Antipode*, 26(2), 135-40.
Mrazek R. (Ed) (1994). *Alternative Paradigms in Environmental Education Research*, Monographs in Environmental Education and Environmental Studies, Volume VIII, Troy: The North American Association of Environmental Education.

Payne, P. (1999). Postmodern Challenges and Modern Horizons: education for 'being for the environment', *Environmental Education Research*, 5(1), 5-34.

Pepper, D. (1993). *Ecosocialism: from deep ecology to social justice*, London: Routledge.

Plant, M. (1998). *Education for the Environment: stimulating practice*, Dereham: Peter Francis Publishers.

Plant, M. (2001). *Developing and Evaluating a Socially Critical Approach to Environmental Education at Philosophical and Methodological Levels in Higher Education*. Unpublished PhD thesis. Nottingham: Nottingham Trent University.

Posey, D. (1994). Environmental and Social Implications of Pre- and Post-contact Situations on Brazilian Indians: The Kayapo and a New Amazonian Synresearch, in Roosevelt, A. *Amazonian Indians from Prehistory to the Present*, Tucson: The University of Arizona Press, 271-286.

Robottom, I & Hart, P. (1993). Towards a Meta-Research Agenda in Science and Environmental Education, *International Journal of Science Education*, 15(5), 591-605.

Soper, K. (1995). Feminism and Ecology: realism and rhetoric in the discourses of nature, *Science, Technology, & Human Values*, 20(3).

Smyth, J. (1989). When Teachers Theorize their Practice: a reflexive approach to a distance education course. In: T. Evans & D. Nation (Eds.) *Critical Reflections on Distance Education*, London: The Falmer Press.

Williams, M. (Ed.) (1996). *Understanding Geographical and Environmental Education*, London: Cassell

Worster, D. (1994). Nature and the Disorder of History. In: M. E. Soule, and G. Lease (Eds.), *Reinventing Nature: responses to postmodern deconstruction*, Washington DC: Island Press, 65-86.

BIOGRAPHY

Dr Malcolm Plant is Principal Lecturer and Honorary Fellow in the Faculty of Education at Nottingham Trent University, Nottingham, UK. His recent teaching responsibilities include cross-curricular undergraduate environmental education programmes, as well as a postgraduate physics course in Initial Teacher Education. His main research interests lie at the crossing points between science, technology and society, inspiring him to co-write and tutor an *MA in Environmental Education* course by distance learning at Nottingham Trent University. This course was the subject of a book, *Education for the Environment: stimulating practice*, he wrote in 1998. Using the MA course as an educational setting for his PhD study, Malcolm examined the philosophical and methodological issues underpinning the mediation of this course to a culturally diverse group of students. As his case study shows, he is interested in the potential of the philosophy of critical realism for developing and evaluating the 'depth' of student-tutor learning experiences, and in critical action research as an approach to students' professional development related to their cultural contexts. As tutor and writer, he is contributing to the *MSc in Environment and Development Education* at London South Bank University; and he is External Examiner for a distance education Lifelong Learning programme in Environmental Studies at the University of Exeter.

CHAPTER 23

THE PEDAGOGY OF PLACE: THE ENVIRONMENTAL TECHNOLOGY CENTER AT SONOMA STATE UNIVERSITY

Rocky Rohwedder

INTRODUCTION

While those of us involved in the education profession pay great attention to the content and methodology of our lessons, rarely do we pay as much attention to the message or "lessons" of the physical spaces or environment in which we teach. We can probably agree that the place where we teach has tremendous pedagogical power, yet what are the lessons communicated by the structures and grounds of our academic institutions?

I would like to begin by offering a set of fundamental questions about the pedagogy of place. While these questions are focused on institutions of higher education, they apply to any type of educational setting, as well as to any grade level. These questions address some significant challenges for those of us who consider ourselves to be involved in education for sustainability.

Next, I have outlined our efforts at Sonoma State University to utilize the pedagogy of place, in part by creating a new facility -- known as the Environmental Technology Center (ETC). The ETC opened a few years ago and we believe that it embodies a new vision for a university campus, based on the integration of sustainable planning, design, operation and maintenance. After setting context, I have provided some specific examples of where we have so far succeeded and failed, providing essential lessons we have learned along the way.

The point of this chapter is to first assist others in conceptualizing the physical place we call campus by seeing it as a teaching tool for sustainability and then share one small story from our campus. We are one of a growing number of schools around the world who are actively engaged in demonstrating civic responsibility, saving significant amounts money and energy, reducing greenhouse gases, and teaching students an important message about our willingness as educators to invest in their future (Sturgeon, 2001).

Peter Blaze Corcoran & Arjen E.J. Wals (Editors), Higher Education and the Challenge of Sustainability: Problematics, Promise and Practice, 293-304.

SETTING CONTEXT

Tom Bender, in *Environmental Design Primer* (1976) wrote:

> We have drawn a distinction at our skins that is contrary to the most important relationships and processes that concern us and our well being.
>
> WE ARE OUR ENVIRONMENT -- what lies outside shapes what lies within.
>
> WE ARE OUR ENVIRONMENT -- the environment of our minds brings into existence both the conceptual and physical spaces we inhabit. What we are becomes our world.

This quote, written 25 years ago, had a profound impact on me as a young educator. It's a provocative idea: place shapes mind and mind shapes place. Most of us spend little time wondering or worrying about this dynamic. We might agree that at any campus there is a "built environment" comprised of buildings and landscapes, and there is a "learning environment" comprised of academic departments, classrooms, students and faculty. Bender's quote however speaks to the *relationship* between these two environments. It argues that both the built and learning environments of educational institutions fundamentally shape how students and faculty see themselves and how they see the world around them.

Whether looking from the outside in (landscape shaping mindscape) or the inside out (mindscape shaping landscape), the dynamic between the learning community and the places we call school is powerful, purposeful and pedagogic. Treating this relationship lightly, without careful attention and commitment, is a fundamental mistake that is being repeated each and every day at learned institutions all around the world, by highly respected administrators and educators.

QUESTIONS, QUESTIONS, QUESTIONS

I try to ask and answer several over-arching questions to use as a lens for viewing the pedagogy of our campuses. They address the basic "ecological footprint" (Wackernagel & Rees, 1996) of our campus facilities and grounds, the message communicated by the design and operation of our schools, and the impact that message has on our students. While I don't for a moment pretend to have all the answers to these questions, I do hope that the analysis which follows can serve as a touchstone to help each of us begin to explore how we can redefine or renew the teaching power of place within our own academic context.

The first question simply asks, "What lies outside?" The next two questions address the central notion offered by Bender, that "what lies outside shapes what lies within."

– What are the typical landscapes of higher education institutions (such as buildings, grounds and infrastructure)?
– What ideals and principles do these landscapes teach our students and us? In other words, if schools are models of something, what is it that they stand for,

and how is that vision manifested in the design and operation of all aspects of campus planning, design, and operation?

In the pages ahead I also explore what these landscapes of higher education suggest about the preferred cognitive structures and intellectual paradigms for our interior landscape? In other words, how the form and operation of our teaching and learning spaces influence how we perceive, organize, and evaluate ideas, information, appropriate civic behavior, etc.

– How is the physical (and perhaps the curricular) landscape of our institutions of higher education congruous and incongruous with the basic notion of a sustainable campus? Where it is incongruous, how do we begin to align new practices within a historically entrenched institutional context?

These questions are very board, they overlap, they give rise to other questions, and they have many sets of nested questions within them. Nevertheless, they can serve as a framework within which we can begin to explore the pedagogy of place in higher education today, as well as the tasks that lie ahead. Let's take a look at these questions one at a time.

WHAT ARE THE TYPICAL LANDSCAPES OF HIGHER EDUCATION INSTITUTIONS?

What do our students and faculty see each and every day, whenever and wherever they are on campus? They may see a set of consistent structures: large square buildings that house discrete departments (e.g. math, music) or academic units (e.g., natural science, humanities). These large square buildings hold large and small rectangular rooms in which square chairs are typically organized into rows, all facing one direction. Almost everywhere I have visited universities around the world, the structural landscape appears quite consistent.

What about the processes students see within these landscapes? Whether due to design or operation or both, they often see facilities that routinely are very wasteful of energy and other natural resources. The energy that is wasted is typically produced by fossil fuels, adding to the risks of climate change while depleting available fuels for future generations. In addition, educational facilities often generate tremendous amounts of solid and chemical waste.

It seems ironic that places designed to enlighten the mind are built and operated in a manner that often seems so mindless. David Orr, in *Earth in Mind* (1994), underscores this incongruity when he wrote:

> It is paradoxical that buildings on college and university campuses, places of intellect, characteristically show so little thought, imagination, sense of place, ecological awareness, and relation to any larger pedagogical intent.

WHAT IDEALS AND PRINCIPLES DO THESE HIGHER EDUCATION LANDSCAPES TEACH OUR STUDENTS AND US?

First let's take a look at the lessons taught to us by the structure and layout of most universities, and then we will examine how campuses are typically constructed and operated.

When viewed as a whole, the structure and location of university buildings teach us that disciplines should remain separate. Discrete fields of study housed in discrete buildings suggest that interdisciplinary study is not a valid academic enterprise (because it is not physically represented). Campus design suggests that the most knowledgeable people in our society believe we should primarily specialize within isolated disciplines. The overall physical design of universities "teach" us this preferred epistemology.

The fundamental message of higher education landscapes also suggests that to understand the world, it is necessary to just understand the parts (or a part). In fact, those who teach in higher education typically have a doctorate, which means we learned a tremendous amount about a very small aspect of the world of knowledge. The curricular structure of most institutions suggest that to become more *knowledgeable*, we need to learn more about a specific component of knowledge, not more about the relationship between disciplines or about the paradigms of knowledge. Specialization is modeled and encouraged. Integration is rare or even discouraged. Unfortunately, this message clearly promotes and perpetuates the Cartesian, reductionist worldview that lies at the root of so many of our social and environmental problems today.

Specialization *is* a good thing, but it isn't everything. New integrated and applied structures of teaching and learning are needed. Learning communities, project-based learning, and service learning endeavors, for example, offer students formal academic experiences in which knowledge is primarily interrelated and experiential. When our academic institutions only dabble in interdisciplinary and applied approaches to instruction, and instead consistently reinforce the separation of and specialization within disciplines, then they continue to reinforce the reductionist view of *how we should think.* One can't help but wonder, is the environmental crisis more a crisis of mind than a crisis of behavior — and are institutions of higher education one of the primary perpetrators of this crisis?

What about how universities are constructed and operated? The typical design of university buildings, as well as the technologies and materials chosen to build them, often teach us that energy is cheap and that natural resources are unlimited. Inefficient structures, powered by fossil fuels and built from energy-intensive materials that are harvested and manufactured with little regard for the environment, are far too commonplace on university campuses. Even with the tremendous advances in sustainable architecture and engineering made in the last few decades, campus construction continues to reflect an inefficient, wasteful paradigm of the past. The operation of these facilities, once built, reinforces this paradigm. Energy, natural resources such as water and soil, and *especially* paper, are often treated as if there is no tomorrow.

Both the design and operation of university facilities clearly teach us how the "most educated people" build and run the places in which they live and work. In this way, our educational institutions teach us *how we should act*. The lesson is clear and convincing — albeit profoundly disturbing. Educated and responsible citizens pay far too little attention to their consumption of energy, their generation of waste, or the related impacts these behavioral patterns have on future generations or other living things.

HOW ARE OUR CAMPUSES INCONGRUOUS WITH THE NOTION OF SUSTAINABILITY AND HOW CAN WE FIX IT?

This question brings the previous discussion home to roost. While it is easy to point a finger at other forces within the context of any educational institution, we should all take a hard look at what is happening on our own campus right now. While some university classes may address sustainable techniques and technologies, do buildings and operations demonstrate sustainability? If the design and operation of our facilities don't fully exemplify the fundamental application of the lessons we are seeking to impart, what then is the message that we convey to our students? If our students and fellow colleagues can't see clearly manifested on our campuses the principles of sustainability that we hope to see boldly adopted by all of society, then how can we ever expect our students to carry this vision forward?

A LONG JOURNEY IS MANY SMALL STEPS

Clearly it is time to place much greater effort on understanding how the physical landscapes of our schools impact our ability to promote a sustainable worldview. We must understand how the structure of educational institutions and their dominant paradigms impact the ways in which we see our world and act within it. While in school, the models of thought and action we see around us have a profound impact on our mental landscape.

Our challenge is to reshape or remake the fundamental structure and behavior of learning institutions so as to be in alignment with the lessons and rhythms of the natural world that supports us. And as we do, we need to share our visions, failures and successes, so that we present a bold new landscape for higher education. As the Senior Editor of *Building Design and Construction* (Flynn, 2001) noted:

> The evolution of sustainability from a fringe trend to a mainstream design principle has its challenges. Experts say communication and education are needed to spread the message that change is necessary and that sustainability can be performed affordably and in an integrated fashion.

This is a pivotal time in the (re)design evolution of schools. Over the next five years, the United States alone will spend almost $100 billion to build and renovate public schools (George Lucas Education Foundation, 2002). Fortunately, new ideas and examples, such as those in this book, are helping to bring forward a new vision for institutions of higher education, a vision based on the principles of sustainability. In the U.S., efforts such as LEED (Leadership in Energy and Environmental Design) within the U.S. Green Building Council have provided valuable aids in bringing principles of sustainability to school building design. This "whole-building design" concept now needs to be expanded to whole *campus* design so that our schools exemplify eco-efficient systems of thought and practice.

PRACTICE WHAT YOU PREACH WHERE YOU TEACH: A LOOK AT SONOMA UNIVERSITY

At Sonoma State University, efforts to become more sustainable have taken a variety of forms. While there are dozens of examples I could offer, perhaps the six most significant recent accomplishments have been: our student-led Campus Environmental Audit; the installation of 100 KW of solar electrical generation; a building *retrofit* which dramatically improved the quality of the user experience (students, faculty and staff) while saving over 40% of the historical energy load; a new very "green" student recreation center, and our Environmental Technology Center. I would like to say a bit about the last of these efforts towards creating a more sustainable Sonoma.

THE ENVIRONMENTAL TECHNOLOGY CENTER (ETC) AT SONOMA STATE UNIVERSITY (SSU)

In an attempt to answer some of the questions I have proposed, about ten years ago we set out to envision and then create a place where much of what we *explored in our courses* could be *applied on our campus*. We started by reclaiming a one-acre parking lot that we labeled the EarthLab. Step-by-step, we transformed that parking lot into food, herb and flower gardens, a solar greenhouse, and a compost demonstration area. This volunteer-based project involved university students, elementary-aged youth, "youth-at-risk," faculty from several disciplines, community members, and local businesses. Our latest addition to the EarthLab site is a new "building that teaches" and serves as a model of sustainable building techniques and technologies. We call it the Environmental Technology Center, or ETC. For those of you interested in learning more about any aspects of the building or our educational program offerings, please visit our web site (www.sonoma.edu/ensp/etc).

In the hope that it might be helpful to others considering similar projects, below is our mission statement and some related analysis.

> To design, build and operate a dynamic, interactive and integrative facility where faculty, students and community members from a wide variety of disciplines can work together in applied research training, academic study and collaborative environmental projects.

By *dynamic* we mean a building that is constantly changing so as to accommodate shifts in the focus of curriculum, advances in technology, and changes in research methodologies. We knew we could never build a "state-of-the-art building" because by the time one designs, bids, and builds, it is anything but state-of-the-art. Flexibility was a fundamental design principle.

By *interactive* we mean that the building and its users are involved in a participatory exchange. The building is informed by the weather and it reacts accordingly. We as users are informed by the building, and we in turn interact with it so as to achieve the desired results in heating, cooling, lighting, and function.

By *integrative* we mean that the building is designed to bring together on and off-campus academics and practitioners from different disciplines so we can integrate our knowledge and experience. The point here is that this is a facility designed to connect: discipline with discipline, campus with the community, and sustainability with science and technology.

We also began with a commitment to *applied research*: research that promotes direct experimentation in the context of natural phenomena and contemporary applications. We believe students need to learn how to conduct valid research, but they also need this research to translate directly to what they experience in their everyday world. This approach gives context and meaning to the academic principles introduced in course work while encouraging critical thinking and heuristic problem solving.

Finally, we believe that the ETC should promote *collaborative environmental projects*: projects that bring together students, faculty and community members with diverse interests and backgrounds to focus on projects that can benefit from collaboration.

The objectives of the ETC included:

– to serve as a center for interdisciplinary environmental science education, demonstration and research training;
– to serve as a model for other campuses by addressing campus-related environmental technologies and techniques;
– to assist Sonoma State University in becoming a model of public sector environmental responsibility and sustainability.

When you add up the mission and objectives you can see that from both a philosophical and programmatic basis we are focused on a wide range of education, training and research efforts.

What about the building itself? In thinking about building design and operation, we tried to bring together the ancient wisdom of nature and culture with the latest advances in environmental science and technology, resulting in a building that *renews* lessons that our grandparents tried to teach us. That short cuts don't make it in the long run, treat the land and all living things with respect, don't be wasteful, teach through your actions and not just your words, and strive to make the world a *better* place for future generations. With a focus on science, synergy, and sustainability, we can have beautiful, functional spaces in which to teach and learn that has far less impact on the planet while saving significant amounts of energy, natural resources, and money.

That all sounds fine, but when it comes time to put pencil to paper, what does it really look like and how does it function when it is actually built and operating? When you move from rhetoric to reality, what comes out of these ideals and grand visions? Since we always saw the ETC as potential catalyst in sustainable building on campus, when we had to make choices we were always committed to utilizing technologies and design criteria that other campuses could afford, adapt and adopt. We labeled this approach as "state-of-the-shelf technology, with state-of-the-art design."

The resulting building reflects these broad ideals and has some interesting aspects, including a building that:

- uses 80% less energy to operate than other new university buildings;
- generates, with a solar electric roof, more electricity than is required for all aspects of the building operations;
- was built with a variety of recycled and sustainable materials, including rammed earth walls, recycled plastic lumber, recycled auto glass tiles, cellulose insulation, and seaweed acoustical panels;
- responds to the rhythms of nature and local climate to tap renewable sources of energy for almost all lighting, heating and cooling needs (with no air conditioner);
- utilizes advanced optical coatings such as smart windows and low-e paint;
- has a brain (an interactive computer system) that monitors weather conditions and building demands, and then responds by adjusting building technologies such as shutters, light shelves, windows, and fans (more on this later).

Now that we can look back on our evolution and current endeavors, it is clear that to date the most important impacts of our Environmental Technology Center have nothing to do with technology per se. They are really about the teaching and learning that has occurred along the way.

From the outset, extremely important lessons were taught during the design process. For our students, helping to envision and design the ETC was a hopeful and empowering process. They saw a public institution that was for once not teaching hypocrisy and was instead attempting to "walk the talk." By participating in the process, they also learned directly about the complexity of sustainable design, the importance of knowing their "hard" sciences, and the critical role of economics in decision-making. Today, students take classes in a building that exemplifies the physics, architecture and engineering they are learning in their classes. Some of them are ETC tour guides for visiting schools and community groups, and some are employed through the center to conduct energy service projects in local schools.

We knew the ETC could be a catalyst for others who would like to build similar buildings at their school or community in the future. We have therefore sought to involve future and current building professionals from our region who could translate ideas demonstrated at the ETC into reality in other places. Initially students in the Energy Management and Design Program at Sonoma State University were involved in generating fundamental ideas for a new facility. Towards the end of the schematic design stage graduate students at UC Berkeley helped us analyze the impacts of wind flow on our building. Local architects, engineers, consultants, and manufacturers were used whenever possible throughout the remainder of the design and construction process. To continue in the role of community catalyst, today the ETC offers a wide variety of programs and projects. For example, we currently offer professional seminars and training programs as well as a community lecture series. We also give educational tours for school and community groups and we frequently host visitors from other institutions who are interested in exploring more sustainable design efforts at their own campus. Each June we also offer a major green building expo to showcase the latest tools, materials, and techniques as well as provide important networking opportunities.

One last note about the educational impact the ETC on our own campus and our campus facility personnel. It was extremely important to have the consistent involvement of our on-campus facilities personnel in our design and construction process. Facilities personnel can understandably see new efforts at sustainability to be just more work. By consulting and involving facilities personnel early and often, you'll have a far better chance of long-term support. This was especially true on our campus. Our facilities personnel were reluctant at first to be working on our "think outside the box" project. Today they have become strong advocates of sustainable building design on our campus. They have led several new sustainable building projects around campus that in many ways surpassed the accomplishments of the ETC. All of the campus buildings following the ETC have demonstrated extremely high levels of energy efficiency, renewable power generation, and environmentally-conscious design.

LESSONS LEARNED AND STILL BEING LEARNED

Perhaps there are two major lessons we have learned iin our pioneering efforts to create a learning laboratory model of sustainability on a university campus. The first lesson is that, compared to today's typical campus construction, sustainable building projects can cost more at the outset. A second major lesson is that highly complicated systems have a tendency to break down, while simple ones do not.

First let's look at higher initial costs. There are a variety of reasons for this. The planning and design costs of green buildings, especially in an academic context, can take extra time and money. For our building, we pursued an integrated design process that required additional early involvement of architects, engineers, contractors and end users. Then add in the additional layers of university facilities personnel, sub-contractors, various funding agencies, and other members of the academic community such as other faculty and students. Having all of these levels and layers of involvement in the design process can bring out some excellent ideas, yet it can lengthen the design process and therefore increase initial project costs. It's worth it in the end, but designing complicated green buildings in an academic environment can take more time. This is true not only for designing. It can also apply to rendering, bidding, and building.

Many sustainable materials also cost more. At present, many of the construction materials with lower embodied energy or recycled content are simply more expensive than materials that are not as "green." For example, when we choose shelving made from compressed sunflower seeds, we paid a premium for that. The same is true for most materials that use green or recycled content.

In our case, because we wanted our building to make a contribution to the field of sustainable design, we choose to take on additional initial construction costs through our research into new material options for concrete mixes. We were using slab-on-grade construction and lots of thermal mass, so our project was using a relatively large amount of concrete, which is very energy intensive to manufacture. Because we couldn't find in the industry what we felt was a reasonable effort to develop new concrete mixes, we made the commitment to spend almost $10,000 experimenting with various mix designs to replace energy-intensive cement with the waste products from coal and rice hull power plants. This mix would of course have to meet or exceed all engineering and safety tests for school buildings. The test results were very positive and the best mix was used to construct our building. Three years later, the final product has worked very well. The cost of this success was incurred early in the design process, but in the end it removed almost 50% of the cement from our building and the related CO_2 emissions from the air and provided stronger concrete while utilizing a current waste product.

This new mix design presented construction challenges as well. In order to properly cure this new mix design, it needed to set several days longer than what was conventional. Unfortunately, the contractor cutting the expansion joints in the slab didn't know that, so when they cut on a typical schedule they cut into concrete that wasn't fully cured, resulting in a jagged instead of a clean cut. New materials may require changes to traditional construction practices.

The ETC also utilizes roof-integrated photovoltaics to generate electricity with the sun. This is a new "peel and stick" technology that adheres directly to a standing seam metal roof. No racks or panels are needed and the manufacturer guarantees the electrical output for 20 years. The materials, installation, and control systems for this solar power system were not cheap. Initially we also had big problems with the inverters. Currently the system is performing well and we are pumping electricity into the grid. Most buildings don't factor in the cost of their future energy bills. Because our building produces more energy than it uses, it runs the meter backwards and saves money every month on the campus utility bill, especially on summer afternoons when electricity is most expensive. In the end, we know that it costs more up front to do all of this, but that the returns on our investment will come in many ways, only some of which are measured by energy savings and a lower utility bill.

The second biggest lesson we have learned is that complicated systems are hard to develop and sometimes easy to compromise. It is clear in our case that the current significant technical failure of the ETC is the highly computerized building management system or BMS. Because it was designed in part to be a laboratory for building energy analysis, the ETC is extensively wired with various measurement devices (such as light levels, air and mass temperatures, wind speed and direction). Unfortunately, the computerized system that takes in the data from these measurement devices has rarely worked. Our so-called "smart" building has been mostly brain dead for two years. We've never really had the ability to analyze the quantitative performance of the building. We can only gather the most basic of thermal performance data with the BMS. The relationships of these data points, which are primarily not being read, are suppose to determine much of how the building's air and light levels are adjusted — by ceiling fans and the opening and closing of windows, shades, and light shelves. Manual adjustments are all we have been able to do to date. We finally have a budget and plan in place to rewrite the software and hopefully begin to address this long-standing problem. When we do, we can begin to fulfill more directly the mission of this facility to serve as a building that teaches.

I would like to add that it is important to note that in spite of this limitation in building analysis and control (as well as having no air conditioner), we have maintained excellent thermal comfort and daylight levels throughout all seasons. This is a testimony to the capabilities of basic passive solar design, which has performed superbly without the aid of computer-controlled devices. Our window orientation, thermal mass, and building shell have worked so well that we have met most of our light and thermal comfort needs without the aid of a building management system.

As we continue to educate, demonstrate and research there will no doubt be many more lessons to learn. Public, working models of education for sustainability, such as the ETC, are instrumental in presenting new ideas and establishing new patterns.

CONCLUSION

Our academic campuses serve as a model of our human societies highest ideals. Students learn lessons from us not only though our lectures, texts and websites. They see what we believe and honor in the design and operation of our places of practices. That's the pedagogy of place that I hope we will turn our attention to in the years ahead. Since so many new school facilities are being planned, built, and retrofitted, this is an important time to redefine what is possible and permissible for an institution that has such an important impact on creating a more sustainable world.

The challenges ahead are many. Pioneering efforts often have to swim upstream and break through a variety of obstacles along the way. If a campus effort towards sustainability impacts multiple academic units (which it logically would), then the isolated, compartmentalized structure of most academic thinking and budgeting can be a significant problem. Integrative efforts are sometimes counter current in academic settings. Even if a building has no direct connection with particular academic units, the initial cost estimates are often higher for more sustainable projects. Unfortunately, they will be unevenly compared with conventional school buildings with smaller initial buildings costs, yet much higher environmental, social, maintenance, and operational costs. Conceptually limited and short-term fiscal accounting has to be addressed and overcome. We also learned that high technology is almost always promoted as the answer, yet simple, elegant design is often the most cost effective way to achieve energy savings and occupant satisfaction. In our case, smart design was far more important than computer-based energy management systems.

Every institution will of course have its own unique challenges and opportunities. Budget, administrative perception, academic priorities, bureaucratic reluctance, and so on can all get in the way of any new effort. Fortunately, many of the avenues to a more sustainable campus can save money, look good for administrators, advance academic endeavors in many disciplines, and even inspire some bureaucrats. Every campus will also have its' own set of opportunities. Some campuses may succeed with campus-wide initiatives, some will focus on the efforts of a particular department or student organization, some will have building projects such as our Environmental Technology Center. Some will start small, some will think grand. Regardless of the specific approaches we take towards campus sustainability, we have to accept the responsibility that schools are models of how we think, behave, live. Further, it's our responsibility to promote and demonstrate a new form of design and operation at our schools that symbolizes a commitment to sustainability — a form that sends a clear message to our students about a commitment to their future through an investment in it today.

REFERENCES

Bateson, G. (1979). *Mind and Nature*. New York: E.P. Dutton.
Bender, T. (1976). *Environmental Design Primer: A Book of Meditations on Ecological Consciousness*. New York: Schocken Books.
Flynn, L. (2001). Sustainability Takes Root. *Building Design and Construction*, April.
George Lucas Educational Foundation (2002). Edutopia: (Re)designing Learning Environments. Fall, 3.
Orr, D. (1994). *Earth in Mind*. Washington, D.C.: Island Press, 112.
Sturgeon, J.(2001). The Green Schools Revolution. *College Planning and Management*, March. 22-29.
Wackernagel, M. & Rees, W. (1996). *Our Ecological Footprint*. New York: New Society Publishers.

BIOGRAPHY

W.J. "Rocky" Rohwedder is a Professor and past-Chair of the Department of Environmental Studies and Planning at Sonoma State University. His primary teaching and research areas are environmental science education, environmental technology, and computer-aided communications. Rocky has a B.A. in Social Ecology (UC Irvine), an M.S. in Resource Policy and Management (University of Michigan), and a Ph.D. in Environmental Planning (UC Berkeley). His recent keynote speeches have focused on the pedagogy of place as well as how we can transform the place we call school into teaching laboratories for sustainability. He has been an educational consultant for numerous organizations, including the World Resources Institute, U.S. Environmental Protection Agency, U.S. Agency for International Development, U.S. Peace Corps, President's Council on Sustainable Development, The Energy Foundation, and the California Department of Education. He was a facilitator in the Middle East Peace Process, a Visiting Scholar at the George Lucas Educational Foundation, a co-founder of the EcoNet computer network, and co-founder of the Environmental Technology Center at Sonoma State University.

CHAPTER 24

DEVELOPING SUSTAINABILITY IN HIGHER EDUCATION USING AISHE

Niko Roorda

INTRODUCTION

In the Netherlands, the so-called "Stichting Duurzaam Hoger Onderwijs" (Dutch Foundation for Sustainable Higher Education) is working on several projects to strengthen the role of sustainability in the Dutch universities. One of those projects, now completed, was the development of an instrument for the investigation of the status of sustainability within a university or a university faculty or department. This instrument, called AISHE (short for "Auditing Instrument for Sustainability in Higher Education"), is now used for sustainability audits in many universities.

The instrument is built around a list of 20 criteria, divided into three groups, "Plan", "Do" and "Act", corresponding to three of the four parts of a quality circle, also known as the "Deming Wheel" or "PDCA" (see Deming, 1986). The 20 criteria are shown in see Table 1.

For each of these 20 criteria, a five-point ordinal scale is designed. The characteristics of these scales are shown in Table 2. This is based on an earlier model for general quality management, the EFQM model (EFQM, 1991) which INK, a Dutch organisation for quality management, converted into a "Five Stages Model" (INK, 2000). A higher education expert group on quality management (HBO Expert Group, 1999) developed a higher education version of this model (Van Schaik et al., 1998). AISHE was tested in 2001 in a number of universities in the Netherlands and in Sweden. At the end of that year, it was first published (Roorda, 2001). AISHE was compared with other assessment tools for sustainability in higher education by Shriberg (2002).

Facilitation, advise, and training

In 2002, a follow-up project was started in which AISHE was used as a tool for the development of a policy for the integration of sustainability in university settings. Furthermore, a training programme was (and still is) offered to (future) sustainability co-ordinators in universities, in order to enlarge the number of people able to perform AISHE audits. From 2003 on, this training is offered outside the Netherlands as well.

Peter Blaze Corcoran & Arjen E.J. Wals (Editors), Higher Education and the Challenge of Sustainability: Problematics, Promise and Practice, 305-318.

Table 1. The criteria list.

Plan	1.	Vision and policy	2.	Expertise
		1.1 Vision		2.1 Network
		1.2 Policy		2.2 Expert group
		1.3 Communication		2.3 Staff development plan
		1.4 Internal environmental management		2.4 Research and external services
Do	3.	Educational goals and methodology	4.	Education contents
		3.1 Profile of the graduate		4.1 Curriculum
		3.2 Educational methodology		4.2 Integrated Problem Handling
		3.3 Role of the teacher		4.3 Traineeships, graduation
		3.4 Student examination		4.4 Speciality
Check	5.	Result assessment		
		5.1 Staff		
		5.2 Students		
		5.3 Professional field		
		5.4 Society		

Table 2. General description of the five stages.

Stage 1: *Activity oriented*	-	Educational goals are subject oriented.
	-	The processes are based on actions of individual members of the staff.
	-	Decisions are usually made ad hoc.
Stage 2: *Process oriented*	-	Educational goals are related to the educational process as a whole.
	-	Decisions are made by groups of professionals
Stage 3: *System oriented*	-	The goals are student oriented instead of teacher oriented.
	-	There is an organisation policy related to (middle)long-term goals.
	-	Goals are formulated explicitly, are measured and evaluated. There is feedback from the results.
Stage 4: *Chain oriented*	-	The educational process is seen as part of a chain.
	-	There is a network of contacts with secondary education and with the companies where the graduates find their jobs.
	-	The curriculum is based on formulated qualifications of professionals.
Stage 5: *Society oriented*	-	There is a long-term strategy. The policy is aiming at constant improvement.
	-	Contacts are maintained, not only with direct customers but also with other stakeholders.
	-	The organisation fulfils a prominent role in society

THE AISHE AUDITING PROCEDURE

In short, the procedure for an audit is as follows:
1. Preparation with the internal assessment leader:
   – Explanation of the method;
   – Discussion of the procedure;
   – Selection of criteria and appendices to be treated;
   – Composition of the group of participants.
2. Written information to the participants.
3. Introduction with the group of participants:
   – Explanation of the AISHE method;
   – Discussion of the procedure.
4. Filling in the criteria list: by the participants individually.
5. Consensus meeting, participants + consultant.
6. Review with internal assessment leader.

Some of these steps will be explained in some more detail.

Participation

In small organisations (up to about 15 staff members) each staff member can participate. In larger organisations a group of 10 to 15 participants is selected. The group has to be representative for the complete teams of the staff members and the students, so there have to be one or more managers, a number of teachers (professors, lecturers, etc.) coming from a wide variety of disciplines and curriculum parts, some students, and perhaps one or more members of the non-teaching staff.

Filling in the criteria list (individually)

After the model has been explained to all participants, they are asked to read the part of the *AISHE* book that contains the descriptions of the five stages for all criteria. While doing this, individually, they compare this to their own organisation (i.e. a degree programme, a faculty or an entire university), and find the stage that most resembles their own situation.

In the end, they write their conclusions down on a form and hand it to the assessment leader, who combines the conclusions of all on one composite form.

Consensus meeting

Next, a meeting takes place in which all of the participants are present. At the beginning (or earlier) the copied composite form is distributed. As before, every participant has the AISHE book, in which their own scores and annotations are written: these are essential for the meeting.

All participants have an equal weight in the discussions and in the decision-making process. If possible, decisions are made based on consensus. If, however, for some criterion no consensus can be reached, the chair will conclude that, of all proposed scores, the *lowest* is the one that is decided upon: this is, because a (higher) score has only definitively been realised if all participants agree with it. In *no* case at all, decisions are made by voting.

Desired situation, priorities, and policies

During the discussion of the criteria, naturally a number of possible improvement points will emerge. This will enable the group to formulate – for each criterion – a *desired* situation. This desired situation is defined, not only in terms of a stage to be reached, but also in terms of a series of concrete targets and associated activities that will lead to the desired stage.

In order to create the necessary sense of relevance and urgency, a decision is made early on about the timeframe in which the necessary actions need to take place. This may, for instance, be a period of one year, starting at the moment of the assessment.

When for all 20 criteria, or a majority of them, policy intentions are defined in, a large list of goals and activities to be worked on can be generated. But then of course the danger is that if this list is rather long and it is a well-known fact that a policy plan with more than three priorities usually has not much chance of success.

This is why the meeting ends with the assignment of those elements in the list of policy ideas that the group considers the most important.

The results

In the end, the audit results consist of:
-   A report containing a description of the *present* situation, in the form of a number (the stage) for each criterion plus a description for each criterion in words;
-   A ditto description of the *desired* situation;
-   A *date* on which this desired situation has to be reached;
-   A list of first *priorities*, that are considered to be crucial in order to be permitted to conclude that the policy will have been successful (see Figure 1).

In the end, this package has the status of "recommendations to the management".

This set of recommendations has a good chance of being accepted by the management and to become a part of a concrete policy plan. This is because the management itself is represented in the group of participants (and that is exactly why that is so vital!). Furthermore, the recommendations have – if all went well – been chosen in consensus by a representative group from the staff and the students, so it is likely that there is support for the conclusions.

For an assessment in which all 20 criteria are investigated, the consensus meeting(s) will probably take four to five hours.

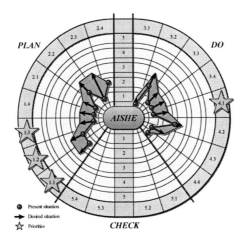

Figure 1. The result of an AISHE audit[1].

AISHE AUDIT AND TOTAL QUALITY MANAGEMENT

One of the results of an AISHE audit will be a list of improvement points leading to a better description of a "desired situation". This description is not yet a complete policy plan, and by far no activity plan on an operational level. But these can be made, using the AISHE-report. In fact, this appears to be the main task of the AISHE consultants. A few interesting cases will be shown below.

Probably, the policy plan will contain a deadline for achieving the desired situation. On that date, AISHE can be used again, in order to evaluate the results of the activities that have taken place. In this way, a quality cycle (*PLAN – DO – CHECK - ACT*) is completed. Next, the results of this second AISHE audit can be used as a starting for a new policy plan, etc.

This is exactly the way in which general quality management usually works. This is no coincidence: in the optimal situation, the sustainability policy is integrated in the total quality management. Or, to put it in a different way: the logical consequence of the implementation of a Total Quality Management System is the integration of sustainability.

This is reflected in the way AISHE can be used in a system for quality management in Higher Education: think of self-evaluations, visitations and accreditation. On several occasions, AISHE has been used as a part of a self-evaluation process in preparation of an external visitation. In other cases, it was the reverse: complaints by an external visitation committee about a lack of sustainability in the curriculum gave rise to a request for an AISHE audit.

---

[1] The balls show the present situation; the arrows indicate the desired situation. The stars on the edge mark the first priorities. The numbers in the outer circle correspond with the criteria listed in Table 1.

At present, the AISHE auditing team has contacts with the designers of the Dutch academic accreditation system, in an attempt to give sustainable development a prominent position in the accreditation system. As it seems, this will result in a situation in which universities or study programmes can adopt sustainable development as a special institutional characteristic.

For Dutch universities for professional education ("hogescholen" or polytechnics), a "Charter for Sustainable Higher Education" has been developed by the Dutch Foundation for Sustainable Higher Education (1999). This Charter differs from the Charters of Talloires (1990), Copernicus (1994) etc., in that it calls for a series of concrete activities and assessable results, as embodied in a number of sustainability levels (also referred to as Protocols for sustainability). The demands are formulated as criteria and stages of AISHE.

More than 60% of the Dutch universities for professional education have signed the Charter. Those who meet the demands are granted the Certificate for Sustainable Higher Education. About ten universities for professional education are in possession of this Certificate

RESULTS TO DATE

A number of interesting conclusions can be drawn from the audits that have been done so far:

- *Communication* about sustainability (criterion 1.3) is, so far without *any* exception, a main point for improvement. Usually, many things are less than optimal, because of a lack of effective communication between the management and the staff, among staff members themselves, with other people or parties involved and especially, between the university and the students. In all investigated cases, it was mutually decided that the improvement of communication should have first priority.
- In most audits, improvements in the *vision* and the *policy* about sustainability (criteria 1.1 and 1.2 in Table 1) have high priority. The vision and the policy often lack an explicit mentioning of sustainable development. In some cases, explicit reference is made to aspects such as ethics and social responsibility, in other cases, those are not mentioned at all. An improvement regarded as vital, is the explicit formulation of sustainability in the mission statement and in policy plans in such a way that there are real implications for all of the university's activities.
- Usually, there is considerable *diversity* of opinions and stakeholders. It is not uncommon that the opinions about a criterion vary from stage one up to stage four. It appears that there are two main causes for this. One cause is a lack of effective communication. The other cause usually is a difference of opinion about the concept of sustainability and the meaning of it in relation to education. Nevertheless, it appears to be possible to find a consensus on all criteria.

- In a number of criteria, the *manager is more optimistic* than the other participants. This, too, is usually caused by a lack of communication: often, the manager knows much more about management processes that are going on, but less about their effectiveness, compared to the staff and (especially) the students.

In the AISHE audit report, a small group of global indicators is calculated:

- The *median* of the 20 scores is, in most audits, stage one. In many of the audits, the participants define a desired situation with a median of two. Usually, the desired situation has a date that is one year from the audit date; sometimes it is one-and-a-half or two years.
- The *"Plan Do Balance"* is simply the difference between the added scores of the "Do" part (Table 1-criteria 3.1 through 4.4) and those of the "Plan" part (Table 1-criteria 1.1 through 2.4). If this indicator is far below 0, this indicates that the university is making a lot of plans and visions, but is not very successful in implementing this in is education. If, on the other hand, the indicator is very high above zero, much has been achieved with respect to its education, but there is not much support from the management, and so there is a risk that the achievements may vanish in the near future as they are not anchored in university policy.
- The *"Policy Ambition"* is calculated by adding all scores of the desired situation, and subtracting the sum of the scores of the present situation. Policy ambitions appear to vary between about five and about twenty. An interesting phenomenon is that usually the ambition is higher when the present situation is higher: it seems that the forerunners like to preserve their front position.
- The *"Distance to Protocol"* is related to the already mentioned Dutch Charter for Professional Higher Education. When this distance is zero, the audit indicates that it is likely that the Certificate will be granted.

With regards to this Certificate for Sustainable Higher Education, some interesting conclusions can be drawn as well:

- In some cases where the Certificate was granted to university departments, a subsequent AISHE audit pointed out that in the present situation the demands for the Certificate were definitely not met. The most likely cause is that the method that is used for the Certificate assessment, mainly based on filling in a series of questionnaires by university staff members themselves, has not a high validity, mainly because the staff is eager to obtain the Certificate. During the AISHE audit, although also being a self-evaluation, the critical role of the AISHE consultant is a guarantee that the validity of the exercise is higher. From the middle of 2003, the tests for this Certificate will be done exclusively through AISHE audits.
- Quite a lot of Dutch colleges for professional education (compare with the former polytechnics in the UK-system) show a real interest in being able to sign the Charter and obtain the Certificate. AISHE audits clearly show that there is a strong positive effect of the existence of the Certificate on the process of developing and implementing sustainability in the education and the university operations. This implies that it is worthwhile to investigate whether such a

Certificate could be a means of strengthening the process of implementing sustainability in universities in an international context.

THE CASE OF AN ECONOMICS STUDY PROGRAMME

In a large university, an AISHE audit was done for a study programme in Economics (Van den Bergh & Withagen, 2001). The median of the present situation was in stage 1; in fact, 70% of all scores were in stage 1 or lower. An interesting set of improvements was suggested for a desired situation, to be reached in one year; the policy ambition was 14, which is rather high. The high priorities were set on the usual criteria 1.1, 1.2 and 1.3 (*vision, policy* and *communication*), as well as on criterion 4.1: *curriculum*.

A small part of the resulting audit report will be described here.

Criterion 1.3. Communication

Present situation: Stage 1
Only a few staff members know that the Copernicus Charter has been signed. Nevertheless, sustainability is a frequent subject in meetings, especially of the management. In several educational projects, sustainable development is present, for instance in the projects of rural renewal and urban renewal. One education development group had the task of implementing sustainability. The manager has asked students to investigate "phase 3".
Desired situation: Stage 2 – *High Priority*
In order to spread the management vision on sustainable development, there must be an intense communication in the near future. A good opportunity is the coming process of curriculum redevelopment. The staff may be involved through e.g. the university magazine and the e-mail news bulletins, and also in meetings. Students may be informed through brochures and on information days.

Criterion 1.4. Internal environmental management

Present situation: Stage 1
Environmental management is not a part of the policy and the management. As a consequence, you can see a lot of polluting processes. Many people are unsatisfied with this.
Desired situation: Stage 2
Start with the main aspects. Within the team, attention will be given to paper waste, printer toner catering waste, use of energy.
Students will be asked to design and perform a quick-scan. They could make use of the ecological footprint, see www.novib.nl).
A problem is that the department that is now investigated has no own authority regarding many environmental problems. Therefore, the situation must be discussed with the utilities department. The manager will take this initiative.

Criterion 2.1. Network

Present situation: Stage 1

> We keep regular contacts with companies in the professional field. But sustainable development is not an important aspect in these contacts. In one large student project, there is a relation with the environment department of the local government.

Desired situation: Stage 2

> Sustainable development as an aspect of the contacts with the companies has to be given a high priority: it should be anchored within the university department. A small number of partners in industry (e.g. 5) will be selected with which state of the art expertise about sustainable development will be exchanged intensively. These will be used for guest professors, for traineeships, and for curriculum development.

The manager of the study programme was optimistic. He thought it was a good set of intentions: ambitious, coherent and realistic at the same time. Together with his co-ordinating team, he designed an activity plan to realise all intentions within the chosen time period of one year. In this stage, no AISHE consultant was involved.

Half a year later, part of the intentions had been realised. A basic module in sustainable development for the propaedeutic year was made, and was about to be used in practice. An educational project for the students was designed and already used once. From a methodological viewpoint, much had been realised. So, the work on criterion 4.1 (curriculum), which had a high priority, was rather successful. On the other hand, sustainability had not yet been integrated into the curriculum in a systematic way: the sustainable elements were not logically connected as a thread throughout the curriculum.

At the same time, a Mission Statement had been made for the entire university. The team of the Economics study programme had had a role in it. However, although this Statement contained a number of elements that were sustainability-related (e.g. ethics, professional responsibility), the concept of sustainability itself was not mentioned explicitly, and the text was rather abstract, so it was difficult to draw conclusions from it with respect to a policy or to concrete activities. The most important problem, according to the manager, was the definition of the professional profile of the future graduates (criterion 3.1): he and his team experienced a gap between the university vision, as formulated in the Mission Statement, and the professional profile of the Economics programme. If it would be possible to make the vision more explicit, i.e. to operationalise it, then it could be used to formulate the professional profile, and next to redesign the curriculum in such a way that sustainability could be integrated systematically.

On the subject of communication (criterion 1.3), some achievements were made, but again not in a systematic way. Some communication on a university level about the Mission Statement had taken place, but since sustainability was not made explicit, this was not fully successful. In the university magazine, some attention had been given to sustainability in the education and the university operations, but here too, the risk existed that this was not going to be repeated. Besides, sustainability had been on the agenda of some meetings.

All in all, in half a year a rather good job had been done, but there was still much more to be done. At that moment, the manager and his team were somewhat confused about the sustainability policy as a whole. The AISHE team was consulted, and it had two meetings with the manager and his co-ordinating team. During the first meeting, the situation was analysed. It appeared that the main problem at that moment had to do with communication. As a result of the AISHE audit, a necessity was felt to intensify the communication about sustainable development, and so, this communication had become a target on it's own, somewhat neglecting the reasons why communication was important. So, all kinds of communication had been used in the last half-year, and now they did not know what to do next.

At the same time, it was important to revive the involvement of the staff and the students, which had faded away a little bit in the six months after the AISHE audit. So, a necessity was felt to find out a way to systematise the way in which the communication with all kinds of stakeholders was made. However, the team could not think of a way to do this. A simple scheme for the communication system about sustainability was suggested to the co-ordinating team, in the form of a matrix. On one axis of this matrix, a variety of kinds of communication are set out, like: "Give information to", "Receive information from", "Generate information together", "Create support", and so on.

On the other axis, a list of possible stakeholders was put, being the result of a stakeholder analysis, for instance:

Teaching staff - Students - Management - PR department - Professional field - Public media – Government, and so on.

In the cells of this matrix, it was possible to fill in two kinds of things:

1. the reasons for communication; and
2. suitable communication tools.

In this way, it was possible to discern all kinds of reasons for communication, and for each of them think of suitable tools to realise this communication systematically and periodically.

The manager and the team considered this a realistic way to invent the communication system they needed. Besides, they judged that it could be a good starting point for a systematic development of sustainability within the university. Based on good communications, it would be possible to revise the Mission Statement in co-operation with staff members, students and of course the central university board. Next, based on an operationalised university policy for sustainability, a policy plan could be designed, aiming at several things, among which the definition of the profile of the graduate, to be formulated as a set of professional competencies. This looked like Figure 2, also showing a possible way to enhance this scheme with some next steps.

The Economics team is working along the lines of this scheme at present. Some time ago, a Certificate for Sustainable Higher Education has been awarded to them.

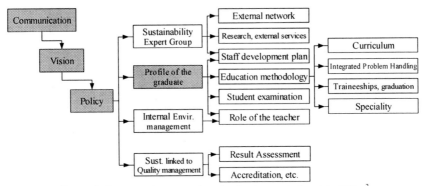

Figure 2. A systematic approach to the development of sustainability.[2]

THE CASE OF AN ENVIRONMENTAL TECHNOLOGY STUDY PROGRAMME

In the Netherlands, for almost all of the university programmes in environmental science and -technology, the number of students is decreasing strongly. At the same time, investigations in the professional field indicate that the need for environmental experts will diminish in the coming years. Because of this situation, several studies were performed. Dröge and Schoot Uiterkamp (2000) looked at the future needs of the professional field for environmentalists, and attempted to redefine the professional competencies they will need. In another investigation a commission of the Dutch Association of Universities for Professional Education ("HBO-Raad") looked at the question what the relation should be between the environmental study programmes and sustainability, regarding the fact that more and more non-environmental university programmes are integrating aspects of sustainability in the curriculum: the environmental programmes are "loosing territory" (HBO-Raad, 2000).

In the final report of this latter investigation, it was recommended that three major profiles are to be discerned for the future environmental experts: the *consultant*; the *researcher*; and the *process manager*. For all of those profiles, an interdisciplinary role as part of a team of various disciplines will be vital.

After the report was published, many of the universities with environmental programmes were searching for a new definition of this programme, a new "*raison-d'etre*".

In this context, an AISHE audit was done in one of those environmental programmes. Not surprisingly, the results showed an emphasis on the need for the development of a new vision. The high priorities for improvement were criteria:

1.1 - vision
1.2 - policy

---

[2] Shaded boxes suggest high priority areas.

1.3 - communication
2.3 - staff development plan
3.1 - profile of the graduate
4.1 - curriculum

The problems investigated in the above mentioned studies were reflected in the discussions during the consensus meeting. A sample of the audit report reflects this clearly. The present situation was described as follows:

> A "kind of a" vision exists, but the contents are not formulated very explicitly. There is much emphasis on environmental subjects, and not enough attention to sustainable development in general. That is to say, sustainability is interpreted too narrowly as "mainly environmental matters".

> It is virtually impossible to check whether the students acquire the right and enough professional competencies, because the staff team hardly has an idea about what kind of professional competencies related to sustainable development they should be taught.

The ambiguity regarding the role of the environmental professional, appearing during the audit, was formulated even stronger when, a month after the audit, a meeting took place of the co-ordinating team of the study programme. There, it appeared that there existed a lot of confusion about a mixture of subjects, all related to vision, policy and the profile of the graduate.

The discussions had been complicated by an attempt to interpret the recommendations of the HBO-Raad report. The emphasis in this report on interdisciplinarity had been interpreted by some team members as a recommendation to see the environmental expert as specialising in interdisciplinarity, as a "spider in a web", as the one who was going to connect all kinds of other specialists with each other. This seemed as an impossible task, because in this vision, the environmentalist almost had to be an expert in all kinds of specialities. In this vision, the environmentalist was to be seen as an "interdisciplinarity specialist".

In contrast, some other team members thought of quite another interdisciplinary role, where the environmentalist still is a specialist in his own field, and functions as just one of the members of an interdisciplinary team. Figure 3 shows the distinction between the two visions.

It took a lot of discussions, before this distinction was made explicit; at the start, it all seemed like a diffused set of opinions. After this distinction was discovered, clarified and understood by all, the team concluded that it was possible to structure the decision process in a step-by-step approach.

First, decisions about the profile of the graduate should be made: especially, a fundamental choice between the two possible roles of the environmentalist should be made. From that, a vision about the relation with sustainability could be developed, followed by a policy plan leading to a curriculum and to a staff development plan for sustainability subjects.

Even before that, it was vital to develop a good plan for communication with all kinds of stakeholders. Only if there was a solid communication structure, guaranteeing that all interests of the professional field, of NGO's, of governments and of other stakeholders would get the right attention, it was to be expected that a valid and durable profile of the graduate could be developed.

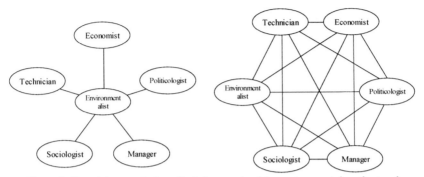

Figure 3. Two visions on the interdisciplinary role of the environmental professional.

As a consequence, a development scheme was designed which, superficially, resembles the one shown in the earlier case of the economical programme (section 5), but in reality differs fundamentally. This is shown most clearly by the different position of the "profile of the graduate" (see Figure 4). Before it is possible to discuss the vision, the profile of the graduate has to be made clear.

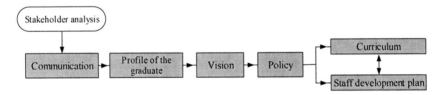

Figure 4. The development scheme for the environmental programme.

In terms of quality management: the environmental programme development is in a quality circle, a "Deming wheel", at the moment. The AISHE audit functioned as a "Check", testing the achievements of the years before. The discussions after the audits can be seen as the "Act" phase. The development scheme that resulted was the start of the "Plan" phase, which went on afterwards as the designing of a policy plan for the coming years. And at this moment, the staff is performing this policy plan: "Do". In one or two years, a new AISHE audit will be done, closing the quality circle and assessing the results.

CONCLUSIONS

The cases described above show how the implementation of sustainable development in a university, i.e. in the vision, the policy, the organisation and the education, can be treated as a part of the general quality management, and how AISHE can be of help therein.

At the moment, there are no examples yet of university departments where the whole Deming wheel has been completed, that is, where an AISHE audit has been done twice. This will be the challenge for the coming years: to investigate what the results will be of a time period after an AISHE audit, and to investigate through a second audit what the effects are of an approach towards sustainability in higher education in a quality management style.

REFERENCES

Copernicus Charter (1994). The University Charter of Sustainable Development of the Conference of European Rectors (CRE), Geneva 1994. See http://www.copernicus-campus.org.

Deming, W.E. (1986). *Out of the crisis.* Cambridge: MIT Press.

Dröge, F. and Schoot Uiterkamp, T. (2000). Higher environmental education and the environmental labour market in the Netherlands – a survey of the influence of internal and external factors on higher education environmental programmes and the labour market for environmental professionals in the countries of the European Union. Essence Programme, 2000.

Dutch Foundation for Sustainable Higher Education (1999). *Charter for Sustainable Higher Education.* See http://www.dho21.nl/hbohandvest.

EFQM (1991). *EFQM Model.* European Foundation for Quality Management. http://www.efqm.org.

HBO Expert Group (1999*). Method for improving the quality of higher education based on the EFQM model.* 3rd version, Hanzehogeschool (representative), Groningen, Netherlands. Translation of: Expertgroep HBO.

HBO-Raad (2000). *Van milieu tot duurzaamheid.* Eindrapport van de Verkenningscommissie Milieuopleidingen. Den Haag: HBO-Raad.

INK (2000). *Gids voor toepassing van het INK-managementmodel.* INK, 's Hertogenbosch, Netherlands

Roorda, N. (2001). *AISHE – Auditing Instrument for Sustainability in Higher Education.* (Available in English and in Dutch). Dutch Committee for Sustainable Higher Education (CDHO), Amsterdam.

Van den Bergh, J. & Withagen, C. (2001). *Economie en Duurzame Ontwikkeling.* Netwerk Duurzaam Hoger Onderwijs en UCM/Katholieke Universiteit Nijmegen, Netherlands.

Van Schaik, M., Van Kemenade, E., Hengeveld, F. and Inklaar, Y. (1998). *The EFQM based method for continuous quality improvement adapted to higher education.* Proceedings of the EAIR Forum, San Sebastian, Spain.

Shriberg, M. (2002). Institutional assessment tools for sustainability in higher education. *International Journal of Sustainability in Higher Education,* (3)3, 254-270.

Talloires Declaration (1990). *The Presidents Conference, University Presidents for a Sustainable Future - The Talloires Declaration.* Talloires 1990. See: http://www.ulsf.org.

BIOGRAPHY

Niko Roorda, works as a consultant for universities on the implementation of sustainable development in higher education. He is a member of the Dutch Committee for Sustainable Development (CDHO).

He studied theoretical physics at the universities of Leiden and Utrecht. After working as a teacher for some years, he developed a study-programme for Sustainable Technology in the Brabant University for Professional Education, After functioning as the manager of this programme for a number of years, he developed and managed the Cirrus Project, which worked on the implementation of sustainable development in the curricula of the study programmes of the Faculty of Technology of the same university. This project was awarded the Dutch National Award for Innovation and Sustainable Development in 2001.

CHAPTER 25

CURRICULUM DELIBERATION AMONGST ADULT LEARNERS IN SOUTH AFRICAN COMMUNITY CONTEXTS AT RHODES UNIVERSITY[1]

Heila Lotz-Sisitka

Visvanathan has argued that the university is a futuristic institution that makes innovative use of the past. But at the same time, as one of the last surviving of medieval institutions, in fact the only one of its guilds to adapt and survive in modern society, the university has still remained a microcosm of the walled city. Today the wall may not exist, but the separation between the university and society is real. It is a source of tension, but also a source of creativity (Odora Hoppers, 2002, p. 22).

INTRODUCTION

Since the 1992 Earth Summit in Brazil, many African countries have signed a number of multi-lateral environmental agreements, including Agenda 21 and numerous environmental conventions, which influenced local policies and practice, including education. South Africa has a history of socially unjust conservation laws and protection of the land for the benefit of the few. This has been to the detriment of the majority of South Africans, most of whom were disproportionately affected by socio-ecological impacts of apartheid legislation. These impacts include environmental degradation such as soil and water pollution and poorly serviced living areas and workplaces, leading to health risks. Post apartheid South Africa has emphasized the relationship between social justice and ecological sustainability, with environmental issues and risks being closely linked to human rights and social responsibilities in numerous policies, including the Constitution of 1996.

While these policies signal a transformational intent, debates surrounding the World Summit on Sustainable Development (WSSD) held in Johannesburg in 2002 indicated that environmental issues and risks are increasingly complex and contested (Beck, 1992; 1999; 2000). They require deep-seated social transformation and reflexive responses which include challenges to epistemological frameworks and the

---

[1] The full name of these courses is the Rhodes University / Gold Fields Participatory Certificate course in environmental education. For purposes of readability, I use the abbreviated version: Rhodes University Participatory Courses. Over the years the course has expanded into different countries and contexts, and have come to be known as: The RU/ Swaziland Participatory course or the RU/SADC Participatory course etc.; depending on the partner group involved in implementing the course.

Peter Blaze Corcoran & Arjen E.J. Wals (Eds.), Higher Education and the Challenge of Sustainability: Problematics, Promise and Practice, 319-333.

ability to engage with the processing of 'unawareness' (Beck 1999; Raven 2003; see also Popkewitz & Brennan; 1998). Paradoxically, in spite of being 'flooded' with transformational policy and intent, the current context in South Africa reflects painfully slow socio-economic development and redistribution, along with an increase in the range and extent of socio-ecological issues and risks that have their roots in inappropriate development frameworks (Odora Hoppers, 2002)[2] have also increased. Many of these issues and risks manifest at a local level, affecting the livelihoods and sustainable living options of local people. The education system is directly affected and involved, and Higher Education Institutions are being called upon to draw on their research and teaching resources to assist society in responding to these sustainability issues. Many of the problems outlined above are prominent in the Grahamstown area, a small university town in the Eastern Cape Province, where Rhodes University is situated.

HIGHER EDUCATION – COMMUNITY RELATIONSHIPS: ESTABLISHING THE RHODES UNIVERSITY PARTICIPATORY COURSES

In 1990 Rhodes University (RU) recognized the need to respond to emerging socio-ecological issues and risks through education, with the establishment of the Murray & Roberts Chair of Environmental Education in the Faculty of Education[3]. A key focus of the Chair has been to respond to socio-ecological issues and risks (sustainability issues) through the provision of education and training programmes and educational research. One of the early programmes launched by the Murray & Roberts Chair in 1992 was a community-capacity building programme, which aimed to support environmental educators working in community-based and other settings. Influenced by international developments at the time (post Rio), the changing political economy in South African society, and recognition of a need for professional development of environmental educators (the field was relatively new at the time), a participatory ethos characterized the first community capacity building course run in partnership with the Chair in 1992. The course typically involves one year of semi-distance study with three national workshops and an average of ten regional tutorials. The course is run nationally on an annual basis.

In the mid 1990's the course started expanding to other countries in the southern African region (including Zanzibar, Zimbabwe, Swaziland, Malawi, Zambia, Angola, Namibia) and into diverse community and institutional contexts (including industry, conservation and teacher education) as more students became tutors and / or course co-ordinators. The participatory ethos, the workplace-based and community oriented focus of the course; and the tangible outcomes appeared to appeal to many environmental education practitioners in southern Africa. To support the rapid expansion of this community oriented programme, Rhodes University

---

[2] The crux of the problem of development theory has been that of how to overcome its Eurocentric bias. By Eurocentric bias is meant development theories and models rooted in Western economic history and consequently structured by that provincial, unique, though historically important experience (Odora Hoppers, 2002).
[3] The Murray & Roberts Chair of Environmental Education is the only Chair of Environmental Education in Africa.

established the Gold Fields Environmental Education Service Centre in 1997; with the explicit aim of extending these community capacity building programmes started by the Murray & Roberts Chair in 1992. Expansion of this programme reflects current sustainability education thinking which emphasises that education for a sustainable future should engage a wide spectrum of institutions and sectors, including but not limited to business/industry, international organisations, youth, professional organisations, non-governmental organisations, higher education, government, educators and learners in schools (UNESCO, 1997). Participants on the courses are associated with all of the above institutions and community sectors, enabling Rhodes University to support community developments in a range of contexts. To date more than 700 environmental educators from a wide range of sectors have participated in the RU participatory course programme in the southern African region.

Contrary to most courses in the University, the participatory course was established as an 'open entry, open exit' course, and requirements for certification are linked to evidence of professional development and participation. In 1996 the Rhodes University senate agreed to certify the course and since then participants on the course have been awarded a Rhodes University Certificate in Environmental Education[4]. Community-based socio-ecological issues and risks, and the work of course participants in responding to these, forms the core of the course. All assignment work involves review of work-in-context, thus fostering immediate linkages with community contexts and issues. Participants are encouraged to become tutors, as a way of deepening professional competence. The participatory ethos involves both students and tutors in constructing and learning on the programme, as both 'learners and educators' (ICAE, 1993).

The course has always been characterized by reflexivity and change. Early on, course developers deliberated critically on dominant educational practice and trends in environmental education. These deliberations illuminated that environments and environmental issues (at the time) were mainly associated with biophysical problems.

Assumptions about education were often linked to linear models of awareness raising and behaviour change[5]. Through ongoing reflexive deliberations and research, a stronger focus on long term *social processes* (O'Donoghue, 1997), *social critique* (Huckle, 1991; Fien, 1993) and *social change* (Popkewitz, 1991; Janse van Rensburg 1995) emerged, introducing a concern for *history, context and socially*

---

[4] Initially the course was only certified with a certificate of participation. With participants wanting formal recognition for their engagement with the challenging course processes, Rhodes University agreed to offer a certificate in Environmental Education. In 2002 new qualifications were passed by the South African Qualifications Authority, and the course has been re-designed to enable participants to gain 24 credits towards a fourth year Advanced Certificate in Environmental Education, a nationally recognized qualification. This has required numerous changes to the course, particularly relating to assessment.
[5] Janse van Rensburg (1995), identified and critiqued some of the more apparent views of what change in environmental education entails in a South African context. She identified "linear, rationalist models of change" in environmental education. These include: Change as restoring order (centre to periphery or managerial orientations to change); Change as the resolution of practical problems (community problem solving orientations to change which are often underpinned by liberalist ideologies); and Change as reconstruction (a critical orientation to change).

critical perspectives in course processes. Course participants are encouraged to probe the history of environmental issues and risks: globally through critiquing modernist trends and development models; and locally through considering the nature, causes, and impacts of issues and risks in their contexts. They are also encouraged to consider social processes associated with the environmental and educational problems they experience, and thus to take a 'deeper', more critical view of human-environment relationships and educational practice in their community and work-related contexts.

From the start, the course emphasized issues of *environment and development,* and the close relationship between these, thus reflecting many of the tenets of what has come to be labelled as *'Education for Sustainability'* in recent years[6]. The course is characterized by environmental education processes that involve both tutors and learners in "...promoting sustainable development and improving the capacity of people to address environment and development issues" (UNCED, 1992, Chapter 36, p. 2). The course supports a perspective that sees environmental education processes as processes of *social transformation and change* (Janse van Rensburg, 1995). Following earlier deliberations on linear rationalist models of change, Janse van Rensburg (1995), recommended a 'reflexive perspective' or orientation to change, which emphasises process rather than product, which is not concerned with a linear model of change, and which does not rely on doctrines, 'tools' or 'methods' to bring about change. She sees environmental education as a '...responsive process of change', and sees reflexivity in environmental education as a tentative engagement with change processes through "... collaboratively developing capabilities (tools, resources, action competencies) to deal with and encourage change in local contexts" (ibid, p. 168). This became a key focus of the course assignments and associated support processes as tutors and course participants worked with each other in developing the tools, resources and action competencies to respond to sustainability issues and risks in diverse community settings.

RE-ORIENTATION OF PEDAGOGY AND TEACHING PRACTICE: ARTICULATING FEATURES OF THE COURSES

In 1997, UNESCO published a document titled *"Education for a Sustainable Future: A Transdisciplinary Vision for Concerted Action"*, which indicated that 'too little has been achieved' in the environmental arena. It re-emphasised the importance of life-long learning; curriculum reform and highlighted the need for a fundamental re-orientation of education and training (including Higher Education) towards sustainability. An important facet of this re-orientation involves how learners are viewed in Higher Education and community-oriented settings; and how this influences curriculum design. Higher Education (and adult education more broadly)

---

[6] See for example UNESCO-UNEP, 1996; UNESCO, 1997; Huckle & Stirling, 1996; BGCI, 2000. While many educators agree that the agenda of sustainability should be furthered by education, some educators have begun to question the instrumental rationality adopted by much of the EfS 'doctrine', and the assumptions that 'sustainability' provides an adequate conceptual framework for education (for example Jickling, 1999).

in South Africa has been influenced by the institutionally located concept of 'andragogy' (normally used in opposition to pedagogy to indicate that adults learn differently to children). This has led to wide-ranging technological assumptions influencing curriculum and teaching practice in adult education. Andragogy, has, however, been critiqued for being an 'abstract form of individualism' (Hanson, 1996, p. 98), which assumes general characteristics and similar life experience amongst a group of learners, leading to de-contextualised curricula and teaching practice (see also Usher et al., 1997; Edwards, 1997). These critiques of dominant forms of adult education emphasise the need for contextually located engagements with learners themselves within their real life situations, which address the context and learners' active experience of the world.

In an evaluative review of the Rhodes University participatory course in 1998, Janse van Rensburg and Le Roux (1998) identified key aspects of the course orientation, which have come to shape the pedagogy and teaching practice in the course. Further deliberations on the outcomes of this evaluative review led to a clarification of 'key features' of the course curriculum. It also involved clarification of processes of *curriculum deliberation*, as a way of orienting community-based adult education courses in Higher Education (see Lotz, 1999). This led to an exploration of trends in adult learning, particularly as these related to our interest in enabling environmental learning through higher education – community links.

Trends in adult education and environmental education professional development

The need to involve adults from all walks of life in addressing environmental issues has led to the expansion of adult education programmes aimed at responding to sustainability issues. This is consonant with a more general trend towards life-long learning and the 'formalising' of adult learning, with consequent growth in the number and range of professional development courses and programmes for adults (as is evident in the expansion of the RU participatory courses). From another perspective, Usher et al. (1997) reflect on the changing landscape of adult education and note that changes in adult education signal a dissatisfaction with the dominance of external control within technical-rational models of practice, which have, until very recently characterized course and curriculum development in adult education and training world wide (see also Robottom, 1987). These approaches are, unfortunately, still relatively dominant in South African Higher Education settings. Changes in adult education have occurred alongside a broad leveling of the power gradient in society (globally and nationally)[7]. These trends affirm the significant place of *the learner in learning*, rather than the significance of the institution or traditional practices of teaching or learning. With affirmation of the significance of the learner's role in the learning process, comes a recognition that educational goals, forms and practices are shaped by diverse cultural and socio-ecological contexts, not by universal norms or institutional models and goals which are implicit in these. Thus, learners (or participants) become involved in defining what constitutes a learning opportunity and are enabled to provide and shape the course design. We

---

[7] See e.g. Usher et al. (1997, p. 26); Doll (1989); Popkewitz & Brennan (1998) and Muller (1997).

found that a review of trends in adult education, together with the articulation of the orientation and key features of the course (see Lotz, 1999), helped to provide the re-orientation necessary to find new ways of working *with* course participants in community contexts. Following this review, we were more able to develop course frameworks and models of process which are *deliberated with participants* in order to enable responses to diverse and complex environmental issues and risks in different contexts.

Key features of the course

In the review five 'key features' of the course curriculum were identified, and articulated in the form of open-ended questions to guide curriculum deliberations (see Figure 1). All are key to affirming the learner's role in defining learning (Lotz, 1999). These were articulated in relation to the course orientation, which in turn is based on epistemological considerations which guide curriculum decisions. As noted above, the course orientation includes a consideration of *history and context; reflexivity; social construction of meaning; social critique* and *social transformation.* We have found many of these curriculum features to be important in fostering more sustainable living practices and better environmental management through education; although application of these key features in curriculum processes have not been without their challenges.

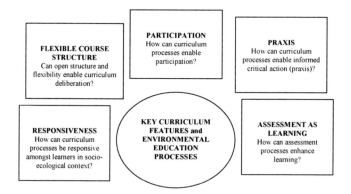

Figure 1. Key features of the course curriculum

These key features include:[8]

1. *Responsiveness:* In the RU participatory course tutorial group in the Grahamstown area in the Eastern Cape (there are tutorial groups in different provinces) participants have developed tools and materials to support educational activities in the following projects or programmes (amongst others): an environmental health project in an early childhood education school; a waste management project amongst unemployed citizens; a tree planting project to 'green' degraded urban areas; a river audit to establish levels of pollution; a project involving the documenting of indigenous knowledge of water conservation practices; and an HIV/AIDs education project at the local clinic (similar examples are evident in all tutorial groups). Significant to the defining of these (and other) projects and programmes developed by participants, is the deliberation of curriculum processes that allow for responsiveness to the needs of the participants (and their community-related challenges); *and* the complexity of the socio-ecological issues they engage with. Responsiveness in curriculum processes, however, requires a consideration of the context and constraints of the course tutors, as well as the context and constraints of the participants. This would, for example, involve material / technical aspects such as whether tutors are able to interact with participants from far flung areas in the same way that they can if learners are geographically close. Language, widely recognized as crucial to the fostering of learning, has also emerged as a complex issue in the context of the course (the course invariably takes place in multilingual settings in which English is normally the second or third language spoken by course participants). Heylings (2000), for example, noted that the language in which the course is offered may create 'distances' between learners and the course, which reduces its responsive capacity. We have found, for example that having two tutors in each region, at least one of whom is able to converse in the local dialect, increases responsiveness within the course. Responsiveness also requires insight into the often complex contexts in which adult learners work and live. For example, in one case a course participant was unable to respond adequately to the problem she had identified as a result of the gender-specific cultural customs that prevented her from interviewing 'senior' male figures in the community (in this case the village chief). Responsiveness also requires in-depth insight into *social habitus* (Bourdieu, 1998), or those socially embedded factors that constrain agency (which are often 'hidden' or even 'unconscious'); and embedded power relations that operate on, or 'beneath' the surface in different social contexts. The complexity of contextually located issues and risks can therefore, in itself, be a constraining factor in enabling responsiveness in courses such as the RU participatory course.

2. *Flexible course structure:* In the RU participatory course, the course curriculum is negotiated with participants within a framework of open-ended themes.

---

[8] See Lotz (1999) for a more in-depth review of these features.

Participants are encouraged to present their expectations and 'stories' of practice at the start of the course, and through this a range of issues that course participants want to address (environmental, educational and professional) are articulated. These requirements become the building blocks of the curriculum plan for the course. For this to work, however, much work needs to go into the framing of the open-ended themes. The framework then provides the structure, which allows the open-processes to work. In recent years, the RU participatory course has been using a set of open-ended outcomes to guide the curriculum deliberations. These outcomes are reflexively reviewed with course participants as the course unfolds, to avoid the narrowing of learning opportunity through pre-determined outcomes (see for example, Barnett, 1994; Harley & Parker, 1999). One of the key issues we have confronted, is the difficulties many course participants experience with open processes. Most are interested in the 'different' way of approaching course design, but resulting from the predominantly technicist orientation of previous course experiences, are uncertain of how to respond within a more open-process framework. This issue points to some of the complexities associated with changing the culture of learning in Higher Education / community settings. We found this to be a particular issue when working with participants from the business and industry environment. Not only did they struggle to orient to an open-process approach to course design, but they found it difficult to apply the learning in context during the course, due mainly to the dominance of technicist approaches that characterize the culture of business and industry training (Jenkin, 2000).

3. *Participation:* Participatory orientations to curriculum development have been influenced by global trends towards the democratization of institutional and social life; and by constructivist learning theories, which recognize prior knowledge and experience of learners in the learning process. Critical theory propositions of 'empowerment' and 'emancipation'; and post modern recognition of the role of language, social interaction and diversity of cultural and symbolic systems in learning processes have further popularized participatory orientations. As mentioned above, the RU participatory courses were established with a 'participatory ethos'. It took a number of years, however, to clarify the meaning of this participatory ethos in the context of the course. For example, one course tutor noted that it is not useful to place 'preaching' and 'participatory' educational orientations in opposition, and that curriculum deliberation requires adult educators to make decisions about which processes are most useful in particular contexts and situations. This, however, is not as easy as it sounds, as it requires maintaining a certain 'ambivalence' in relation to participatory orientations (Bauman, 1991) if one is to avoid participation being co-opted as a 'moral imperative'; replacing previous determinist approaches to teaching and learning with equally deterministic participatory approaches (see O'Donoghue, 1999). Encouraging participation by drawing on participant experience involves processes of encouraging learners to problematise and interrogate their experience, as much as to access and validate it (Usher et al., 1997). Workplace based assignments encourage participants to

work with others in context, thus extending participation beyond the course; which again leads to ambivalence, as participants are required to draw the line between their work in the course and their work with colleagues (often similar but different processes). Participation in the course is also multi-levelled and multi-facetted, in that participants on the course participate in different ways, for example by negotiating assessment criteria; by deliberating and working with others on group tasks; by contributing readings and other materials that may be useful to other participants; and through their assignments. Janse van Rensburg and Le Roux (1998) reflect on the epistemological relationship between tutors and participants, and note this to be significant dimension of the participatory ethos of the course. While central to the course and its orientation, we have found that participatory orientations are not without their problems, and in the RU participatory courses, the tensions that exist between facilitation which has the potential to reduce educators to 'stage workers' that are 'robbed of their pedagogical authority' (Shalem, 1997, p. 2) and critically engaged mediation processes are always at play. Janse van Rensburg and Le Roux (1998) argue that participatory learning processes require a crucial shift towards an understanding of meaning as something co-constructed. Harley and Parker (1999) note that this requires significant role changes for both educators and learners, particularly in a society that has been ideologically dominated by repressive forms of colonialism and apartheid for centuries, and where 'formal education / learning' as experienced in modern day societies in Africa is primarily a western / colonially inspired construct.

4. *Praxis:* Praxis involves a conscious recognition of the relationship that exists between practice and its rationale(s). Praxis, in the RU participatory courses constitutes deliberation on the 'why' question. It involves "…asking why we do things in certain ways, and this questioning affects what we do next" (Janse van Rensburg & Le Roux, 1998, p. 104). Assignments in the RU participatory courses are all praxiological. They require participants to draw on work or community experience and reflect in writing on their action(s) in practice. Another key dimension of the praxiological nature of these assignments is their basis in real experience, not in hypothetical examples. Recent research by Raven (2003) has highlighted one of the dangers of practical, applied approaches to assignments. She notes that participants have a tendency to 'avoid' specific engagement with theoretical perspectives introduced in the course (particularly when these are challenging), thus narrowing or reducing the praxiological orientation of the course. Constant attention is required in the narration of practice to ongoing engagement with theoretical perspectives, and the way in which these enable participants to inform their practice. Raven (2003) identifies tutorial support and assessment as two further crucial features of enabling praxis. We have also found that praxis, in itself, is not an unproblematic steering idea, given that its roots lie in modernist dialectical reason. Perhaps the ongoing dialectic between theory and practice that exists in the course (despite attempts to present the theory-practice relationship as non-dialectical), paradoxically, arises from attempts to do away with the dialectic

through a focus on praxis? This question is currently the subject of a research project within the course.

5. *Assessment as learning:* Environmental issues and risks are not neutral, but of a socio-cultural and political nature and thus heavily value laden. In the RU participatory courses, which have been designed to respond to the diversity and contextuality of environmental learning, course developers have struggled to define an appropriate assessment and accreditation process, which does not conflict with the open-ended, participatory and reflexive orientation of the course. The advent of an outcomes-based education system in South Africa (the governments preferred model for enabling educational transformation), requires that pre-determined outcomes and assessment frameworks be established for accreditation purposes. This introduces a governmentality into the course as assessment frameworks of this nature are, more often than not, devised within a control framework. In defining an assessment strategy that is consistent with the course orientation, course participants and tutors have, in the past few years, been involved in negotiating assessment criteria for each of the themes, and assignments. A mix of assessment strategies are used, including self and peer assessment. Participants are encouraged to draft at least three drafts of their assignments, each of which is shared with tutors. Tutors pose critical questions of the participants, who are then required to respond to the questions, after deliberations with the tutor. In this way, assessment becomes more a learning process, than a process of judgment. In her research into reflexive competence in the context of the course, Raven (2003) identified assessment *as* learning, as one of the key course processes that enable reflexivity and change.

As indicated above, these features are underpinned by a conceptual framework, which involves epistemological considerations; social change theories and a consideration of curriculum theory and practice. These features of the RU participatory course may well have relevance in other Higher Education settings. With the recent establishment of a course developers network, established in partnership with the SADC[9] Regional Environmental Education Programme, these course curriculum features are being considered by a range of Higher Education institutions in the region, as course developers meet to work on course design. In conclusion, I briefly review the epistemological considerations that have shaped the RU participatory courses; by way of indicating some of the implications for curriculum development.

IMPLICATIONS FOR CURRICULUM DEVELOPMENT IN HIGHER EDUCATION INSTITUTIONS

---

[9] The Southern African Development Community involves 14 member stakes.

Meaning making and social change

Karembu and Kinyanjui (1997) indicate that, as environmental educators, we need to continuously reflect on the way we think about and perform educational practice. The case example shared above draws on wide ranging experience of environmental educators in the SADC region, as they have, through educational processes, attempted to confront issues of ecological safety and risk, increasing poverty and inequitable patterns of economic growth, unsustainable consumption patterns and changes in organizational and social life in different community contexts. Drawing on and reflexively reviewing applications of social theory has assisted in the clarifying of appropriate responses to these issues through educational practice, and has been an essential feature of the course development process over the past ten years.

Out of ten years of deliberation on an appropriate curriculum framework for environmental education processes that are responsive to sustainability issues in community contexts, has come a recognition that:

> ... the knowledge that shapes our educational practice and our actions in the environment is socially constructed, and hence open to review; ... such review needs to be informed by ongoing contextual evaluation of what a better environment for all (or sustainable living/ social justice etc) would entail; ... such a review may also involve a re-thinking of educational practices (Janse van Rensburg & Le Roux, 1998, p. 40).

The social construction of meaning (see Berger & Luckmann, 1966; Gergen 2001) has influenced the developments associated with the RU participatory courses over the past ten years. This has challenged course developers, tutors and participants equally, as education and training in South/southern Africa is historically rooted in positivist views of learning and knowledge creation. A re-orientation of curriculum processes in Higher Education would appear to involve deep-seated epistemological challenges and ruptures. In an African context, Odora Hoppers (2002) argues strongly for a recognition of the social construction of *knowledges* and *inter-epistemological discourse* in re-orienting Higher Education in post colonial Africa. She notes that this recognition is intimately linked to issues of social justice and states that "The call to *social justice* is not a quick ticket handed out by a machine. It represents different kinds of creativity and calls for *radical innovations in pedagogy*" (ibid, p. 2, my emphasis).

In the RU participatory course knowledge has been presented as historically shaped and open to change through reflexive review. This understanding of knowledge has permeated the way in which the course materials were designed, which present knowledge as contested, historically influenced and socially constructed. Part of this involves creating spaces in the course materials for the stories that participants have to tell, and enabling these stories to become the basis of their understanding of the course content. Kinyanjui (1995, p. 2) emphasizes the significance of this perspective for a re-thinking and re-defining of educational practice in an African context when he states that "... any serious environmental education *has to be rooted in local communities*, bring with it *local participation, local knowledge*, orientation and *be geared towards dealing with concrete realities of daily life*" (emphasis mine). In this sense, curriculum development processes amongst adult educators are presented, through the RU participatory course case, as *'processes of open-ended inquiry within and around local, regional and global environmental issues and risks'* (Lotz, 1999, emphasis original). As indicated above, this presents a range of challenges to course developers, not least is that of needing to disrupt the dominance of narrow forms of rationalism and positivist views of knowledge in Higher Education settings.

Curriculum as open-ended inquiry

The above discussion of the RU participatory courses indicates that curriculum is an ongoing process, which is best refined and reflexively reviewed from within the course process over time. Thus, after ten years of involvement with the curriculum development associated with the RU participatory courses, I[10] am unable to provide a model curriculum or a model for curriculum development for fostering community partnerships and sustainability in Higher Education contexts. The case story described above, has attempted to point to different possibilities for curriculum processes amongst adult learners in community contexts, and to offer some insights into key elements and struggles within a process that remains a developing story. Thus the case story, in the words of Patti Lather (1991, p. 159) is presented *"... not as a set of answers, but making possible a different practice"* (emphasis mine).

CONCLUSION

Curriculum deliberations amongst adult learners, as outlined in the case example above involve deliberations about learning and social change; an understanding of changing political economies (global and national) that influence educational practice; a deeper understanding of trends in adult education; a recognition that meaning is socially constructed and clarification of key features of courses. A consideration of these key features (and related questions, see Figure 1) identified in the RU participatory courses has enabled hands-on, practical engagement with sustainability issues in a range of community contexts, as evidenced by the kinds of projects participants engage in, and in the expansion of the RU participatory courses into different countries and contexts. The key features also indicate that curriculum deliberation involves processes that are *deliberately focused* on achieving sustainability goals (for example the setting of praxiological assignments; reading core texts and prescribed readings); and *deliberative* (for example enabling participation in the design of the curriculum around a structure of key themes; negotiating assessment criteria; and responding to tutor comments and questions on assignment work).

   The RU participatory course case indicates that curriculum deliberation amongst adults in community-based education programmes involves achieving a good

---

[10] I have been extensively involved in the RU/ Participatory courses over the past ten years. I have been a course participant, course tutor, materials writer, researcher, course co-ordinator and have supported the expansion of the courses in other countries and contexts, and have supported research associated with the courses.

balance between the deliberate and essential from the outside and insights and practices in context. A tentative 'middle ground' has been opened between prescriptive narrowing interventionist perspectives and liberal individualized perspectives that locate everything in the freedom of the individual to choose (O'Donoghue & Lotz-Sisitka, 2002). This middle ground has created the space for challenging learning in context (including ongoing learning amongst those of us involved in course development – community relationships in Higher Education). What seems to have developed in the context of this case, is an African ethos of open consensus seeking in community (ibid). As noted above, this ethos has, over the past few years, started to permeate a number of other environmental education programmes in the Faculty at Rhodes University, and in other institutions in southern Africa. It has also helped us to establish ongoing relationships with a number of community-based projects in Grahamstown and in other centers around the country. Ongoing reflexive engagement in Higher Education-community relationships through educative responses to environmental issues and risks over the past ten years, has become a creative space for re-orienting pedagogy and practice.

REFERENCES

Bauman, Z. (1991). *Modernity and Ambivalence*. Cambridge: Polity Press.
Barnett, 1994. *The Limits of Competence: Knowledge, Higher Education and Society*. Buckingham: Open University Press.
Beck, U. (1992). *Risk Society: Towards a new Modernity*. London: Sage Publications.
Beck U. (1999). *World Risk Society*. Oxford: Blackwell Publishers Ltd.
Beck U. (2000). *What is Globalization?* Cambridge: Polity Press.
Berger, P. & Luckman, T. (1996). *The Social Construction of Reality*. Harmondsworth: Penguin.
BGCI. (2000). Guidelines for Botanic Gardens in Education for Sustainability. Unpublished Draft Document. Surrey. UK.
Bourdieu, P. (1998). *Practical Reason*. Cambridge: Polity Press.
Delanty, G. (1999). *Social Theory in a Changing World: Conceptions of Modernity*. Cambridge: Polity Press.
Doll, W. (1993). *A Postmodern Perspective on Curriculum*. New York: Teachers College Press, New York.
Edwards, R. (1997). *Changing Places? Flexibility, Lifelong Learning and a Learning Society*. London: Routledge.
Fien, J. (1993). *Education for the Environment: Critical Curriculum Theorising in Environmental Education*. Victoria, Australia: Deakin University Press.
Fien, J. (1998). Re-orienting teacher education towards sustainability: An Action Research Network Approach. UNESCO-EPD Demonstration Projects in Teacher Education for Sustainability. Paris: UNESCO.
Gergen, K. (2001). *Social Construction in Context*. London: Sage Publications.
Hanson, A. (1996). The search for a separate theory of adult learning: Does anyone really need andragogy? In Edwards, R., Hanson, A., Raggart, P. (Eds) *Boundaries of Adult Learning*. London: Routledge.
Harley, K. & Parker, B. (1999). Integrating differences: Implications of an Outcomes-based National Qualifications Framework for the Roles and Competencies of Teachers. In Jansen J. & Christie, P. (Eds). 1999. *Changing Curriculum: Studies of Outcomes-based Education in South Africa*. Cape Town: Juta & Co. Ltd.
Heylings, P. (1999). Professional development in environmental education in Zanzibar, Tanzania: Distances encountered in a semi-distance learning course. Unpublished M.Ed thesis. Department of Education, Rhodes University, Grahamstown, South Africa.

Huckle, J. (1991). Education for Sustainability: Assessing pathways to the future. *Australian Journal of Environmental Education*, 7, 43-62.

Huckle, J. & Sterling, S. (1996). *Education for Sustainability*. London: Earthscan.

ICAE, 1993. Treaty on environmental education for sustainable societies and global responsibility. *Adult Education and Development*, 40.

Janse van Rensburg, E. (1995). Environmental education and research in southern Africa: A landscape of shifting priorities. Unpublished Ph.D thesis, Department of Education, Rhodes University, Grahamstown.

Janse van Rensburg, E. & Le Roux, K. (1998). Gold Fields Participatory Course in Environmental Education: An Evaluation in Process. Sharenet. Howick.

Jenkin, N. (2000). Exploring the Making of Meaning: Environmental Education and Training for Industry, Business and Local Government. Unpublished M.Ed thesis, Department of Education, Rhodes University, Grahamstown.

Jickling, B. (1999). Beyond sustainabity? Should we expect more from education? *Southern African Journal of Environmental Education*. 19, 60-67

Karembu, M. & Kinyanjui, K. (1997). *Co-ordination, Emerging Needs and Actions in Environmental Education*. Report of the Regional Workshop for Eastern and Southern Africa held in Nairobi, Kenya, 2-8 November, (1997). IDRC.

Kinyanjui, K. (1995). Research Agenda in Environmental Education in Africa. In: *Integrating Environment, Social and Economic Policies*. Nairobi. IDRC.

Lather, P. (1991). *Getting Smart*. Routledge, New York.

Lotz, H. (Ed) 1999. *Developing Curriculum Frameworks: A Sourcebook on Environmental Education amongst Adult Learners*. SADC Regional Environmental Education Centre, Howick.

Muller, J. (1997). A harmonized qualifications framework and the well tempered learner: Pedagogial models and teacher education policy. In: Bensusan, D. (Ed.). *W(h)ither the University?* Proceedings of the Kenton Education Association Annual Conference 1996. Juta & Co. Ltd. Cape Town.

O'Donoghue, R. (1997). Detached Harmonies: A study in/on Developing Processes of Environmental Education. Unpublished PhD thesis, Rhodes University, Grahamstown, South Africa.

O'Donoghue, R. (1999). Participation: An under theorized icon in research and curriculum development. *Southern African Journal of Environmental Education*. Vol 19, 14-27.

O'Donoghue, R. & Lotz-Sisitka, H. (2002). *Special ten year report: 1992-2002. Gold Fields Participatory course and Gold Fields Environmental Education Service Centre*. Grahamstown: Rhodes University Environmental Education Unit.

Odora Hoppers, C. (2002). *Higher Education, Sustainable Development, and the Imperative of Social Responsiveness*. Concept paper prepared for the Human Sciences Research Council, the University of Fort Hare and the University of the North. Paper presented at the Environmental Management for Sustainable Universities Conference (EMSU 2003). Rhodes University. South Africa. 11-13, September 2002

Popkewitz, T. (1991). A Political Sociology of Educational Reform. Power/Knowledge in Teaching, Teacher Education and Research. New York: Teachers College Press.

Popkewitz, T. & Brennan, M. (Eds.) (1998). *Foucault's Challenge: Discourse, Knowledge, and Power in Education*. New York: Teachers College Press.

Raven, G. (2003). *Course processes that enable the development of reflexive competence: A case study of an environmental education professional development course*. Unpublished Ph.D study (final draft). Department of Education, Rhodes University. Grahamstown.

Robottom, I. (1987). Towards inquiry-based professional development in environmental education. In Robottom, I. (Ed). *Environmental Education: Practice and Possibility*. Geelong, Victoria: Deakin University Press.

Shalem, Y. (1997). *Epistemological labour: The way to significant pedagogical authority*. Paper presented at Kenton-at-the-Gap, Hermanus, October 1997.

UNCED (1992). *Earth Summit '92*. London: The Regency Press

UNESCO (1997). *Educating for A Sustainable Future: A transdisciplinary Vision for Concerted Action*. UNESCO. November 1997.

UNESCO-UNEP (1996). Teaching for a sustainable world: Environmental Education for a New Century. *Connect*. Vol. XXI, No. 4, December 1996.

UNESCO-UNEP (1996). Education for Sustainable Development: A priority for the world community. *Connect*. Vol. XXI, No. 2, June 1996.

Usher, R., Bryant, R. & Johnson, R. (1997). *Adult Education and the Postmodern Challenge: Learning beyond the Limits*. London: Routledge.
Visvanathan S. (2000). Democracy, Plurality and the Indian University. *Economic and Political Weekly*. September 30, 2000. pp: 3598

BIOGRAPHY

Heila Lotz-Sisitka (Associate Professor) holds the Murray & Roberts Chair of Environmental Education (Africa's only Chair of Environmental Education). Her research interests include: curriculum and professional development; participatory learning processes; industry environmental education and training and research methodology. Professional contributions include the articulation of deliberative orientations to professional development in environmental education in southern Africa and participatory articulations of an environmental and social justice focus in national curriculum policy transformation in South Africa.

CHAPTER 26

INCORPORATING SUSTAINABILITY IN THE EDUCATION OF NATURAL RESOURCE MANAGERS: CURRICULUM INNOVATION AT THE ROYAL VETERINARY AND AGRICULTURAL UNIVERSITY OF DENMARK

Susanne Leth & Nadarajah Sriskandarajah

INTRODUCTION

For the last 250 years, sustainability has been seen as a conceptual foundation for Danish forestry, but seen in the sense of sustained tree production, described by Larsen (1997) as a *classical* paradigm of sustainability. During the last couple of decades, societal discourse about environmental issues has moved further on to place the focus not only on silvicultural[1] issues and ways of growing and producing trees, but also on the function of the forests. In the Brundtland version of sustainability, sustainable development is described, as "one that meets the needs of the present without compromising the ability of future generations to meet their own needs" (World Commission of Environment and Development 1987), relating sustainability in a more global context. In forestry, this demand for sustainability can be seen as an agreement across generations about the multifunctional use of forests and, therefore, as indicating the need for balancing not only economic and ecologic dimensions, but also the social dimension of practicing forestry (Larsen 1997, Larsen & Emborg 2002). Today, this latter interpretation of sustainability has a profound impact on the policies dealing with Danish forestry and therefore also on Danish forestry education.

The highest level of forestry education is offered in Denmark at The Royal Veterinary and Agricultural University (KVL) which offers a broad range of five-year Masters degree programs within veterinary, food and agricultural sciences. For many years, forestry education had a small intake of students (approximately 12 a year) and most of the graduates were employed in the traditional forestry sector. In recent decades, the graduates have also been successful candidates for employment in other sectors dealing with broader aspects of environment and natural resource

---

[1] Silviculture is the art and science of controlling the establishment, growth, composition, health and quality of forest stands.

Peter Blaze Corcoran & Arjen E.J. Wals (Editors), Higher Education and the Challenge of Sustainability: Problematics, Promise and Practice, 335-345.

management. KVL has, therefore, raised the intake of students to approximately 50 students per year in 2002. In line with the above development concerning sustainability, KVL has also been working at including ecological, economic, social, recreational and technical aspects in the recent forestry curriculum.

Even though social and recreational aspects are mentioned as issues to be dealt with in the education, the curriculum still primarily focuses on aspects of natural science and economics. In year 2000, teachers of a course titled Silviculture, put forward the argument that this course was suitable for incorporating the social aspects of sustainability in line with the Brundtland interpretation of the concept. The course was at the time a two-semester unit in the Masters program, offered in the eighth and ninth semesters, including excursions and field trips during the semesters and a week's summer excursion placed between the semesters. The course had previously emphasised giving the students a foundation on silvicultural issues through lectures, assignments and smaller projects. This time, the teachers attempted to integrate a new paradigm of sustainability, and paid special attention to the introduction of a social dimension. In order to achieve the latter objective, the silviculture course was merged with a course in forest and natural resource planning, for offer during the second of the two semesters. Conflict management became the theme of the combined course and this formed the basis of a workshop with the students during the summer excursion in the inter-semester break. The workshop was designed in a way that students and teachers would work towards agreeing on the content and form of the second semester and through this, the students would also get experiences in conflict management.

This chapter describes the teachers' effort to integrate the new interpretation of sustainability into the forestry curriculum with the particular aim of uncovering actual implications of converting the idea of sustainability into educational practice.

THE CASE STUDY

Integrating new concepts raises questions about how to change, reformulate and rewrite curriculum and the study and course plans. Besides this, students' aspirations and expectations can have a profound impact on the practical context of their education thus becoming an essential factor for success when integrating new concepts. In the case study presented here, we focus on how the teachers conceptualised the new idea of sustainability into a framework of the silvicultural subject area to present to the students during lectures in the first semester. We also focus on the students' reactions to this framework and the related discourses in the classroom. The findings are illuminated and interpreted through the uncovering of the students aspirations and expectations to their education and the silviculture course. These findings from the first semester are supplemented by the students' reaction to the participatory curriculum development workshop at the end of the first semester.

The discourses in the classroom setting were followed by participant observation. Students' aspirations and expectations were uncovered by a set of open-ended questions given to them as a questionnaire in the beginning of the course. The outcomes of the participatory curriculum development workshop were also evaluated using three questionnaires given at the beginning, during and at the end of the workshop.

The data presented here is part of a larger educational research project, some aspects of which have been published already (Leth et al., 2002). What has been selected for discussion in this paper relates to the following:

–   Students' aspirations and expectations in order to uncover the paradigm of sustainability they relate to;
–   Subject matter presented in the course in order to characterise the framework used by the teachers to integrate the new paradigm of sustainability;
–   Interaction and discourses between teachers and students in order to uncover the students' reactions to the presented framework;
–   Outcomes of holding a participatory curriculum development workshop in order to uncover the students' reactions to this particular innovation.

ASPIRATIONS AND EXPECTATIONS OF STUDENTS

The questionnaire given to the students in the beginning of the course revealed that interest in ecology was the strongest argument for the students to take up this education, though a lot of them were also interested in management combined with ecology. It was also revealed that the students' more personal aspirations and dreams about their future life were also important reasons such as, for example, wishes for combining professional life with a hobby. The students valued the acquiring of competencies in biology and ecology very highly and often in combination with achievement of competencies in management. They also valued the more personal competencies such as the ability to lead, plan and negotiate.

Concerning the silviculture course, the students expected basic knowledge from previous courses to be synthesised and to learn how to convert scientific, ecological knowledge into the practice of growing trees. The students found the course extremely relevant and, not surprisingly, necessary to hold a Masters degree in forestry.

In the questionnaire, the students were asked what they understood sustainable forestry to be. A little less than half of the students described an understanding, which fitted very well with the Brundtland definition of the concept (see examples in Table 1). An equal number gave answers, which related to an understanding of sustainability in the three common dimensions, namely, economic, ecologic and social, but where the economic dimension was considered most important, in line with the earlier description of sustainability in the classical sense. The rest of the students expressed deep scepticism towards the concept.

Table 1. Examples of students'different perceptions of sustainability.

Sustainability in relation to the Brundtland definition: "To grow forest in such a way that the biological, the economic and the social sides are considered and taken into account. You can talk about sustainability in many ways, but which one to be put on the top or how to list them is a society/political issue. Sustainability is also not to diminish the foundation for growing in such a way that the future generations will be worse off than ourselves when they are to use the resource."   *Sustainability in relation to Brundtland, but prioritising economics:*   "I understand silviculture that lasts in the long run. Silviculture shall be economically profitable, but also consider ecology and being accepted by the surrounding society."   *Classical sustainability:*   "A sustainable silviculture is both economically and ecologically sustainable."   *Scepticism:*   "A raped concept that loose more and more dignity. I don't think it will be found in places other than literature in 10 years!"

Though approximately half of the students seemed to have adopted the Brundtland definition of sustainability, their understanding and expectations of their education and the competencies it should provide them with, still seemed to be grounded in the classical understanding of sustainable forestry. The forest was seen as a production unit somewhat detached from landscape and society.

THE FRAMEWORK USED TO PRESENT THE NEW CONCEPT OF SUSTAINABILITY

Observations of the lectures uncovered how the teachers understood sustainability in a forestry context, as well as how this perception influenced the choices of actual content in the teaching situation. At the outset was an overall approach to forestry, which can be characterised as holistic. Focus on multi-functionality was weighted, relating to a balancing of economic, ecologic and social dimensions concerning the needs of the present society and a concern for the possibility of resource use by future generations. In this perception, the forest is not considered as an isolated unit of production with only the interest of the stakeholders, but as part of an overall landscape management and of the interest of society. The specific locality used for forestry was seen as offering an opportunity for natural resource management, and silvicultural questions become questions regarding which functions are desired for the particular locality, a desire that is negotiable with society. In dealing with these functions, sustainability is brought into practice in silviculture as a weighting between the three dimensions to be taken into account in formulations of management plans. The actual choice of tree species and provenances then becomes closely related to the function of the forest.

The approach was seen to be successful in the interaction with the students when dealing with the more general forestry issues as, for example, locality functions. One example concerning the effort to introduce and weigh the importance of the different functions of the forest was a slide show session in the third lecture of the semester. Through showing the different forest settings and thereby discussing the functions obtained by the way silviculture was practiced, it was possible to discuss sustainability in the new understanding in the classroom. The students participated in this session and the related discussion with enthusiasm. They willingly challenged their own understanding of practicing forestry, grounded in the classical understanding, and also the new paradigm and how to understand it. This session could be considered a success because the students easily seemed to adapt the idea of focusing on the different functions of the forest.

Moving the focus to the more specific issues such as choice of tree species and provenances, the approach came under challenge. All through the semester, the students tended to ask very specific questions related to the practice of silviculture, such as the tree species and provenances that had to be chosen. These questions could be seen as reflecting the students' need for knowledge regarding forestry and silviculture, linked to the single function of productivity. Though the intention of the teacher was to maintain the discussion about the specific silvicultural issues and the choices to be made in relation to multiple functions of the forest, the discourse between teacher and students in fact remained locked to dealing predominantly with production as the important issue. In those situations, it seemed impossible to move beyond the pre-existing paradigm.

PARTICIPATORY CURRICULUM DEVELOPMENT – INTRODUCING THE SOCIAL DIMENSION

In their effort to integrate the new concept of sustainability, the teachers paid specific attention to the introduction of the social dimension in forestry. As conflicts were becoming an important aspect to be managed in modern forestry, it was decided to work with conflict management as a theme in a workshop, drawing on the theories of participation and learning. As participation was seen central in the new paradigm, it was decided to involve the students in the process and engage them in a workshop designed to achieve participatory curriculum development. The structure and content of their own course to be run during the second half of the revised curriculum in the following autumn semester became the focus of this workshop. Through the workshop, it was intended to expose the students to current thinking in conflict management and the tools available for it. By using curriculum development as a subject matter to work with, it was also intended to place the new paradigm and the required changes in educational practice on the agenda and to make the outcomes of participation more immediately relevant.

The objectives of the workshop can be summarised as follows:

- To introduce students to methods of problem solving and the learning process;
- To allow students to gain practical experience in problem solving and negotiation;

- To engage students in participatory curriculum development by having them define objectives, formulate course content, develop a time schedule and activity plan, and if possible reach consensus on this course plan;
- To enable the development of personal competencies.

Students formed six groups of 6-8 persons each and engaged themselves in problem identification and structuring, using one of six established 'soft' operation research methods described in a textbook for the course (Sørensen and Vidal 1999). Six volunteers from the class were instructed in the methods and prepared beforehand to act as facilitators for each of the groups. Prior to the group work, all students were introduced to Kolb's experiential learning theory (Kolb, 1984) with an emphasis on problem solving as a learning process.

EXPERIENCES FROM THE PARTICIPATORY CURRICULUM DEVELOPMENT WORKSHOP

The students were informed about the workshop and introduced to its aims and content during a plenary session. They reacted with great surprise and uncertainty to this introduction and the information about merging the rest of the course in silviculture with forest and natural resource planning. The students expressed scepticism and reluctance concerning the change of course content and raised questions regarding the feasibility of such a change within the existing curriculum, course descriptions and rules of the study board. They were concerned about the possible disparity between what they chose from the study handbook and what they were in fact going to receive. Seen as most problematic was the impression that these changes would lower the importance of biological and ecological content in the Silviculture course, and therefore a possibility of not achieving enough silvicultural knowledge.

Immediately after the plenary introduction of the workshop, the students were given the first questionnaire. Contrary to what might have been expected from the students' initial reactions, the picture that emerged here was much more positive. The students valued most the achievement of personal competencies such as the ability to co-operate, communicate, give and receive critique and to learn about own strengths and weaknesses. They expected this workshop to give them further tools and experiences to improve on those competencies. The importance of an integrated viewpoint was expressed in their comments (Table 2).

Table 2. Examples of students'comments on the relevance of the workshop in their education.

"Silviculture can not be seen isolated"

"In today's forestry/everyday life conflict management is needed"

"Important to have influence on own education. Problem solving is an important discipline"

"I do not expect an occupation where silviculture will be a part. Conflict management is extremely important in all contexts, especially in working with developing countries which is of my interest"

"Leading and problem solving is underestimated in the present study plan – therefore relevant to spend time on, especially at master level"

"Silvicultural problems are complex and demand insight and methods to give solutions"

During the workshop, the students continued to express a positive attitude as evident in their responses to subsequent questionnaires. A very valued aspect of the final process of negotiation was the ability to reach consensus about the objectives for the autumn semester. The students also reacted very positively to their involvement in the planning process. The workshop was rated as highly relevant, because of the opportunity to influence their own education and the possibility to achieve personal competencies.

In relation to changing the content of the two courses, the students' main concern was to do with the silviculture course specifically. They felt that the new direction would imply a trade-off between the new focus on conflict management/problem solving and the usual focus of the course on more detailed knowledge of silvicultural practices. Furthermore, their concern about limited acquisition of specific silvicultural knowledge from the course turned into expression of a sense of insecurity about what this change would mean for their examination, their performance in it as well as for their future employment.

The final course plan agreed on through consensus reflected the views of the students and attempted to clarify the distinction between silvicultural issues and social issues. The key elements included in the course were:

- Introduction to theories of conflict management, negotiation, systems thinking, qualitative methods, stakeholder democracy, power and participation.
- Workshops in conflict management and negotiation with involvement of different types of stakeholders.
- Introduction to different tree species (ecology and growth) and silvicultural systems appropriate to Masters level education.

PARADIGM TRANSITION

Until the close of the nineteenth century, forests in Europe were managed according to a 'natural' approach, whereby silviculture was based on the ecology of natural indigenous forests, and it emphasised an adjustment to rather than radical changes in natural forms of the ecosystems (Matthews, 1999). The more modern approach to silviculture in many countries retained the concept of the forest as an ecosystem but had the emphasis shifted from mixed and uneven aged stands of trees towards more pure and even aged stands of plantations and towards higher productivity. Today, there is a return to more 'nature-based' thinking in silviculture and to conversion of mono-specific plantations to silviculturally flexible, ecologically stable and biologically diverse forests (Larsen, 2000). If this 'ecologisation' of silvicultural methods is already a shift from the classical paradigm of sustainability with its focus only on sustained yield of wood, then the incorporation of biological, aesthetic and social values and viewing of landscape and forests in a multi-functional way amounts to much further broadening of the meaning of sustainability. Inclusion of this social dimension in the sustainability discourse in the classroom was exactly what the revised course was attempting to do in our case study.

It was clear from our study that students generally subscribed to the classical paradigm of sustainability referred to earlier, emphasizing sustained yield as the outcome of managing silvicultural systems. Many had difficulty in seeing sustainability as an inherent property of that system in its social dimension as well as the ecologic and economic ones. Furthermore, in their quest for the required knowledge of the specifics of silviculture according to their paradigm, they appeared to be not willing, initially at least, to trade off that knowledge of content for a learning process that emphasised the management of social conflicts. However, over time, it seemed that they were willing to acknowledge the importance of learning to deal with the social aspects of management of silvicultural systems, but only after they were confident that their basic silvicultural knowledge from a biological and ecological perspective would be catered for.

The students demonstrated their views about the validity of one form of knowledge relative to another. They showed an ability to separate learning needs associated with professional competencies, which in their view were essential to secure employment, and personal competencies, which they saw as useful to perform well in their jobs. This indicates that, at least for some of them, it was possible to not only distinguish the two paradigms of sustainability but also to hold on to different paradigms at different times. How incommensurable are the two paradigms and to what extent were the differences enacted in the classroom? The demands of moving between different set of assumptions associated with different paradigms can be heavy but are not insurmountable, especially in a learning situation. Appreciation of multiple perspectives and acquiring pluralistic methodologies has a special place in sustainability education. The philosophical, cultural and psychological resistance to transforming a long-held paradigm and broadening one's ontologic framework can be overcome through a learning process which emphasises effective facilitation of the learners' engagement in dialogue and self-reflection and their reaching of a critical attitude (Midgley, 2000).

SYSTEMIC THINKING AS BASIS FOR A FRAMEWORK

The specific situation of dealing with silvicultural systems in this study also presented the opportunity to incorporate systemic thinking more explicitly and to assist in the student's epistemic development. The pattern described in the classroom regarding forest management could be depicted as several inter-linked levels through which the practice of forestry and silviculture could be approached with an overview of sustainability in mind (Figure 1). The demands for multi-functionality could indeed be viewed at the level of the landscape, which would then set the approach down to the levels of the forest and the individual species of trees that make up the forest.

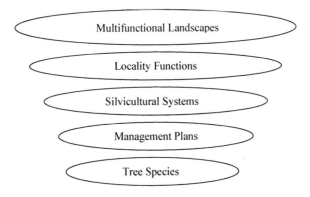

Figure 1. A framework for approaching the practice of forestry incorporating sustainability concerns.

The knowledge content associated with the lower levels of this hierarchy, which the students appeared to value greatly, may be offered by the usual didactic means. Epistemologically speaking, the basic information processing, which goes with that level of learning should leave the students in a relatively comfortable state of no confusion or ambiguity, the stage of dualism according to Salner (1986). The more systemic orientation called for in the learner when dealing with the grey areas, conflicts and subjectivity that are associated with the higher order issues, in the upper levels of our framework, cannot be achieved by didactic instruction. Adopting Salner's terminology for the stages of cognitive processing further, learners have to make the transition from dualism to multiplicity, the stage of accepting that there are many truths rather than the single absolute one, and then on to the stage of contextual relativism where they become aware of the importance of contexts in defining truth and value. Exposure to meaningful contexts in real life settings and broadening of perspectives through such learning experiences should be aimed for not just in one area of education, such as silviculture in this case, but in as many instances as feasible within the whole curriculum.

CONCLUSION

In an attempt to incorporate sustainability in its broadest meaning in forestry education, the teachers brought the social dimension into classroom discourse in this case study. The students seemed willing to engage in that discourse but only after their need for the specifics of silvicultural knowledge were met. We have argued that through effective facilitation learners' resistance to paradigm transition can be overcome. This, in conjunction with the adoption of systemic thinking as a framework for the curriculum, should lead to better appreciation of multiple perspectives and a shift towards higher levels of epistemic development amongst learners. These transitions are equally challenging to students and to the teachers. Grappling with concepts such as sustainability and learning from the multiple realities of people is provided for if the students get the opportunity to experience the real life dilemmas more fully, while being supported in the reflective processes and the synthesis of arguments in support of their own value positions.

REFERENCES

Kolb, D. A. (1984). *Experiential learning: Experience as the Source of Learning and Development.* Englewood Cliffs, NJ: Prentice-Hall.
Larsen, J. B. (1997). *Skovbruget ved en skillevej – teknologisk rationalisering eller biologisk optimering?* (Forestry at crossroads – technological rationalisation or biological optimising?) Særtryk af Dansk Skovbrugs Tidsskrift, 82.årgang.
Larsen, J. B. (2000). *From plantation management to nature-based silviculture – a Danish perspective.* Congress Report, 3rd International Congress: Sustainability in Time and Space. Fallingbostel, Germany, July 7, 2000.
Larsen, J. B. and Emborg J. (2002). *Fremtidens skovbrug i Danmark (Forestry in Denmark in the future).* In: Jensen, E. S., Vejre, H., Bügel, S. H., and Emanuelsson, J. (Eds.). Visioner for fremtidens jordbrug (Visions for the agriculture of the future). Gads Forlag.
Leth, S., Hjortsø, N. and Sriskandarajah, N. (2002). Making the Move: A Case Study in Participatory Curriculum Development in Danish Forestry Education. *Journal of Agricultural Education and Extension,* 8(2), 63-73.
Matthews, J. D. (1999). *Silvicultural Systems.* 2nd edition, Oxford Science Publications. Oxford: Clarendon Press.
Midgley, G. (2000). *Systemic intervention: Philosophy, methodology and practice.* Dordrecht: Kluwer Academic/Plenum Publishers.
Salner, M. (1986). Adult cognitive and epistemological development in systems education. *Systems Research,* 3(4), 225-232.
Sørensen, L. and Vidal, V. (1999). *Strategi og planlægning som læreproces* (Strategy and planning as a learning process). Copenhagen: Copenhagen Business School Press.
World Commission of Environment and Development (1987). *Our Common Future.* Oxford: Oxford University Press.

BIOGRAPHY

Susanne Leth is currently the Administrative and Pedagogical Head of Postgraduate Education in Orthodontics at the School of Dentistry, University of Copenhagen. She is completing a PhD thesis at the Royal Veterinary and Agricultural University working with science education, curriculum development and educational sociology at university level. In her research, she has been focusing on natural resource management education with forestry education as a case study, studying the implications of converting a vision about sustainability into a practical educational context. She holds an MSc in geography and has several years of teaching experience within natural resource management education at university level in Denmark.

Dr Nadarajah Sriskandarajah is Associate Professor at the Unit for Learning and Interdisciplinary Methods, Royal Veterinary and Agricultural University, Copenhagen, Denmark. Prior to moving to this position, he was with the School of Environmental Management and Agriculture at the University of Western Sydney, Hawkesbury in Australia, as an active member of the group which led many innovations in agricultural education. His main interests have been in systemic and learning approaches to education and research within agriculture and rural development.

CHAPTER 27

THE PRACTICE OF SUSTAINABILITY IN HIGHER EDUCATION: A SYNTHESIS

Arjen E.J. Wals, Kim E. Walker & Peter Blaze Corcoran

Case studies may provide ideas, suggestions, or imagery that might sensitize outsiders to issues they may have not considered, particularly with regard to the process of institutional change. The lessons to be learned from the case studies presented in Part Three will partly depend on the reader's own background and experience. It is not only difficult for us to distill lessons that are relevant or worthwhile for all, it is also undesirable to suggest that based on the case studies. As we wrote elsewhere (Corcoran et al., 2004), a study is more transformative when it challenges the reader and sets challenges for the writer. The development of sustainability in higher education has both personal and shared elements to it. The dialectic between the text and the reader allows one to relate her or his ideas, insights, experiences, and feelings to those expressed in the case studies. The result is likely to be a mixture of consonance and dissonance. Both are important elements of so-called 'case-inspired self-generalization' (Wals & Alblas, 1997). While the consonance fosters reassurance and confidence, the dissonance generates the reconsideration of current practice in light of contesting viewpoints.

Elsewhere we have suggested that four broad areas of concern need to be considered when conducting case study research: purpose, roles, tensions, and challenge (Corcoran, et al., 2004). The examplary practices presented in Part Three not only reaffirm the value of these areas for case study research itself, but also as area's of concern to be addressed when changing institutional practice.

The first area of concern, purpose, stresses the importance of having a clear aim for taking on sustainability as an institutional challenge.

The second area of concern, the role of the players, emphasizes the importance of involving a variety of key stakeholders in the change process, particularly when representing potentially diverging interests. This also suggest that the power base needs to be explored in institutional change processes, in other words, whose interests and goals were being served in the initiative.

The third area, tensions between the universal and the contextual, requires sensitivity towards the different ways in which sustainability issues are dealt with in situ. It is important that a institutional innovation benefits from well-documented experiences elsewhere, not by blind adoption but by critical adaptation.

Peter Blaze Corcoran & Arjen E.J. Wals (Editors), Higher Education and the Challenge of Sustainability: Problematics, Promise and Practice, 347-348.

The fourth area, challenge, stresses the importance of contestation of the various perspectives the different stakeholders bring to the institutional change process. This will create the dissonance that is so crucial in social learning processes that ultimately might lead to a shared framework of institutional sustainability.

Having included case studies from three continents, means that there are inevitably cultural differences that make a comparative analysis difficult if not impossible. Meeting the challenge of sustainability in higher education is culturally embedded. At the same time, this challenge is closely tied to the academic history and curricular tradition of the institution concerned. Clearly, there is no panacea for the failure of institutions to take on this challenge. Some institutions will choose to add on to existing programmes, others will opt for a more transformative approach. The decision about the most desirable reform approach is highly contextualized but appears to have more impact when resulting from an open and communicative process in which all stakeholders play their own, respected roles.

However contextualized the cases presented in Part Three may be, they do suggest that sustainability can be a catalyst for institutional change and for a transition towards higher learning and new ways of knowing. Serious attempts to integrate sustainability into higher education, like the ones described in this book, introduce teachers, students and administrators alike to a new pedagogical world that opens up promising avenues for both institutional and individual practice.

REFERENCES

Corcoran, P.B., Walker, K.E. and A.E.J. Wals (2004). Case Studies, Make-Our-Case Studies, and Case Stories: A Critique of Case Study Methodology in Sustainability in Higher Education. *Environmental Education Research*, 10 (1).

Wals, A.E.J. & A.H. Alblas (1997). School-based research and development of environmental education: a case study. *Environmental Education Research*, 3 (3), 253-269.

RESOURCE LINKS

Maintained by Rogier van Mansvelt

In an experiment with our publisher, Kluwer Academic Press and the Dutch Foundation for Sustainability in Higher Education (St-DHO), we have organized the resource section as a web-based extension of the book. You are invited to the web-site for Higher Education and Challenge of Sustainability, Problematics, Promise, and Practice. There you will find resource links on many dimensions of the content of this book.

Information and links are available on:
- Professional, and non-governmental organizations;
- Academic journals;
- Declarations related to sustainability in Higher Education;
- Selected electronic resources;
- Assessment processes.

The web-site will be maintained by Rogier van Mansvelt. Depending upon use, the site may grow to include extended topics such as scholars active in the field and upcoming events. Certain key links will also be updated on an ongoing basis, such as the Selected Bibliography on Higher Education and Sustainable Development of the International Association of Universities. This is an exceptionally diverse and comprehensive listing of books, documents, periodicals, monographs, chapters, and declarations.

You are invited to send suggested additions to the web-site and comments on its usefulness to us. We especially invite suggestions to broaden the site across languages, cultures, and types of Higher Education. Please email the Dutch Foundation for Sustainability in Higher Education at info@dho.nl. The resource link for Higher Education and the Challenge of Sustainability: Problematic, Promise, and Practice may be accessed directly at: www.dho.nl/SHE-resources .

BIOGRAPHY

Drs. E.R. van Mansvelt is working as a teacher at the University of Amsterdam. Together with his colleagues he started the Expertise Centre for Sustainable Development (ECDO) at the faculty of natural science. Students from all faculties participate in interdisciplinary courses. The students work on real-time challenges in society concerning sustainable development. Besides teaching Van Mansvelt participate in the Dutch national network for sustainable development in higher education curricula. He is editor of the COPERNICUS-news letter and is AISHE consultant (Auditing Instrument for Sustainability in Higher Education).

AFTERWORD

Sustainability is becoming an integral part of university life: wishful thinking? or reality? The editors – themselves trailblazers in the field of education for sustainable development – have brought together an impressive range of experts to answer this question. This has led to a challenging and diverse tapestry of contributions of both a more theoretical and more practical nature.

We are, indeed, living in a time of profound changes characterized by, among other things, an increasingly globalized and knowledge-based world. These changes inevitably impact the way we organize and disseminate knowledge, curriculum development, teaching and learning. The rapid development of improved systems of communication and transport has changed our world from a complex, and sometimes, chaotic blanket of territories and borders, to a hierarchical system of nodes and channels. Our society is becoming more complex and heterogeneous, consisting of individuals characterized by intriguing sets of multiple identities. And..., it is undeniable, together we are set on an unsustainable course, using so many of our planet's natural resources, that the future of younger generations is jeopardized.

As several authors in this book have illustrated so powerfully, we can no longer ignore the interlinkages between globalization~trade~poverty~development, and environment. Perhaps that is what education for sustainable development is all about: to learn to understand the whole, complex reality and to act in adequate and informed ways. That is where education comes in: to learn to know, to do, to live together - with others - and to be (UNESCO, 1996); to become aware of our individual responsibilities to contribute, to make responsible choices, to respect other people, nature, and diversity.

This type of learning is truly life-long: it never stops and most of it is non-formal and incidental. In fact regular, formal education only prepares us to enter this challenging trajectory with, hopefully, more competence and confidence. Education, understood broadly as an ongoing process, including both formal and informal modes of teaching and learning, plays a crucial role in preparing and updating people for their future in a highly connected, interlinked, rapidly changing, and globalized world. Higher education, in particular, has an important role to play:

"Universities and equivalent institutions of higher education train the coming generations of citizens and have expertise in all fields of research, both in technology as well as in the natural, human and social sciences. It is consequently their duty to propagate environmental literacy and to promote the practice of environmental ethics in society."(CRE-COPERNICUS, 1994).

Although this was formulated with regard to the environment, it is equally true for sustainable development. Higher education prepares an important portion of the population for their entry into the labour market, including in most cases, the teachers that are responsible for education at the primary and secondary levels. Here, universities are called upon to teach not only the skills required to advance successfully in a globalized world, but also to nourish in their students, faculty and staff a positive attitude towards environmental issues and cultural diversity; to help them understand how richness of both nature and cultures can benefit all peoples, and can contribute to a better life, in a safer world, for all.

Since the Earth Summit in 1992, sustainable development has been high on the political agenda. Quite surprisingly, though, the role of education was not very well articulated. Neither was education defined as one of the stakeholder groups. Nine stake-holder groups were identified, among others Youth, Science and Business, but not education. Within the World Conference on Higher Education, however, a thematic debate was organized on "Higher Education and Sustainable (Human) Development". This debate brought together fourteen different organizations, in particular, university organizations, but also student groups and the World Business Council for Sustainable Development. In a follow-up, the Global Higher Education for Sustainability Partnership (GHESP) was formed in 2000. During the 2002 World Summit on Sustainable Development (WSSD) in Johannesburg, GHESP was launched as a type II partnership in which the International Association of Universities, the University Leaders for a Sustainable Future, COPERNICUS-CAMPUS and UNESCO co-operate. Through these organizations, over one thousand universities have vowed to promote sustainability in higher education.

Many other initiatives have followed the WSSD. Particularly important was the Ubuntu Declaration (2002): the UNU Institute for Advanced Studies (UNU-IAS) took the lead in bringing together eleven strong partners, including the International Council of Science (ICSU), the Third World Academy of Science (TWAS), the African Academy of Science, the Science Council of Asia, the World Federation of Engineering Organizations (WFEO), major university organisations, GHESP and the United Nations University. These organisations recognized "that integrated solutions for sustainable development depend on the continued and effective application of Science and Technology and that education is critical in galvanizing the approach to the challenges of sustainable development". They called upon higher education to respond to these challenges.

The Japanese Government chose education for sustainable development as a spear point for its contributions to the WSSD. Education for sustainable development was given an important place in the Johannesburg Plan of Implementation (JPOI). Then at the proposal of, among others, Japan and Sweden, the United Nations in its General Assembly in the autumn of 2003 decided to designate 2005 as the Year and 2005-2014 as the Decade of Education for Sustainable Development. UNESCO was invited to take the responsibility to be the lead agency. The Year and the Decade will bring -indeed- global attention to the critical movement for education for sustainable development.

The path for universities and disciplines to take is not clear, however, even though quite a few exemplary courses and case studies have been developed, as Part

Three of this book has shown. After all: it does involve a challenging and dynamic process of educational change and profound institutional innovation. It is evident, however, that a decisive start has been made. Part One of this book raises some of the many problematics involved in this process and offers helpful and critical thinking for those engaged in the transition. Delineation of promising new avenues of intellectual work is outlined in Part Two. They are an important contribution to those fields, but also to university administrators, researchers, teachers, and students moving to work across disciplines. Part Three provides well-documented cases from a range of the practice of sustainability. The cases represent a useful variety of geographies, cultures, and types of institutions. This collection, by leading practitioners, critics, and researchers, is a valuable educational tool. It has the potential to advance significantly the movement towards education for sustainability.

Together these three parts give a thought-provoking overview of some important work already done and which will contribute to the further development of education for sustainable development. Yes, indeed, sustainability is increasingly becoming an integral part of university life.

Hans van Ginkel
President, International Association of Universities
Rector, United Nations University, Tokyo, Japan

REFERENCES

CRE-COPERNICUS. (1994). *CRE-COPERNICUS Declaration.* Geneva: CRE-COPERNICUS Secretariat.
Ubuntu Declaration (2002). *On Education and Science and Technology for Sustainable Development,* September 2002. UNESCO-UNEP.
UNESCO (1996). *Commission Delors: Learning the Treasure from Within,* Paris: UNESCO.

ABOUT THE EDITORS

Peter Blaze Corcoran is Professor of Environmental Studies and Environmental Education at Florida Gulf Coast University, US, where he is developing The Rachel Carson Center for Environmental and Sustainability Education. He has been a visiting professor in Australia and The Netherlands and works extensively in international environmental education. He is Past President of North American Association for Environmental Education. Corcoran serves on the editorial boards of *International Journal of Sustainability in Higher Education, International Research in Geographical and Environmental Education, and Environmental Education, Ethics and Action in Southern Africa*. He is also Senior Fellow in Education for Sustainability at University Leaders for a Sustainable Future in Washington DC and is Senior Advisor to the Earth Charter Initiative in San Jose, Costa Rica. Research interests include the significant life experiences that lead to environmental concern, assessment of sustainability in higher education, professional development and teacher education in environmental education, nature study, and environmental ethics.

Arjen Wals is an Associate Professor within the Education & Competence Studies Group of the Department of Social Sciences of the Wageningen University in the Netherlands (http://www.wur.nl). He specializes in the areas of environmental education and participation, and social learning in the context of sustainable living. His PhD, obtained from the University of Michigan in Ann Arbor, U.S.A., under the guidance of Bill Stapp, focused on young adolescents' perceptions of nature and environmental issues and their implications for environmental education. He is a past-president of the Special Interest Group on Ecological & Environmental Education of the American Educational Research Association (AERA) and of Caretakers of the Environment International. He is the (co)author of over 100 publications on environmental education related issues and serves on the editorial board of four research journals: the *Canadian Journal of Environmental Education, Environmental Education Research, Environmental Education, Ethics and Action in Southern Africa* and *Tópicos en Educación Ambiental*. Email: arjen.wals@wur.nl